Geothermal Engineering

Arnold Watson

Geothermal Engineering

Fundamentals and Applications

Arnold Watson
51 Ash Grove
Te Awamutu 3800
New Zealand

ISBN 978-1-4614-8568-1 ISBN 978-1-4614-8569-8 (eBook)
DOI 10.1007/978-1-4614-8569-8
Springer New York Heidelberg Dordrecht London

Library of Congress Control Number: 2013946231

Printed on acid-free paper

Springer is part of Springer Science+Business Media (www.springer.com)

Preface

There are very many specialist geothermal journal and conference papers, but papers are mostly written to advance a topic incrementally and are no way to learn the fundamentals of the subject. So at the outset, the aim of this book was to provide a basic understanding of the most important aspects of geothermal engineering—a framework on which to build. However, that caused me to examine the fundamentals on which my own understanding relied. Geothermal engineering is multidisciplinary, and almost every main topic is a slight variation of subject matter developed in connection with industries that figure more prominently on the global energy scene. It is hard work to pick up specialist books, adapt to their wavelength, and gain the understanding that one is after. The result has been the inclusion of the basic ideas of thermo-fluids in this book, with explanations sufficiently detailed to perhaps assist earth scientists as well as engineers. However, the book is about engineering and makes no attempt to cover the earth science topics that geothermal engineers need to be familiar with—a route through the engineering material exists without needing the full earth science coverage.

Energy technologies wax and wane. We do not know whether we are in the early phase of geothermal heat use that will develop smoothly into a major energy industry or whether the geothermal future is just a short period of more of the same style. The text incorporates historical perspectives partly for this reason.

I have included references to geothermal conference papers where possible. The International Geothermal Association is to be congratulated for making these freely available on the Web, to which most of us have access.

I came into geothermal engineering from a half career in nuclear thermo-fluids, by joining the Auckland consulting firm of KRTA early in its geothermal history. A fine company spirit was engendered which carried us all along during my 16 years there. A shorter period in the Geothermal Institute of the University of Auckland then followed, during which I was lucky enough to work with the New Zealand Regional Authorities on assessing resource use applications.

I am grateful to the University of Auckland for the library access granted to me. Also to Springer and the anonymous reviewers engaged for help in improving the manuscript. I have learned a great deal from my clients, colleagues, students, and friends in the geothermal world and dare not risk mentioning names for fear of omitting anyone. However, I can say with certainty that I would not have been able to write this book without the support and encouragement of my wife, Kathy, who even turned her fine art talents to reproducing my sketches.

Te Awamutu, New Zealand Arnold Watson

Contents

1 Introduction 1
1.1 Background 1
1.2 Scope of This Book and Order of Presentation 3
1.3 A Typical Geothermal Project 4
1.4 Power from Heat 6
References 9

2 Sources of Geothermal Heat 11
2.1 The Structure of the Earth 11
2.2 Processes Taking Place in the Crust 13
2.2.1 Collision Boundaries 14
2.2.2 Subduction Boundaries 14
2.2.3 Evidence of Magma Bodies at Drillable Depths 17
2.3 Geothermal Surface Discharges 20
References 22

3 Thermodynamics Background and the Properties of Water 25
3.1 Definitions and the First Law of Thermodynamics 25
3.2 The Steady Flow Energy Equation and the Definition of Specific Enthalpy 29
3.2.1 The Steady Flow Energy Equation 30
3.2.2 Specific Enthalpy 32
3.2.3 The Kinetic and Potential Energy Terms in the SFEE 33
3.2.4 Adiabatic Pressure Drop or Throttling 33
3.3 Entropy and Absolute Temperature 34
3.4 The Thermodynamic and Transport Properties of Water 38
3.4.1 Phase Change, Clapeyron's Equation and the Saturation Line 38
3.4.2 Properties of Water 39
3.5 Hydrostatic Pressure and the Boiling Point for Depth Curve 42
References 45

4 The Equations Governing Heat and Single-Phase Fluid Flow and Their Simplification for Particular Applications 47
4.1 Introduction 47
4.2 The Equations Governing the Flow of Fluid and Heat 49
4.2.1 The Continuity Equation 49
4.2.2 The Equations of Conservation of Momentum 51
4.2.3 The Equation of Continuity of Energy 55
4.2.4 Bernoulli's Equation 57
4.3 Flow in Pipes and Engineered Equipment 58
4.3.1 The Equations for Steady Laminar Flow in a Pipe 58
4.3.2 Turbulent Flow 61
4.3.3 Making the Governing Equations Dimensionless to Reveal the Controlling Parameters 61
4.3.4 Presenting the Experimental Results for Use in Design 64
4.4 Natural Convection 67
4.5 Thermal Conduction in Solids 69
4.6 Flow in Permeable (Porous) Media 71
4.6.1 Darcy's Law 71
4.6.2 Reducing the Set of Continuity and Momentum Equations to a Single Equation 72
4.6.3 If the Fluid Is Incompressible 73
4.6.4 If the Fluid Has Small Constant Compressibility c 73
References 75

5 Geothermal Drilling and Well Design 77
5.1 Introduction 77
5.2 Well Construction 78
5.3 The Drilling Process 81
5.3.1 Drilling a Hole 81
5.3.2 The Drilling Rig 81
5.3.3 Drilling Fluids 85
5.3.4 Drilling Site Requirements 86
5.3.5 Blowout Prevention 87
5.4 Well Design 87
5.4.1 The Properties, Strength and Failure Criteria of Steel 88
5.4.2 Failure Due to Instability: Buckling and Collapse 90
5.4.3 The Properties and Failure of Rock 91
5.4.4 The Properties of Casing Steel and Casing Specification 91
5.4.5 Selection of Casing Depths 92
5.5 Summary of Modes of Well Failure 93
5.5.1 Drilling Operations That Could Cause Failure 93

5.5.2 Possible Modes of Failure of a Completed Well 94
5.5.3 Inspection Methods 94
5.6 The Completed Wellhead 95
References 97

6 Well Measurements from Completion Tests to the First Discharge 99
6.1 Internal Flow in Wells 99
6.2 A Typical Measurement Programme for a New Well 102
6.3 The Interpretation of Temperature and Pressure Surveys 103
6.3.1 Finding the Water Level 105
6.3.2 A Well with a Downflow 107
6.3.3 A Well with the bpd Temperature Distribution 108
6.3.4 Water Loss Surveys 109
6.3.5 Heat-Up Surveys 111
6.3.6 Pressure Surveys During Heat-Up and the Pivot Point 111
6.4 Resource Pressure and Pressure Gradient Estimates from Individual Wells 114
6.5 Making a Well Discharge 116
6.5.1 Gas Lifting 118
6.5.2 Lowering the Water Level by Air Compression 119
6.5.3 Heating the Casing with Steam 120
References 121

7 Phase-Change Phenomena and Two-Phase Flow 123
7.1 Background 123
7.2 Surface Tension 124
7.3 Boiling 126
7.3.1 Nucleate Boiling 126
7.3.2 Homogenous Nucleation 127
7.4 Flashing 128
7.5 Condensation 129
7.6 Thermal Explosions and Hydrothermal Eruptions 131
7.7 Two-Phase Flow 132
7.7.1 The Approach to Analysing a Two-Phase Flow 132
7.7.2 The New Variables 134
7.7.3 The Governing Equations for Homogenous Flow 135
7.7.4 Correlations for Obtaining Flow Parameters 136
7.7.5 Two-Phase Flow in Permeable Formations 137
7.8 Geothermal Liquids with Dissolved Gases 138
References 139

8 The Discharging Well 141
8.1 The Discharge Characteristic 141
8.1.1 The Form of the Discharge Characteristic 141
8.1.2 Interpretation of Resource Behaviour from the Discharge Characteristics 143

8.2 Measuring the Discharge Characteristics ... 145
8.2.1 The James Lip Pressure Pipe ... 145
8.2.2 The Available Methods of Measurement ... 146
8.2.3 Route A: A Well Discharging Steam Only (Dry Steam) ... 147
8.2.4 Route B: A Well Discharging a Two-Phase Mixture ... 148
8.2.5 Route C: An Alternative for a Well Discharging a Two-Phase Mixture ... 150
8.3 Further Details of the Measurement Equipment ... 151
8.3.1 The Single-Phase Orifice Plate ... 151
8.3.2 The Two-Phase Orifice Plate ... 154
8.3.3 The Thin-Plate Sharp-Edged Weir ... 154
8.3.4 The Separator ... 155
8.3.5 The Silencer ... 156
8.4 Chemical Measurements During Discharge ... 157
8.4.1 Sampling Arrangements During Discharge Measurements ... 157
8.4.2 Discharge from a Well Producing from Several Formations containing Chemically Different Fluids ... 158
8.4.3 Changes in Concentration of Dissolved Species as a Result of Flashing ... 158
8.4.4 Mass Flow Rate Measurement by Chemical Tracers ... 159
8.5 Surveying Wells During Steady-State Discharge and Predicting Pressure and Temperature Distributions ... 160
8.6 Transient Discharge Measurements and Predictions ... 163
References ... 166

9 The Transient Response of Formations to Flow in a Well: Transient Pressure Well Testing ... 169
9.1 Introduction ... 169
9.2 The Governing Equation in Axially Symmetric Coordinates and Its Solution and Application ... 172
9.2.1 The Preferred Solution ... 172
9.2.2 Conversion of the Equation and Solutions to Dimensionless Variables ... 174
9.2.3 Pressure Distributions in the Formation During a Simple Drawdown Test ... 176
9.3 Superposition and Its Use in Designing Tests ... 177
9.3.1 The Pressure Buildup Test ... 177
9.3.2 The Pressure Falloff Test ... 179
9.3.3 The Detection of an Outer Formation Boundary ... 179
9.4 Formation Testing in the Presence of Real Effects ... 180
9.4.1 The Skin Effect ... 180
9.4.2 Wellbore Storage and the Introduction of Type Curves ... 182

9.4.3 Departure from Infinite Uniform Thickness Formation Geometry ... 188
9.4.4 The Influence of Fractures in the Formation ... 189
9.5 Transient Pressure Testing of Formations Filled with Steam ... 191
9.6 Two-Phase Flow in the Formation ... 192
9.7 Tests Using More than One Well-Interference Testing ... 194
9.8 Concluding Remarks ... 197
References ... 197

10 Economic Issues Relating to Geothermal Energy Use ... 199
10.1 Introduction ... 199
10.2 A Single Isolated Project ... 201
10.3 Economic Evaluation of Projects ... 204
10.3.1 Discounting the Cash Flow ... 204
10.3.2 Net Present Value ... 205
10.3.3 Payback Period ... 207
10.3.4 Internal Rate of Return ... 207
10.3.5 Levelised Cost of Electricity ... 208
10.4 Factors Affecting Project Life ... 209
10.5 Other Economic Aspects: Sensitivity Analysis and Risk ... 209
10.6 Consideration of a Geothermal Station in a Network ... 210
10.6.1 Marginal Cost and Load Following ... 211
10.6.2 Base Load Service ... 211
10.7 Energy Analysis ... 212
10.8 The Steam Sales Contract ... 213
References ... 216

11 The Power Station ... 217
11.1 Introduction ... 217
11.2 Historical Trends in Thermal Power Stations ... 219
11.3 Generation of Electricity Using Steam as the Working Fluid in a Cycle ... 220
11.3.1 The Rankine Cycle as a Representation of the Carnot Cycle ... 220
11.3.2 Steam Turbines in Practice ... 227
11.3.3 Transient Performance of Steam Turbines ... 233
11.4 Using Geothermal Steam ... 235
11.4.1 Flashing the Well Discharge at Several Pressures ... 235
11.4.2 Geothermal Steam Turbines ... 237
11.4.3 The Possibility of Superheating Using an External Heat Source ... 238
11.4.4 Heat Rejection at Geothermal Steam Power Stations: Condensers and Related Equipment ... 238

11.5 Plant Using Working Fluids Other Than Water 240
11.5.1 Organic Rankine Cycles 240
11.5.2 An Example of Dual Working Fluid Plant 244
11.5.3 The Kalina Cycle 245
11.5.4 The Trilateral Flash Cycle and Two-Phase Prime Movers 246
References 248

12 The Steamfield 251
12.1 Overall Design Considerations 251
12.1.1 Silica Deposition 251
12.1.2 Overall Steamfield Layout 254
12.2 Pipeline Design 256
12.2.1 Steam Pipeline Design 256
12.2.2 Two-Phase Pipeline Design 260
12.2.3 Design of Pipes Carrying Water, Including Injection Wells 262
12.3 Steamfield Equipment Other Than Pipes, Including Separators and Silencers 263
12.3.1 Separators 263
12.3.2 Silencers 264
12.3.3 Steam Discharge Silencers 265
12.4 Transient Performance of the System of Wells, Steamfield and Power Station 265
12.4.1 Water Hammer 265
12.4.2 The Potential for Flashing in Separated Water Pipelines as a Result of Fault Conditions 267
References 271

13 The Resource Development Plan 273
13.1 The First Stages of Planning 273
13.2 A Stored Heat Estimate 274
13.3 Numerical Reservoir Simulation 278
13.3.1 The Mathematical Basis of a Geothermal Reservoir Simulator 279
13.3.2 Reservoir Simulation in Practice 284
13.4 The Overall Project Plan 287
13.4.1 The Parties Involved 287
13.4.2 Stages of the Project 288
13.5 Environmental Impact Assessment 289
13.5.1 The Effect on Geothermal Surface Features 289
13.5.2 Effect on Global and Local Air Quality 290
13.5.3 Ground Subsidence 291
13.5.4 Induced Seismicity 292
13.5.5 Effects on Local Groundwater Resources 292

13.5.6 Ecological Effects 293
13.5.7 The Potential for Significant Effects Due to Noise, Social Disturbance, Traffic and Landscape Issues 294
References 294

14 Struggles Between Commercial Use and Conservation: Examples from New Zealand 297
14.1 Background to New Zealand Legislation Governing Geothermal Resource Use 297
14.2 Acts of Parliament Relating to Geothermal Energy 299
14.2.1 The Geothermal Steam Act 1952 299
14.2.2 The Geothermal Energy Act 1953 299
14.2.3 Environmental Protection Versus Development 300
14.2.4 The Resource Management Act 1991 301
14.3 Rotorua 302
14.4 Wairakei 305
14.4.1 The Original Development 1956–2001 305
14.4.2 The Poihipi Development 1988–1997 306
14.4.3 The First Proposed Tauhara Development 308
14.4.4 The Renewed Wairakei Consents and the Te Mihi Power Station 310
14.4.5 The Tauhara II Proposal 313
14.5 Ngawha 315
References 318

Appendix A: Saturation Properties of Water from the Triple Point to the Critical Point 321

Appendix B: Compressibility of Water from 0 to 100°C and 0–1,000 bar 325

Appendix C: The Boiling-Point-for-Depth Curve 327

Index 331

Nomenclature

Variables

A	Area of cross section
A_{m}	Mass of fluid per unit volume of rock
a	Pipe radius
B	Slope of the Willans line
c	Compressibility
C	Discharge coefficient or chemical species concentration or wellbore storage coefficient
C_M	Molar fraction
C_p	Specific heat at constant pressure
C_v	Specific heat at constant volume
D	Diameter or signifying total derivative D/Dt
d	Diameter
E	Energy within a control volume or Young's modulus
E_I	Related to Ei
Ei	Exponential integral
F	Force or Fourier number $\kappa t/L^2$
f	Friction factor
G	Mass velocity
Gr	Grashof number $g\beta\Delta TL^3/\nu^2$
g	Gravitational acceleration
$\dot{H}$	Rate of internal heat generation
h	Specific enthalpy or formation thickness
$\hbar$	Heat transfer coefficient
I	Second moment of area or injectivity

j	Volumetric flux
k	Permeability
K_H	Henry's constant
k_{rf}	Relative permeability of liquid phase
k_{rg}	Relative permeability of gas phase
L	Distance or electrical load
m	Mass or slope of a straight line
$\dot{m}$	Mass flow rate
Nu	Nusselt number $\hbar d/\lambda$
P	Pressure or selling price of electricity
P_{pp}	Partial pressure
Pr	Prandtl number v/κ or $C_p\mu/\lambda$
P_{wet}	Wetted perimeter (perimeter of a duct in contact with the fluid)
PV	Present value
Q	Heat quantity or volumetric flow rate or a sum of money
$\dot{q}$	Heat flux
$\dot{q}_m$	Mass flux
$\dot{q}_e$	A heat source term
Q_p	Volumetric flow rate
q_v	Volumetric flux
r	Radius or interest rate
Re	Reynolds number $\rho\bar{u}d/\mu$
r_e	Outer radius of a formation
r_p	Distance between two wells
s	Specific entropy or dimensionless skin factor
S	Saturation
T	Temperature (°C)
t	Time or wall thickness
$\mathrm{t_p}$	A defined time
U	Specific internal energy
u	Velocity in the x direction
$\bar{u}$	Mean velocity
v	Velocity in the y or r direction or specific volume
V	Specific volume or velocity in a turbine blade
V_w	Volume of a well
W	Work quantity
$\dot{W}$	Rate of working
w	Velocity in the z direction

X	Dryness fraction
x	Coordinate direction
x_m	Mass quality
y	Coordinate direction
z	Coordinate direction

Symbols and Greek Letters

σ	Stress
∇	Differential operator
∇^2	$\frac{\partial^2}{\partial x^2}+\frac{\partial^2}{\partial y^2}+\frac{\partial^2}{\partial z^2}$ or $\frac{\partial^2}{\partial z^2}+\frac{1}{r}\frac{\partial}{\partial r}\left(r\frac{\partial}{\partial r}\right)$
Ø	Dissipation function
α	Void fraction
β	Coefficient of thermal expansion of a fluid
γ	Shear rate of strain component
Γ	Locally defined function, (7.1)
Δ	Indicating a difference, e.g. ΔP
ε	Strain or normal rate of strain component
ζ	Surface tension
η_C	Carnot efficiency
Θ	Absolute temperature
κ	Thermal diffusivity $\lambda/\rho C_p$
λ	Thermal conductivity
μ	Dynamic viscosity
ρ	Density
Λ	Dimensionless group
ν	Kinematic viscosity
τ	Shear stress or characteristic time
ϕ	Porosity

Suffixes

∞	At infinity
a	Axial
accel	Due to acceleration
atmos	At atmospheric pressure
c	Critical
D	Dimensionless or discharge
e	Equivalent or electrical
eff	Effective

f	Liquid state or failure stress
fg	Difference in property between liquid and gas phases
fric	Due to friction
g	Gas or steam
grav	Due to gravity
h	Hoop
hp	High pressure
i	Undisturbed
in	Flow entering the well from a formation
lip	Referring to lip pressure
lp	Low pressure
m	Mass
max	Maximum
min	Minimum
pp	Partial pressure
pump	Injected flow
R	Separated water
ref	Reference value
s	Saturation
sep	Separator
sil	Silencer
t	Total
Theis	Refers to Theis solution
turb	Relating to a turbine
w	In the well, at the sandface or at a wall
weir	At the weir
wet	Wetted surface perimeter
wh	Wellhead

Superscripts

-	Average
.	Rate

Introduction

1

This chapter begins by showing the growth of installed geothermal electricity generation equipment since 1950 and discusses the multidisciplinary nature of the industry and the difficulties of obtaining predictable outcomes from a natural system. The scope and order of presentation of the book is explained, and a typical geothermal project is described. A historical review illustrating the reasons for the growth in demand for heat, which began in the eighteenth century, concludes the chapter.

1.1 Background

Central power stations generating electricity in bulk are a product of the twentieth century. Until the late 1940s, their energy sources were almost exclusively river flows and fossil fuels, driving water and steam turbines, respectively; nuclear energy was added in the 1950s. It was not until the 1970s that a formalised worldwide search for alternatives was recognisable. These alternatives included geothermal energy, on which a start had been made 60 years or more before.

Geothermal energy was first used on an industrial scale in Italy in 1912 and was adopted in New Zealand in the 1950s, in the USA and Japan in the 1960s and in many other countries since then. The rate of growth of installed generating capacity changed in the 1970s and has been constant since then, as shown in Fig. 1.1. The International Geothermal Association [2012] has provided details of installed capacity by country. The average incremental step, that is, the size of power station installed as a single project, increased at about the same time, demonstrating confidence in the technology. It too has remained constant, possibly an indication of typical resource size.

This book describes the engineering of an electricity generating system using a source of geothermal heat; by system is meant the power station itself, the pipelines that deliver the well outputs to it, the wells and the manipulation of the local subsurface environment referred to as the geothermal resource or reservoir. The order in which the investigation and planning for any particular development

A. Watson, *Geothermal Engineering: Fundamentals and Applications*,
DOI 10.1007/978-1-4614-8569-8_1,

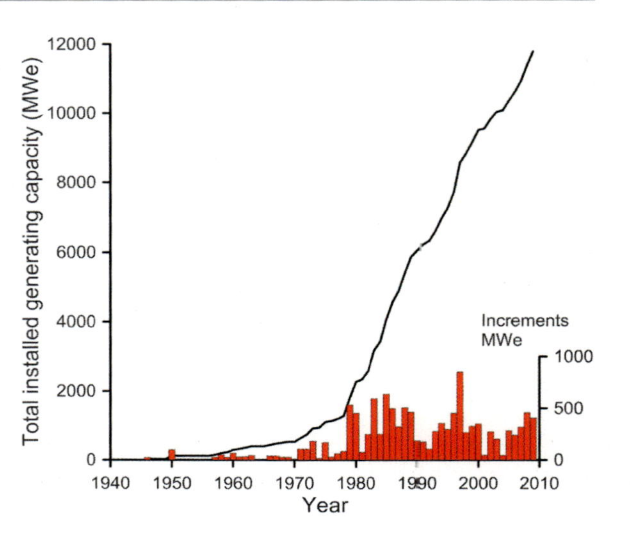

Fig. 1.1 Showing the growth in installed generating capacity and the incremental step size (*Data provided by Rugerro Bertani*)

take place necessarily begins with a scientific exploration of the resource, which guides the drilling, which in turn determines the power plant. The specifications for interconnecting pipelines and other equipment to process the well discharges then follow. The whole operation, of course, is governed by economic considerations; almost everything requires some degree of earth science–engineering–economic optimisation. The availability of suitable machinery with which to generate the electricity from the extracted heat is not guaranteed. Thermal power plant has been under continuous development over a period of almost exactly 300 years at the time of writing, and the manufacturing of heavy power station equipment over approximately the last 120, but they are not usually "off the shelf" items.

The New Zealand government agencies of the 1950s developed methods of drilling and well construction to deal with their subsurface conditions, which differed from those in Italy. Rather than finding steam at the depths drilled, in New Zealand hot water was found virtually from the surface downwards, and new approaches were required for drilling and processing the well outputs. Research in geology, geochemistry and geophysics was necessary to enable the resource characteristics to be deduced from surface measurements prior to drilling. New methods of well measurement and discharge testing followed, together with the topic now referred to as geothermal reservoir engineering, the understanding of natural fluid flow in geothermal areas and how flows can be modified to advantage. The engineering equipment in the power stations and large-scale chemical processing plant (the manufacture of wood pulp to produce paper) presented less fundamental problems.

Using geothermal energy for electricity generation is a multidisciplinary activity. Like petroleum engineering, geothermal engineering is strongly fluid mechanic and heat transfer in bias and requires close working with earth scientists of all disciplines. There is a fundamental difficulty in combining earth sciences with engineering.

Engineering by definition has quantitative outcomes, and bulk electricity generation must be economical on a timescale of 25 years or less to be worth doing—25 years is the nominal life of power station equipment. Like any large-scale engineering enterprise, the outcome must be predictable. This poses difficulties for earth scientists. Subsurface conditions may be typified and categorised, but they include a degree of randomness which leaves earth scientists with no certain means of quantifying outcomes; they must fall back on experience and incompletely supported opinion. The geological processes by which geothermal heat sources at drillable depths are created are dynamic but so slow that to human view they are stationary—the resource must be dealt with as found. Understanding how it was produced over millennia plays only a small part in the business of using it to generate electricity—it supports the detective work that is part of resource development. This fundamental difficulty is simply the outcome of using natural systems as compared to those entirely designed and manufactured, which are much easier to manage. There is no remedy but to have a clear understanding of the various subsurface processes involved and to make the maximum use of field measurements to support analysis.

1.2 Scope of This Book and Order of Presentation

A geothermal resource can be considered as a small part of the earth's crust, extending from the surface to a depth of ten kilometres or more, with a volume of several hundred cubic kilometres. Each one is geologically and geophysically complicated and unique, and all are the result of processes that are of planetary scale and have been taking place over much of earth's history. Resources of the type dealt with in this book are permeated by aqueous solutions reacting with the rocks. Deciding on the best way to use a resource to generate electricity requires the efforts of a team comprising geologists, geophysicists, geochemists (collectively "earth scientists") and engineers. The details of the resource are progressively revealed prior to and throughout the heat extraction process, and the team must interact, so that each member must know something of the other's discipline. However, it is possible to isolate the engineering fundamentals from the complete activity, and this forms the scope of this book.

It is hoped that the explanations will be readable by earth scientists, but on the other hand, no attempt has been made to explain earth sciences to engineers, with the sole exception of Chap. 2, which offers an explanation of the nature of the heat source in terms that engineers can identify with, namely, what processes produce them and how. No attention has been paid to where they can be found, as this information is already available in print and on web pages to be referred to.

This introduction, Chap. 1, ends with a general description of how a geothermal resource is used for electricity generation, followed by a brief historical review to illustrate how society has come to rely on an ever-increasing supply of heat, the reason for geothermal engineering. The origins of geothermal heat are described in Chap. 2. Hot water and steam are discharged from geothermal wells, and phase change must be understood if wells are to be drilled, so the necessary thermodynamics and properties of water are set out in Chap. 3, followed by an explanation of the equations governing fluid flow and heat transfer in Chap. 4. Geothermal drilling and well design is explained next, followed by Chap. 6 describing well measurements prior to discharge and their interpretation, with a section on discharge stimulation. Discharges are often two-phase, so the necessary background in phase change and two-phase flow is provided in Chap. 7. Well discharge measurements follow, Chap. 8, and methods of measuring the properties of the producing formations using transient pressure tests mainly borrowed from petroleum engineering and adapted for geothermal work are presented in Chap. 9.

The topic of economic analysis of projects is often seen by engineers and scientists as a necessary evil, relegated to second-order importance and not at all "fundamental". Economic analysis by scientists and engineers can be helpful in developing project strategy during the exploration stage and is not merely the province of bankers; it is explained in Chap. 10, using simple spreadsheet calculations.

Geothermal power plant is not a specific type separate from all other, but is based on a century of steam turbine development for fossil-fuelled power stations. Chapter 11 first sets out the principles which guided the latter and then their application to geothermal steam, which leads to the reasons for adopting organic Rankine cycle turbines. Steamfield design is then described, followed by Chap. 13 in which project planning, resource assessment and environmental impact assessment are dealt with. Numerical reservoir simulation is included in this chapter. The final chapter consists of examples of the way the competing issues of resource development and environmental conservation have been dealt with, the interaction of the law, science and engineering. The examples are from three New Zealand resources, Rotorua, Wairakei and Ngawha, but the principles are applicable internationally.

1.3 A Typical Geothermal Project

Geothermal resources are sometimes divided into two types according to what the wells produce. Steam-dominated resources produce steam with very little water, and liquid-dominated resources produce the opposite. The resource is often called a reservoir, although this term has no specific definition; reservoir may be more appropriate in petroleum engineering, where it probably originated. Definitions could be suggested, the resource extending to undrillable depths and the reservoir being the volume drilled and within which the flow of fluids can be altered; however, reservoir and resource are used here synonymously. The heat stored in the body of rock has been accumulated over a time period perhaps as short as thousands of years, usually by a flow of hot water from beneath that is small relative to the flow taken from production wells. Thus, geothermal developments depend primarily on stored heat and not on the "recharge" flow, so they are not renewable in the short term if used at the rate and scale necessary to generate electrical outputs of tens to hundreds of MWe. Approximately 90 % of the heat is stored in the rock and

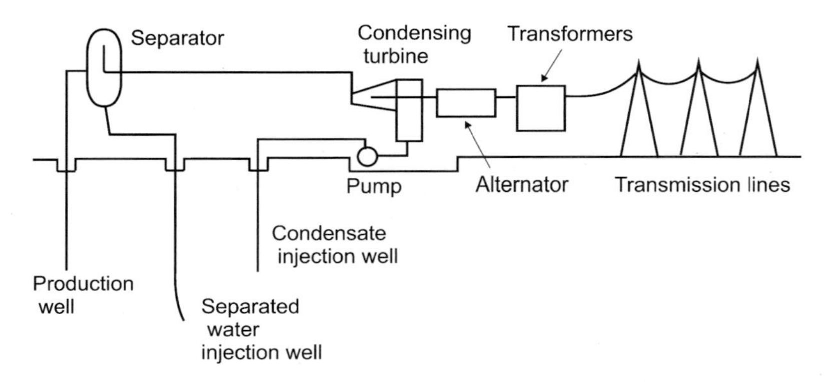

Fig. 1.2 Showing the main components of a geothermal power station development

wells drilled into the reservoir discharge water or steam that holds the balance of the stored heat. Replacement water is essential to extract more heat from the rock.

The natural geothermal water contains dissolved chemical species such as sodium, calcium and many other elements present as chlorides, carbonates and silicates and dissolved gases, particularly H_2S and CO_2. The exact chemical composition is site specific but the fluid is generally damaging to flora and fauna. The production wells usually discharge a mixture of water and steam, and the dissolved species appear in both phases. Figure 1.2 shows the basic elements of a geothermal power station development, with only one of each well type included where in reality there would be several at least.

The output from the production wells is separated into water and steam, the latter being supplied to the power station. Injection (or reinjection) wells are used to return the separated water to the ground. The output from several production wells is often combined and delivered to a separator via a two-phase pipeline (liquid water and steam phases flowing together). The separator performs two functions, it reduces the pressure in the mixture to produce more steam as a result of water evaporation, and it swirls the mixture in a circular path to induce the separation. The turbines require steam as free of water as possible, since water droplets can severely damage the turbine blades. The steam is delivered to the power station through a pipeline. The production wells are usually widely distributed and the distance from separator to the power station, often a kilometre or more, is advantageous in drying the steam still more. The separators operate at several bars above atmospheric pressure so the separated water is able to flow uphill to injection wells if necessary and hence back into the resource.

The power station machinery typically comprises a steam turbine through which the steam passes axially, driving a rotor and blades similar in appearance (but not in engineering) to those visible in the opening to an aircraft jet engine. The steam causes the blades and turbine shaft to rotate at high speed and with it the alternator shaft to which it is coupled. The alternator speed determines the frequency of alternating current generated, so the turbine speed must be precisely controlled if the electricity is to be distributed, to hold the frequency constant. The alternator electrical output passes through transformers which increase the voltage so that large amounts of power can be transmitted by cables with moderate current flow. An electrical switchyard is usually visible where these operations take place. Figure 1.2 shows a condensing steam turbine. The condensate is collected in the condenser at sub-atmospheric pressure, so must be pumped to a disposal well. It is relatively clean of dissolved solids but contains dissolved gases and is disposed of via dedicated injection wells and not mixed with the separated water.

Injection has two aims, disposal of separated water and maintenance of fluid in the resource. Discharging wells without injecting into others results in falling reservoir pressures and smaller discharge rates. However, the separated water is cooler than the fluids remaining in the reservoir so it must not be injected close to production wells; otherwise it may reduce their discharge temperature and hence their electrical generating potential. The distance between production and injection areas must be found by trial and error aided by field measurements. It is possible in principle to inject into formations directly below the production formations, on the grounds that cool water will sink, but this is not a standard procedure and separated water is often injected near the periphery or even outside the resource, away from any hot areas. Both shallow and deep permeable formations (aquifers) will usually exist there; if the shallow ones contain potable water, contamination should be avoided by injecting only into the deeper ones. As a result of the evaporation in the separator, the water for injection has a higher concentration of dissolved solids than the original well discharge; it is more damaging to flora and fauna and can cause deposition of chemical scale in pipelines and wells.

The modern management of a geothermal resource relies on a programme of regular field measurements which are used in conjunction with a numerical reservoir model. Standards of preservation of the environment have risen in recent decades, and determining the nature and extent of environmental effects and how to minimise them is increasingly important.

1.4 Power from Heat

It is now widely accepted that energy exists in several interchangeable forms, and energy has become an everyday media topic. Although this represents an improvement in understanding by the general population, it seems to have masked an appreciation of the extent to which modern human society relies on a continuous supply of heat, which in turn leads to false hopes about the extent to which heat can be replaced by direct mechanical energy sources such as wind, tide and hydro. The fact that not all heat is equally valuable is perhaps also overlooked.

Society's dependence on heat began in 1698 when Savery built and patented a device that converted heat into mechanical energy (Fig. 1.3).

The vessel was filled with steam by manually opening valve V_1. The boiler pressure was little more than atmospheric pressure. Valve V_1 was then closed and valve V_2 opened to spray cold water from a holding tank over the vessel,

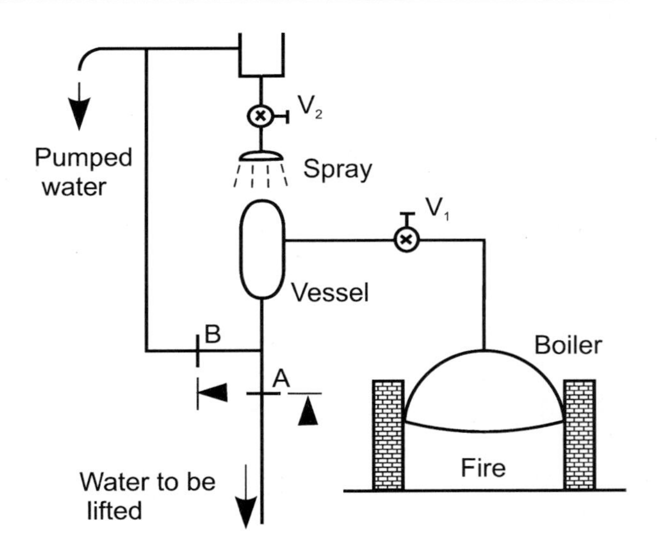

Fig. 1.3 Savery's water pump (1698)

condensing the steam. The valves labelled A and B were non-return valves, so that flow could take place in the direction of the arrow heads when the pressure was higher upstream, but never in the opposite direction. The vacuum created in the vessel brought water from a lower level up through non-return valve A, but no flow passed through B. The spray was now turned off and valve V_1 opened, allowing the steam pressure to drive the water through valve B and empty the vessel, with A allowing no flow. Steam must have appeared from the pumped water exit pipe to show that the vessel had been emptied, at which time valve V_1 was closed and the process repeated. Savery's device worked repeatedly only if someone opened and closed the appropriate valves at the right time, so although not strictly a machine, it was remarkable in its time in that heating the water resulted in a mechanical force to pump water. Before this revolutionary invention, continuous mechanical power could only be produced by animals (including people), windmills and water wheels. The windmills of this period, still to be seen in the Netherlands, were capable of outputs of around 50 kW (Singer and Raper [1978]).

Savery's invention did not come from a sudden flash of inspiration. Experiments with steam had been carried out in Italy by Della Porta in 1606 and Branca in 1629. Torricelli, also Italian, established in 1643 that atmospheric pressure would support a 10 m high column of water, while von Guericke (Germany) demonstrated in 1654 that the evacuated "Magdeburg hemispheres" could not be pulled apart even by a large number of horses. Both von Guericke and the Dutch astronomer Huygens used a piston sliding in a cylinder in their experiments, an item which was essential for the development of engines, but one that was difficult to make with the machine tools of that era. Papin (France), a scientific assistant to Huygens who also worked with Boyle and Hook, carried out experiments with steam, producing the first pressure cooker and also a working model of a pump on the same principle as the full-sized one built later by Savery.

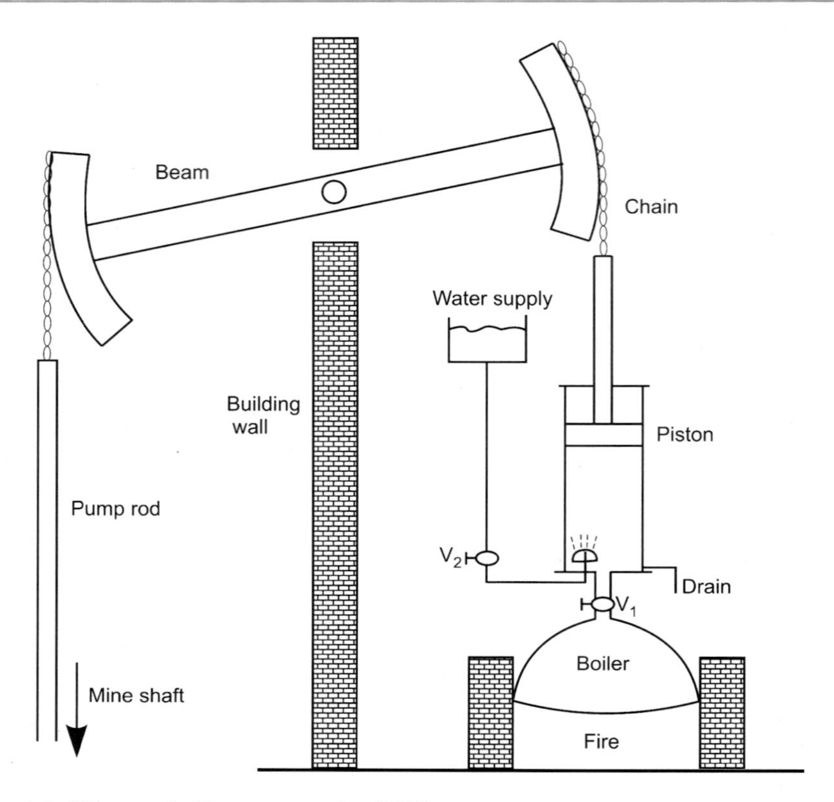

Fig. 1.4 Diagram of a Newcomen engine (1712)

The first recognisable steam engine had a cylinder and piston but a beam instead of a crank, was built in Britain by Newcomen and Calley in 1712 and was also designed to drain mines. It is shown as Fig. 1.4.

It used the same low-pressure steam as Savery's device, with condensation producing the force, but in this case the steam was condensed by a spray inside the cylinder and the force was applied to a separate mechanical pump. This machine also began as a manually operated device before it was realised that the valves could be opened automatically. Many hundreds were built in Europe to this design, and it was almost 70 years before Watt's improvements appeared. More detailed histories of the steam engine and the experiments that led to its development are available in Storer [1969] and van Riemsdyjk and Brown [1980].

Trade was the driving influence in the development of heat engines and of a quite separate demand for heat. By the late 1600s the manufacture of iron goods as trading commodities was well established. This called for a supply of fuel to generate the heat to smelt the ore and then to refine the metal from cast to wrought iron. The forests of Europe were becoming depleted due to the demand for charcoal, the only fuel then known for metal production. In England, Abraham Darby experimented with the use of coal and in 1710 discovered that it was suitable for smelting iron if it was first converted to coke. Cast iron began to be used for the construction of large items such as bridges, the beams of some Newcomen engines, and large machine tools. Manufacturing industries developed in parallel, each having an effect on the other and all requiring fuel and mechanical effort. The British demand for coal by the late 1700s was sufficient to encourage the development of mines further away from markets—in direct analogy with today's development of deep offshore oil as onshore resources diminish. The canal system was in some areas specifically designed to bring coal to market, and it was at this time that the science of geology had its beginnings. William Smith (1769–1839), who went on to set out the principles of stratigraphy and produced the first geological surface map, began as a mapper of coal seams and surveyor of British canal routes (Winchester [2002]).

Taking the widest possible view of human energy consumption, a 70 kg man in a state of complete mental and physical rest requires an energy supply of 80 W, rising to 240 W when "hunter-gathering" or engaged in primitive agricultural work (Alexander [1999]). In Europe the population density 30,000 years ago when mankind lived in this manner is estimated to have been 0.3/100 sq km (Phillips [1980]) compared to 100/sq km at present. The per capita rate of energy use or "consumption" worldwide is now 2.5 kW (Energy Bulletin [2012]) by a population of 7×10^9 (US Dept of Commerce [2012]) with a very wide range of rates across different parts of the world. This indicates that mankind has a problem in population numbers and modus operandi, which is primarily trade driven, an activity which has continued since prehistoric times. The extent to which we should modify the rate of arrival of subterranean heat to the surface in the long term is open to question—a great deal of heat arrives at the surface from the sun—however, changing our modus operandi at the necessary rate requires that geothermal energy development should continue at present, as its currently predominant form is both economically and environmentally profitable.

References

Alexander RM (1999) Energy for animal life. Oxford University Press, New York

Energy Bulletin (2012). http://www.energybulletin.net

International Geothermal Association (2012). http://www.geothermal-energy.org/226,installed_generating_capacity.html

Phillips P (1980) The prehistory of Europe. Penguin Books, New York

Singer CJ, Raper R (1978) A history of technology: vol IV, The industrial revolution. In: Ritson JAS (ed) Metal and coal mining. Clarendon, Oxford, pp 1750–1850, Chapter 3

Storer JD (1969) A simple history of the steam engine. John Baker, London

US Dept of Commerce (2012). http://www.census.gov

van Riemsdyjk JT, Brown K (1980) The pictorial history of steam power. Octopus Books Ltd, London

Winchester S (2002) The map that changed the world. Penguin Books, London

2 Sources of Geothermal Heat

The main purpose of this chapter is to explain how a relatively shallow geothermal resource from which heat is to be extracted relates to the dynamics of the earth. The structure of the interior of the earth is described in general terms. Tectonic plates and events taking place at their boundaries are explained, elaborating on subduction boundaries and discussing the origins of heat sources in the form of intrusive bodies of magma. Examples of these at drillable depths are given. The chapter ends with a brief discussion of geothermal surface discharges.

2.1 The Structure of the Earth

The earth is a sphere 6,400 km in radius which is believed to have formed about 4,500 Ma ago. At present it has a metal core of radius 3,500 km, of either iron or an iron–nickel alloy, which has separated from the surrounding material under the gravitational field. It follows that the core radius must have been increasing over earth's lifetime and the chemical constituents of the surrounding material have been changing. The pressure and temperature at the centre of the core, where the metal is thought to be solid, are estimated to be about 1.4 million bars and 5,000 °C, respectively. The temperature at the outer radius of the core is estimated to be between 3,500 and 4,500 °C. At some radius within the core, the metal is thought to become liquid, although this implies properties like those of common liquids, and "plastic" might be a more appropriate description.

The material surrounding the core is an entirely molten (plastic) layer of oxides of silica and other elements which is referred to as the mantle (literally meaning a cloak or garment) and is 2,900 km thick; the mantle extends essentially to the surface, although not as a homogenous material. It is thought possible that the Moon is made up of mantle material which was separated off at an early stage in earth's development by an asteroid impact. The properties of the mantle have been the subject of considerable research to examine this and other theories about the

A. Watson, *Geothermal Engineering: Fundamentals and Applications*,
DOI 10.1007/978-1-4614-8569-8_2,

early development of the planet (see, e.g. Ohtani [2009]), and a striking feature is the complexity of the mineral mixture and the need for phase diagrams reaching to pressures of 200,000 bars and 5,000 °C if any quantitative estimates of evolutionary processes are to be made. In broad material terms, the earth is made up only of these two components, the core and the mantle, but it is exposed to low-temperature space through a relatively thin and transparent atmosphere, and as a result the surface temperature is low enough for the material to be solid there—the crust. Being solid and exposed to the atmosphere, it has undergone both chemical and physical changes, making it much more heterogeneous than the plastic mantle on which it floats.

Seismological measurements indicate a change in material at the base of the mantle, 2,900 km below the surface, a change referred to as the Gutenberg discontinuity. Another seismological interface, the Mohorovicic discontinuity, occurs at a depth of about 35 km and divides the mantle from the crust. Accordingly, the crust can be thought of as two layers, a lower solid one of silica- and magnesia-rich material which is described as basalt and an upper one of silica- and alumina-rich material generally described as granite (Whitten and Brookes [1972]). Overlying the crust is a thin layer of sedimentary rocks, the result of the processes taking place at the surface by interaction with the atmosphere. The crust is thinnest beneath the oceans and thickest beneath mountain ranges, and its surface is far from smooth. Taking sea level as the average surface level, the highest mountain is 9 km and the deepest ocean 11 km, and maximum undulations of the surface are within an order of magnitude of the thickness.

The quantity of thermal energy contained in the core cannot be calculated because the material properties are not known; however, that is no detriment here. Humanity has existed for only 2 of the 4,500 Ma since the formation of the earth, so not only is the store of heat of literally astronomical proportions but the state of the interior can be regarded as fixed in human terms. The heat leakage from the surface to the atmosphere is an average of 50 mW/m^2 overall, a very small heat flow rate (flux) compared to the 1.4 kW/m^2 of solar radiation arriving at the outer surface of the atmosphere. The net heat loss from the interior is a negligible proportion of the heat stored. However, very much higher heat fluxes than 50 mW/m^2 escape from the ground surface at some locations, and a figure of 800 mW/m^2 has been estimated by Cole et al. [1995] for the Taupo Volcanic Zone of New Zealand (see also Hochstein and Regenauer-Lieb [1989]).

The rotation of the earth and the plasticity of its two main component layers results in the circulation of both mantle and core. The mass of mantle fluid in motion, its viscosity and internal heat generation all combine to fracture the crust and cause the pieces—tectonic plates—to move. Heat is generated as a result of gravitational work done in draining the core metal from the mantle, and the decay of radioactive isotopes in the mantle provides an additional source to keep it molten. The induced buoyancy forces and the Coriolis forces due to the rotation lead to it having a complicated flow pattern. The solid crust is subjected to forces causing it to break into large slabs which move slowly in different directions as a result of floating on a liquid. The study of the motion is called plate tectonics, and articles conveying the current scientific thinking are available on the Internet, for example, by the Institute of Geological and Nuclear Sciences, NZ (GNS) [2012], the US Geological Survey [2012] and the British Geological Survey [2012], including animations of the plate motion.

2.2 Processes Taking Place in the Crust

Processes within the crust are controlled by the physical and chemical properties of the material, the temperature distribution and the motion of the plates. The plates move relative to one another at an average speed of typically 3–10 mm/year, although rates several times this have been measured. Heat reaches the surface at some plate boundaries. It is much too simplistic to consider this simply as leakage at the cracks, and the purpose of this section is to give some appreciation of the physics involved. It is clear from maps of earth's surface that volcanism is associated with subduction boundaries in a recognisable pattern, and many geothermal resources occur in volcanic terrain. The Pacific "Ring of Fire" is perhaps the prime example—see US Geological Survey [2012] and Schellart and Rawlinson [2010]. Plate boundary processes attract a great deal of academic research, but progress is inevitably slow given the inaccessibility of the regions of interest, which are well below the surface. What is relevant here is a picture of how a geothermal resource gains its heat and what physical form it takes.

The plates may converge by moving towards each other in a direction normal to the boundary between them, or diverge likewise, or interact with a shear component, and combinations of these occur. At some plate margins, the mode of interaction changes locally from one type to another. Convergent plates are associated with heat release at the surface and hence with geothermal resources, and the dynamics of converging plates has recently been reviewed by Schellart and Rawlinson [2010]. They provide an extensive list of references and define two types of convergent interaction, subduction and collision. Before describing the boundary interactions, more of the physical aspects of the crust must be understood. It may be considered to be in two layers, the rigid outer one called the lithosphere and the inner plastic one called the asthenosphere (the root is Greek, meaning weak). The crust is thicker beneath continents than beneath the oceans, so indications of thickness are not precise; however, the lithosphere may be up to 75 km thick and the asthenosphere from the bottom of the lithosphere to as much as 200 km. What is being described in the latter case is the depth over which the temperature variations induce major physical property variations, particularly viscosity. In discussing the laboratory modelling of subduction boundaries where oceanic crust is overridden by continental crust, Shemenda [1994] suggests that the materials in order of depth are a low-strength brittle upper layer (clearly lithospheric), then a layer showing a gradual transition from brittle to plastic material properties but with elasticity (the ability to carry stress without continuous strain) which renders the material stronger than the first layer and then a decrease in strength beyond depths of "a few tens of kilometres" into high-viscosity fluid behaviour (clearly asthenospheric).

Fig. 2.1 Sketch illustrating a tectonic plate collision boundary

2.2.1 Collision Boundaries

Collision boundaries are defined as those where both plates are continental plates at the colliding margins. The result of the collision is the formation of a mountain range generated by buckling and folding, e.g. the Himalayas, as illustrated by the sketch (Fig. 2.1).

The mechanical forces involved generate internal heat in the same way that the repeated bending of a piece of wire or the cutting of metal in a lathe causes the metal to become hot, by plastic deformation. The location is a plate boundary, but the heat released has only indirect connection to the heat of the mantle, being due to plastic deformation as a result of the motion which is enabled by the earth's internal heat. Hochstein and Regenauer-Lieb [1998] produced a numerical model of the heat generated by plastic deformation in the collision boundary of the Himalayas. There exists a belt of geothermal springs parallel to the inferred plate boundary, and they related the results of their calculations to field measurements of the hot springs. In doing so they estimated that the rate of heat release at the surface was 100 MW/100 km over much of the length of the 3,000 km plate boundary on which the Himalayas are situated and up to 300 MW/100 km towards the eastern end. They concluded that heat generation resulting from plastic deformation was a likely cause of the higher than average heat flux leaving the surface along that particular plate boundary. The quoted heat fluxes of less than 1 W/m^2 are not encouraging for large-scale geothermal power development.

2.2.2 Subduction Boundaries

At a subduction boundary, one plate rides over the other, which dips (subducts) down into the mantle (Fig. 2.2). Two important surface modifications often occur adjacent to the plate margins, volcanism along a line generally parallel to the overriding plate edge and faulting and surface subsidence some distance behind the overriding plate edge. Schellart and Rawlinson [2010] note that at some subduction boundaries, there is also evidence of the collision process of Sect. 2.2.1 together with the subduction process, referring in particular to the Andes, which is the result of the convergence of the plate forming the bed of the Pacific Ocean moving eastwards and subducting, and the overriding South American continental plate. Buckling and uplift occur adjacent to the subduction boundary.

Shemenda [1994] identifies the difference in density between the lithospheric and asthenospheric materials as a major factor influencing the subduction process. The subducting plate is subject to a drag force from the circulating mantle, a force

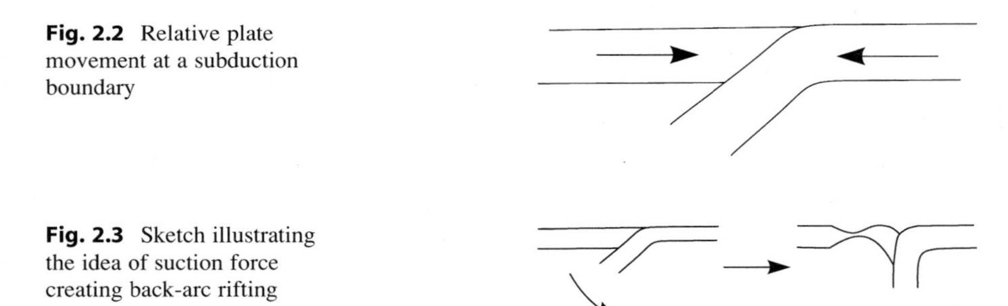

Fig. 2.2 Relative plate movement at a subduction boundary

Fig. 2.3 Sketch illustrating the idea of suction force creating back-arc rifting

acting over its entire area and responsible for its movement. If it is denser than the fluid on which it floats, it will sink as it subducts, producing a force supplementing the fluid drag—both forces are acting normal to the plate boundary and pulling the subducting plate towards the overriding one. The weight of the drooping plate will increase because g increases with depth, although this may be insignificant as it appears to be absent from consideration in the geophysical literature. Given these forces, it comes as a surprise that a major deformation of the overriding plate occurs some distance back but close to its edge and, even more surprising, that the deformation takes the form of a localised stretching consistent with a tensile stress rather than the compression expected from Fig. 2.2. One might be justified in anticipating from Fig. 2.2 that although the overriding plate is driven towards the subduction boundary by fluid drag as a result of the mantle convection current, the obstruction caused by the subducting plate would induce compression immediately up-plate from the boundary. However, within about one thickness distance of its edge, the overriding plate accelerates towards the subducting plate. The stretched, brittle lithosphere is subject to faulting and the thinned crust subsides. In some locations where this subsidence occurs, it is referred to as "back-arc rifting". Shemenda [1994] cites references suggesting that there is a suction force that keeps the falling plate attached to the overriding one, and the sketch of Fig. 2.3 illustrates one way in which this might occur as the subducted plate droops (simply to reinforce the picture of events and not as a proposed phenomenon).

He also reviews evidence that the rifting takes place at an already thinned location of the overriding plate, which may involve the other principal feature of the subduction zones mentioned, volcanism. The volcanoes generally occur along a line parallel to the subduction boundary, but there is variation in the distance of the line from the boundary, and Shemenda [1994] suggests that they may occur along a line of weakness. Alternatively, he suggests, the plate may be eroded on its underside as a result of an eddy in the mantle flow within the vertical wedge formed by the two plates (Fig. 2.4).

The vertical wedge between the lower surface of the overriding plate and the upper surface of the subducting one is the source of the magma for volcanism, which is sometimes loosely attributed to rising magma formed by frictional heating of the moving plate surfaces. A proper explanation would call for considerable

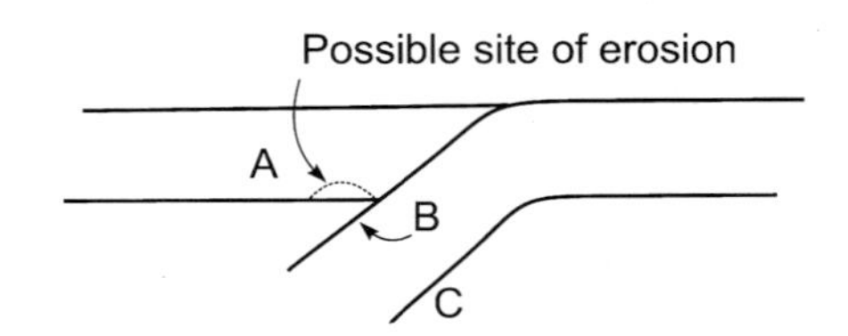

Fig. 2.4 Identification of regions near a subduction zone wedge

analysis, but material properties are poorly known. In Fig. 2.4, the plate lower surface shown at A marks the depth at which the crust changes from solid to plastic—in reality the change is gradual but a simple two-layer picture is sufficient here. Beneath A the mantle material is hot, plastic and in motion. Surface C is similar to A, but surface B is cold, water-saturated crust which will cool the plastic material in the wedge. In Sect. 4.5 it will be shown that the timescale for a slab of thickness L to reach a uniform temperature when one side is suddenly exposed to a heat source is of order L^2/κ where κ is the thermal diffusivity, and knowing the subducting plate speed would allow the time required for melting of the subducted plate to be estimated. If the mantle material beneath A was able to rise towards the surface and intrude or erupt simply as a result of its temperature, then it should have been capable of doing so at any time, regardless of its proximity to the subducting plate. Suction forces, a line of weakness and local erosion are all suggestions as to the mechanism of deformation of the overriding plate. A further suggestion is that release of the water carried down within the subducting plate changes the composition of the asthenospheric material so that it tends to rise by melting its way through the crust. Manning [2004] outlines the relevant chemistry and Gerya and Yuen [2003] illustrate the physics. The wedge material can thus be cooled, intuitively expected to make it solidify, but chemical changes result in it becoming a viscous liquid which rises towards the surface and may reach it and erupt.

The general picture which emerges is of the periodic occurrence of a body of magma sufficiently hot and fluid to rise from within the wedge between the subsiding and overriding plates and penetrate the latter, thinned and weakened as it may be. The periodicity of events is evidently long in human terms but the occurrences are not sparsely distributed, as can be seen from the maps of the "Pacific Ring of Fire" referred to in the introduction to Sect. 2.2. If the penetration is complete, the result is a volcano, and if partial, the body is referred to as a pluton or intrusion. The geological description is unimportant—to bring a large enough quantity of heat into the near-surface crust to create a geothermal resource, the magma body must have a large volume.

Much of the literature on this topic is not written in the language of the mathematical sciences, but focuses on the chemical changes, and the problem is often poorly described in physical–mathematical terms, but there is no point in attempting mathematical analysis before the processes have been adequately described. In what might be viewed as a preliminary skirmish, Norton and Knight [1977] modelled the cooling of a discrete mass of magma as it rose towards the surface by setting out the equations governing heat transfer and fluid flow and solving them for various boundary conditions. The fluid circulating in the permeable rocks around and above the magma body was assumed to have the properties of pure water, and they presented temperature and fluid circulation contours.

The immediate question raised by the above discussion is whether there is any evidence from drilling that magma bodies occur as described.

2.2.3 Evidence of Magma Bodies at Drillable Depths

Drillable depths are far less than the depths involved in the processes discussed above; the limits are set by the temperature encountered and by economics. The result is that wells in most liquid-dominated geothermal resources are less than 4 km deep. Magma bodies have been encountered in several.

Stimac et al. [2008] provide a cross section of the Awibengkok geothermal resource (also known as Salak) in western Java, Indonesia (Fig. 2.5). The data comes from 81 wells, of which several have intersected a geological formation at about 2,000 m below the surface which is interpreted as the top of a large intrusion (of magma). Figure 2.6 is a simplified version of the geological cross section of the Kakkonda resource, Northern Honshu, Japan, produced by Doi et al. [1998]. The figure shows the tracks of six wells which were drilled into the magma body.

Further examples are given by Christenson et al. [1997] at Ngatamariki, New Zealand, although in this case the intrusion is cool because of its age, which they estimate as 0.7 Ma. A magma body referred to as the Geysers felsite has been identified beneath the Geysers geothermal resource in California, USA (Hulen and Nielson [1996]), and a further example occurs in the Tongonan resource in Leyte in the Philippines (Reyes [1990]). In the Tongonan case, it is considered that the outer surface zone of the magma body has increased permeability, perhaps due to fracturing under thermal stresses while cooling, which provides enhanced convection as a plume towards the surface, shown diagrammatically in Fig. 2.7.

The Taupo Volcanic Zone (TVZ), New Zealand, is notable for having many separate geothermal resources close together, that is, separate convective plumes. The usual method of delineating the region permeated by geothermal fluid, where a plume is deflected by the presence of ground surface, for example, is by measuring electrical resistivity. The electrical resistivity of the ground is reduced as a result of chemical alteration and minerals deposition by the convecting geothermal fluid. Individual plumes can be identified by resistivity boundaries and also characterised by the detailed chemical composition of the deep fluid.

Alcaraz et al. [2012] present a model of the TVZ, from the basement upwards, from which Fig. 2.8 is taken. New Zealand is situated along a plate boundary which is only partly a subduction zone—the mode of interaction changes southwest of the TVZ, and to further complicate matters, there is a relative rotational component between the plates where they are subducting. The zone has a high heat output—Hochstein and Regenauer-Lieb [1989] estimated it as 5,000 MW. Various ideas have been proposed for the detailed subduction process resulting in the TVZ, and

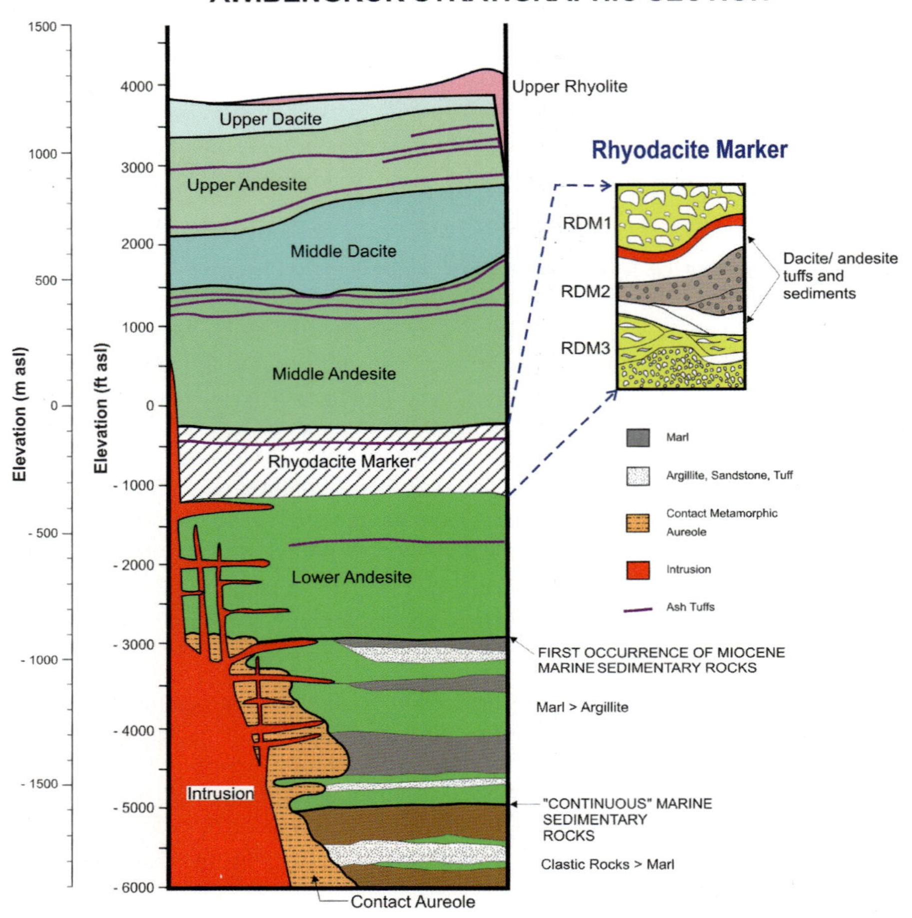

Fig. 2.5 Cross section showing an igneous intrusion at Awibengkok, Indonesia, from Stimac et al. [2008] (*figure provided by J. Stimac and reproduced by permission of Elsevier*)

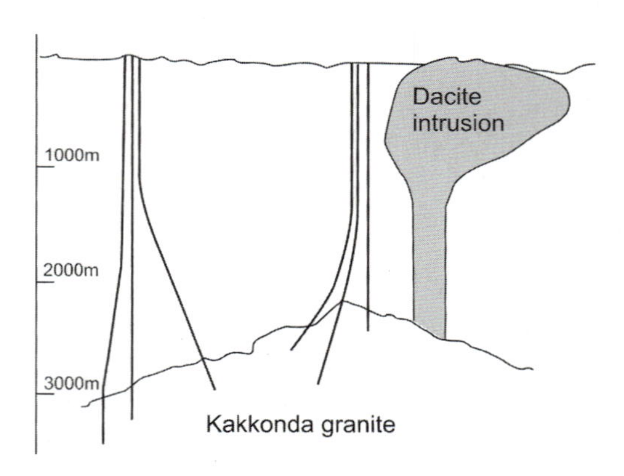

Fig. 2.6 Sketch of the Kakkonda granite body showing the intersection by wells. Based on a cross section from Doi et al. [1998] (*original cross section provided by N. Doi and reproduced by permission of Elsevier*)

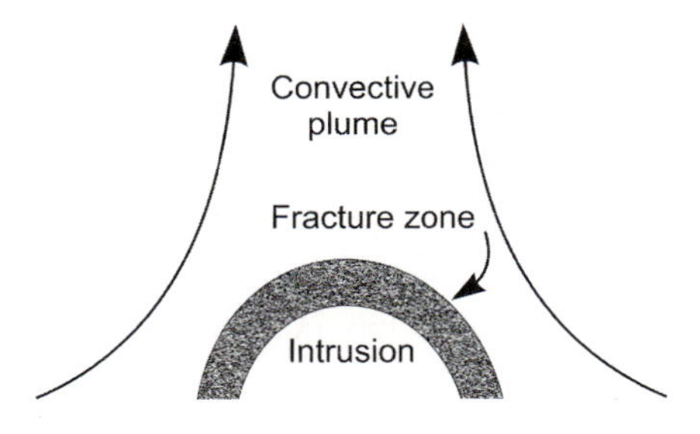

Fig. 2.7 Sketch of a convective plume over a magma body

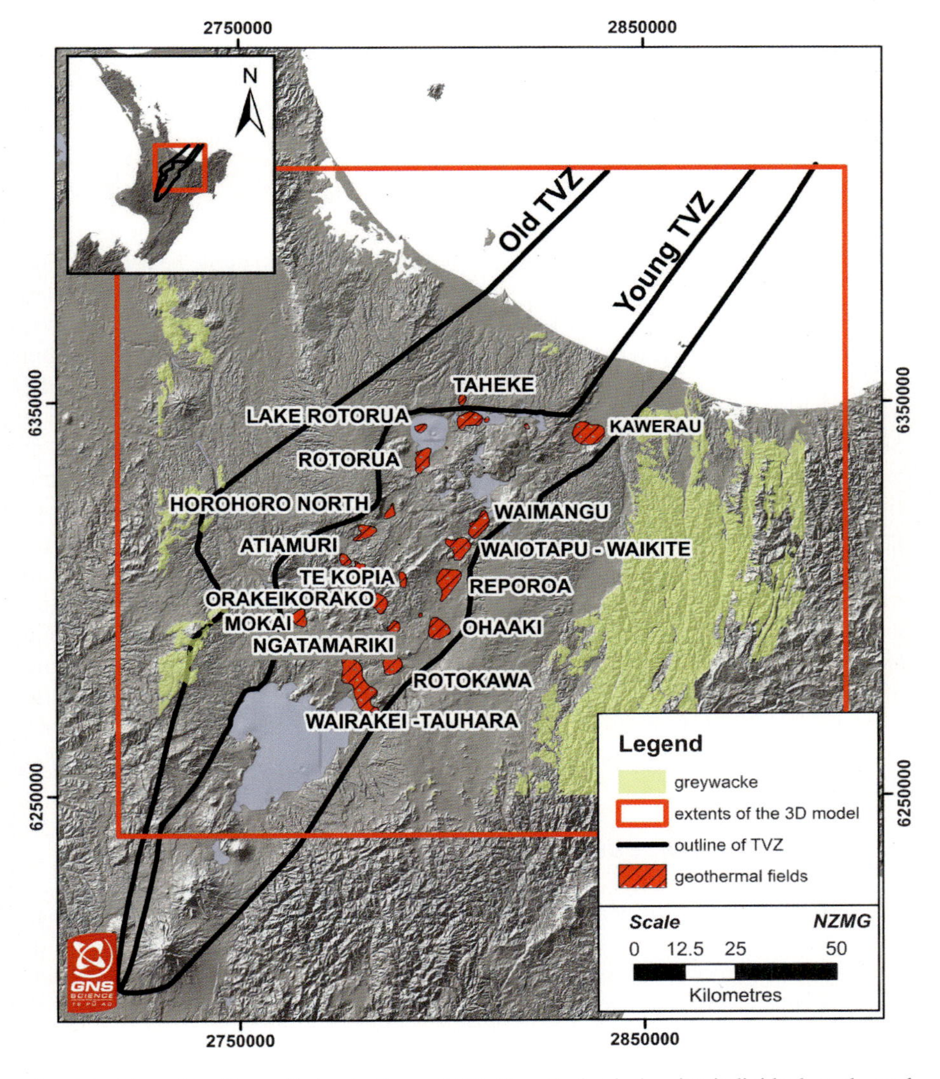

Fig. 2.8 Map of the Taupo Volcanic Zone (TVZ), New Zealand, showing individual geothermal resources as identified by their resistivity boundaries, from Alcaraz et al. [2012] (*figure provided by S. Alcaraz and reproduced by permission of GNS*; *all rights reserved*)

the question appears unresolved at present. McNabb [1992] proposed that the base of the TVZ was in effect a "hot plate", a thin crust heated directly from the mantle beneath. He drew analogy with Benard cellular convection, a laboratory arrangement of a horizontal flat plate held at constant temperature and forming the base of a vessel containing a shallow layer of viscous liquid, which exhibits a pattern of hexagonal natural convection cells known as Benard cells. In fact, Benard cells form in a layer of fluid with a free surface and involve surface tension, but cellular convection in confined layers can occur—see Zarrouk et al. [1999]. Hochstein and Regenauer-Lieb [1989] questioned whether the TVZ is a back-arc rift basin, but Cole et al. [1995] describe it as such, giving the average heat flux in the area as 800 mW/m^2. They note that volcanism in the TVZ has taken place over the last 2 Ma. Dempsey et al. [2012] have produced a model illustrating how a repetitive process of sealing of the upper levels of the convective plume and then faulting could be responsible for the maintenance of the separate geothermal resources observed. The debate continues.

Turning to heat sources in the form of granites within the lithosphere, the decay of radioactive isotopes provides a heat source, particularly thorium and uranium but mainly the latter according to Forster and Forster [2000] who quote heat generation rates of 4–10 $\mu W/m^3$ at maximum. The heat generated within a solid body results in a temperature increase at the location farthest away from the boundary, thus establishing a temperature gradient down which heat conducts. The maximum temperature reached depends on the boundary condition at the surface, the dimensions of the body and its thermal conductivity, as explained later in Sect. 4.5. Enhanced geothermal systems (EGS) are comprised of large bodies of unfractured rock with internal heat generation, the centre temperature of which has risen to well above atmospheric surface temperature. The thermal conductivity of granite is approximately 2 W/mK, and with an outside temperature of 25 °C, a cylindrical body would have a radius of 10 km if the centre temperature was to reach 100 °C by means of the heat generation rate quoted above, neglecting axial variations. The technology proposed to extract the heat is to fracture the rock and circulate water, or as examined by Stacey et al. [2010], carbon dioxide. The capture of carbon dioxide from coal fired power stations is being developed, and trials to store it in stable formations are in progress, so synergy between the two technologies is being examined.

2.3 Geothermal Surface Discharges

The natural discharge of heat at the surface has no doubt always attracted human attention. The discharge produces effects that are rare over most of the globe, ranging from geysers and fumaroles to warm, coloured ground and warm streams growing unusual vegetation. The rising plume of geothermal fluid illustrated by Fig. 2.7 is directed horizontally by the presence of the ground surface, which also adds meteoric water and provides cooling, leading to the wide range of effects mentioned.

Surface geothermal activity is of direct interest to earth scientists, since it provides clues to understanding the nature of the resource and the physical and chemical processes taking place—see, for example, the extensive review and categorisation of the surface discharges from resources with volcanic heat sources (i.e. of the type described in this chapter) by Hochstein and Browne [2000]. Studying it represents a significant part of the role of earth scientists in geothermal development. In contrast, surface activity provides almost no assistance to the problems confronting geothermal engineers. It may create civil engineering problems, if, for example, locations with heat and water or steam discharge have to support buildings or be crossed by pipelines, but these are relatively minor engineering issues. Shallow fluid flow is not amenable to the usual types of engineering analysis, analytic or numerical, primarily because of the random nature of the material through which the heat and fluids flow and the wide variability of material properties that control the resistance to flow. Geysers, a very distinct and infrequent occurrence, are perhaps the sole exception. They arise from flow in natural subsurface tunnels and cavities, as opposed to permeable rock, and exhibit a type of fluid flow that sometimes occurs in wells. They have attracted the attention of engineers and physicists, both experimental and analytical; see, for example, the review by Rinehart [1980] and the work of Lu et al. [2006]. To play their proper role in multidisciplinary activity of resource exploration and development, geothermal engineers must understand surface geothermal activity and its origins, but that information does not form part of the study of geothermal engineering per se.

There is, however, an aspect of the investigation of surface discharges which calls for the direct involvement of engineers, whose duty might be regarded as helping to meet the national demand for electricity as well as helping to develop any particular resource. In a list of the potential effects of geothermal resource use on the natural environment, the effect on geothermal surface features figures prominently. The surface discharges that attract and inform earth scientists of the presence of a usable resource also often have value as tourist attractions and are likely to rank highly as features to be preserved on aesthetic grounds. Despite the conflict between resource development and the preservation of geothermal surface features having first arisen, at least in New Zealand, more than 70 years ago, very little scientific effort has been applied to the problem. Sufficient is understood about the natural discharges to recognise and explain the effects of development by examining changes over a decade or more, but only after capital has been invested to enable the resource use, which is too late, as examples in Chap. 14 will demonstrate. What is required is research directed to achieving the earliest possible identification of effects, which may allow methods of prediction or testing to be developed; a start in this direction was made by Leaver et al. [2000].

References

Alcaraz SA, Rattenbury MS, Soengkono S, Bignall G, Lane R (2012) A 3-D multi-disciplinary interpretation of the basement of the Taupo Volcanic Zone, New Zealand. In: Proceedings, 37th workshop on geothermal reservoir engineering, Stanford, CA
British Geological Survey. http://www.bgs.ac.uk
Christenson BW, Mroczek EK, Wood CP, Arehart GB (1997) Magma-ambient production environments: PTX constraints for paleo-fluids associated with the Ngatamariki diorite intrusion. In: Proceedings of the NZ geothermal workshop, University of Auckland, Auckland
Cole JW, Darby DJ, Stern TA (1995) Taupo Volcanic Zone and Central Volcanic Region: back arc structures of North Island, New Zealand. In: Taylor B (ed) Back arc basins: tectonics and magmatism. Plenum, New York
Dempsey D, Rowland J, Archer R (2012) Modeling geothermal flow and silica deposition along an active fault. In: Proceedings thirty-seventh workshop on geothermal reservoir engineering, Stanford University, Stanford, CA, January 30–February 1, 2012
Doi N, Kato O, Ikeuchi K, Omatsu R, Miyazaki S, Akaku K, Uchida T (1998) Genesis of the plutonic-hydrothermal system around quaternary granite in the Kakkonda geothermal system. Geothermics 27(5–6):663–690
Forster A, Forster HJ (2000) Crustal composition and mantle heat flow: implications from surface heat flow and radiogenic heat production in the Variscan Erzgebirge (Germany). J Geophys Res 105:27917–27938
Gerya TV, Yuen DA (2003) Rayleigh-Taylor instabilities from hydration and melting propel 'cold plumes' at subduction zones. Earth Planet Sci Lett 212:47–62
Hochstein MP, Browne PRL (2000) Surface manifestations of geothermal systems with volcanic heat sources. In: Sigurdsson H (ed) Encyclopedia of volcanoes. Academic, San Diego, CA
Hochstein MP, Regenauer-Lieb K (1989) Heat transfer in the Taupo Volcanic Zone (NZ): role of volcanism and heating by plastic deformation. In: Proceedings of 11th New Zealand geothermal workshop
Hochstein MP, Regenauer-Lieb K (1998) Heat generation associated with collision of two plates: the Himalayan geothermal belt. J Volcanol Geoth Res 83:75–92
Hulen JB, Nielson DL (1996) The Geysers felsite. Geoth Res Countc Trans 20:295–306
Institute of Geological and Nuclear Sciences, New Zeal. http://www.gns.cri.govt.nz
Leaver JD, Watson A, Timpany G, Ding J (2000) An examination of Signal Processing Methods for monitoring undisturbed geothermal resources. 25th Stanford University geothermal reservoir engineering workshop, January 2000, Stanford, CA
Lu X, Watson A, Gorin AV, Deans J (2006) Experimental investigation and numerical modeling of transient two-phase flow in a geysering geothermal well. Geothermics 35:409–427
Manning CE (2004) The chemistry of subduction-zone fluids. Earth Planet Sci Lett 223:1–6
McNabb A (1992) The Taupo-Rotorua hot plate. In: Proceedings of the 14th New Zealand geothermal workshop, University of Auckland, Auckland
Norton D, Knight J (1977) Transport phenomena in hydrothermal systems – cooling plutons. Am J Sci 277:937–981
Ohtani E (2009) Melting relations and the equation of state of magmas at high pressure: application to geodynamics. Chem Geol 265:279–288
Reyes AG (1990) Petrology of Philippines geothermal systems and he application of alteration mineralogy to their assessment. J Volcanol Geoth Res 43:279–309
Rinehart JS (1980) Geysers and geothermal energy. Springer, New York
Schellart WP, Rawlinson N (2010) Convergent plate margin dynamics: new perspectives from structural geology, geophysics and geodynamic modeling. Tectonophysics 483:4–19
Shemenda AI (1994) Subduction: insights from physical modeling. Kluwer, New York
Stacey R, Pistone S, Horne R (2010) CO2 as an EGS working fluid- the effects of dynamic dissolution on CO2-water multiphase flow. Geoth Res Council Trans 34:443–450

Stimac J, Nordquist G, Suminar A, Sirad-Azwar L (2008) An overview of the Awibengkok geothermal system, Indonesia. Geothermics 37(3):300–331
US Geological Survey. http://www.usgs.cgov
Whitten DGA, Brookes JRV (1972) The penguin dictionary of geology. Penguin, London
Zarrouk SF, Watson A, Richards PJ (1999) The use of computational fluid dynamics (CFD) in the study of transport phenomena in porous media. In: Proceedings 21st geothermal workshop, The University of Auckland, Auckland, NZ

3 Thermodynamics Background and the Properties of Water

This chapter has the aim of explaining the thermodynamic principles and calculation processes used in the remainder of the book. Parameter definitions and the first law of thermodynamics are introduced, followed by the Steady Flow Energy Equation, from which the definition of specific enthalpy emerges. Entropy and absolute temperature are introduced next, based on an explanation of Carnot's work. The thermodynamic and transport properties of water are described, focusing mainly on properties along the saturation line, in particular, formulations produced by the International Association for the Properties of Water and Steam (IAPWS). The chapter ends by examining hydrostatic pressure and the boiling point for depth curve.

3.1 Definitions and the First Law of Thermodynamics

Engineering thermodynamics tends to focus on the conversion of heat to work, a process which was a mystery when heat was regarded as some sort of fluid contained in limited amounts within a substance. Rumford found that boring bronze cannon with a blunt drill appeared to release an unlimited amount of the "fluid", and his observations were a key step in understanding that heat and mechanical work are equivalent. Today the idea that energy comes in a variety of interchangeable forms is commonplace; however, thermodynamic reasoning depends on the use of precisely defined terms which must first be introduced.

The changes that take place in a substance when it exchanges heat and work with its surroundings are the focus of thermodynamics. Some authors (e.g. Pippard [1961]) distinguish between classical and phenomenological thermodynamics, the latter drawing on physical observations such as the molecular properties of a substance in any reasoning, whereas the former depends on the bare thermodynamic laws and postulates only, without reference to the wider physics. No such distinction is required here, and simple molecular models will be appealed to sometimes.

A. Watson, *Geothermal Engineering: Fundamentals and Applications*,
DOI 10.1007/978-1-4614-8569-8_3,

Neither heat nor work has any material form, but both are quantifiable in terms of the effect they have on materials. The logical way to approach any quantitative analysis of the effects of exchanges of heat and mass with some amount of a material is to first define the latter; traditionally, this is referred to as "the system". The system is chosen to suit the analysis to be carried out. All of the analyses in thermodynamics and thermo-fluids are based on conservation rules. If the focus of attention is what is happening within the cylinder of a steam engine, as it usually was in the early days of engineering thermodynamic development, then the system is defined as the mass of fluid contained by the cylinder and its piston. For the study of fluid inclusions in crystals, the system might be the inclusion itself. If heat and/or work cross the boundary in either direction, a change in the system takes place, recognisable by changes in certain physical parameters.

Historically, the focus was on a closed system containing a given mass of substance (typically a gas such as air or steam) which did not cross the boundaries, although heat and work did. More precisely, the simplest system is a fixed mass of a single-component homogenous fluid which is not undergoing any chemical reaction, contained within a clearly defined boundary. A full gas cylinder with the valve closed matches this description, as does the cylinder of an internal combustion engine with the valves closed but with a moveable piston. In both cases the substance and the system boundaries are defined, and heat exchange to or from the substance must cross the system boundaries. Work exchange with the fluid in the cylinder is possible by means of the moving piston, work being defined as force times distance moved, but it is not possible for a closed gas cylinder to produce any work output by itself.

The state of the substance filling the system can be defined by two parameters, its pressure P and its temperature T. The energy content of the substance is defined as U, the specific internal energy in units of kJ/kg (specific means per unit mass); substances may have additional energy in other forms, but for a fixed mass contained within simple boundaries and with no chemical reaction taking place and no movement (kinetic energy) or gravitation (potential energy), U defines the energy content. If the system has only recently exchanged heat or work, there may be internal temperature gradients and internal motion. When these have died away and the system can be said to be in equilibrium, P and T are sufficient to define U for the given substance. It follows that if U and P are known for a particular case, then T is at least fixed, even if its value has not been measured. For any given case, the state of the system is known if two of these three parameters are known, and it should now be clear that "state" refers to the energy level of the system. There are parameters other than U which together with P or T define the state of the system. They are related to U and are referred to in engineering thermodynamics as "functions of state".

To make a further step towards analysis, temperature and pressure must be defined. Fahrenheit invented the mercury-in-glass thermometer and his temperature scale in the early eighteenth century. Celsius introduced his scale in the same period, his scale being sometimes called "Celsius", but sometimes "Centigrade" because he divided it into 100 units. Both inventors used different scales and datum points, which are still well known. Mercury-in-glass thermometers, thermal

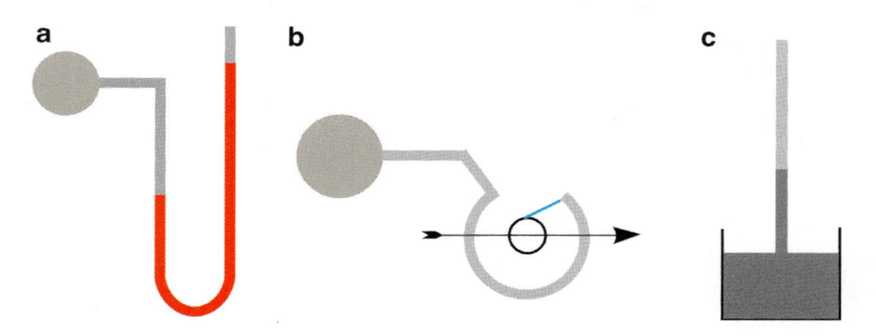

Fig. 3.1 (**a**) A simple manometer and (**b**) a Bourdon pressure gauge, both measuring the pressure in a pipe, and (**c**) a barometer

expansion in solids, change of electrical resistance with temperature and the generation of a voltage between dissimilar metals have all formed the basis of temperature measuring instruments. A loss or gain of heat is usually associated with a temperature change, although an exception discussed later occurs when phase change takes place.

Pressure is defined as the force per unit area exerted by a substance on the walls of its container. If the substance is a solid it requires no container. If its shape is such that one part of its boundary surface is flat so that the solid can stand on a flat plane, then gravity exerts a force that gives the substance weight, and the weight exerts a pressure on the flat plane. Very often the substances requiring thermodynamic analysis are fluids. A fluid may be defined as a substance with no shape of its own—it takes up the shape of its container. No further adjustments to this definition are needed if the fluid is a gas and the system has closed boundaries, such as a gas cylinder. The molecules of the substance fill the container, they have kinetic energy and exert a pressure on the container walls by colliding with it, and the pressure has the units of force/unit area (kgm/s^2 per m^2 or Pa). The classic engineering thermodynamic study is of a gas filling a cylindrical vessel with one end closed and the other formed by a piston capable of axial movement. The pressure of the gas, P, exerts a force F on the piston, which presents an area A to the gas. The magnitude of the force is $F = P.A$ in units of kgm/s^2 which are equivalent to Newtons (N). If the piston is allowed to move an axial distance dx, then the work done by the gas on the piston is $P.A.dx$, or $P.dV$, where dV (m^3) is the increment of volume over which the piston moves. The movement of the piston must be at right angles to the force for this expression to hold. In general the pressure exerted by a gas on its container is normal to the surface, so a similar approach to calculating the exchange of work involved would be applicable to the expansion of a balloon with a flexible envelope and many other situations.

If the fluid is a liquid, its definition differs from that of a gas, because a liquid will always fall to the lowest level in a container which is in a gravitational field. The surface of the liquid at rest will be horizontal. The manometer is a means of pressure measurement which makes use of this phenomenon. The manometer, shown in Fig. 3.1a, automatically measures the pressure relative to atmospheric pressure. The same applies to the Bourdon tube, Fig. 3.1b, which remains a popular instrument and works because the exact shape of a curved tube with a difference in pressure between inside and outside depends on the magnitude of the pressure difference. When the pressure difference changes, the end of the tube changes position; it is attached to a drum by a fine wire, and the drum and pointer rotate accordingly—the drum is spring loaded. It is because the Bourdon tube measures the pressure relative to atmosphere that the term "gauge pressure" was introduced; atmospheric pressure must be added to gauge pressure to obtain absolute pressure. Absolute pressure is used throughout this book and is recommended always. Quartz crystal piezometers are available for electronic recording of pressure; compression of the crystal generates a voltage. Atmospheric pressure is measured using some form of barometer, of which that shown in Fig. 3.1c is perhaps the simplest; it is used on a large scale in creating vacuum in some types of power station condenser (see Fig. 11.16).

Heat is a form of energy and if two systems at different temperatures are placed in contact, heat will flow from the one with the highest temperature into the other. Energy flow is recognisable by the existence of a temperature gradient. This being so, then temperature gradients must occur within a system sometimes rather than just across contacting surfaces—the heat arriving from another system in contact with it establishes a higher temperature in the region where it arrives, which enables its transfer to the rest of the system. In solids, it can be imagined that the molecules vibrate but stay in fixed positions, whereas in fluids they are free to move about. In solids the amplitude of the vibration can be imagined to be directly proportional to temperature, so the condition of thermal equilibrium, at which temperature is everywhere the same, is one of mechanical equilibrium in which all the molecules vibrate to an identical extent. Heat is transferred in solids only by this process, which is known as thermal conduction. Heat transfer by thermal conduction also takes place in fluids but usually accompanied by other transfer processes. Because the molecules of a fluid are free to move, a collection of molecules at a high temperature can migrate as a whole to another part of the system where the fluid is colder, where they share their heat with their new neighbours. This is the process that takes place when fluid is stirred, a highly effective way of producing homogenous and isothermal conditions. Thermal conduction, defined as heat transfer without movement, still takes place in moving fluids but it is no longer the dominant mode of heat transfer. For it to be the only mode of heat transfer in a fluid filled system, the system must be specially designed to prevent fluid movement, as in the case of domestic double glazing, for example. Heat transfer in fluids in the general case where movement takes place is called thermal convection. Earth scientists sometimes use the term "advection" as a general description of transfer of heat and/or mass, but the term is not used by engineers—it has no clear, generally accepted meaning.

The transfer of heat across the boundary of a system containing a single-phase substance is detectable by the measurement of temperature difference; the temperature of a block of metal rises if it is exposed to a flame. The increase in energy level of the system can be quantified by the rise in temperature once equilibrium has been achieved, but not the absolute energy level if the only temperature scales available are Fahrenheit or Celsius, which are arbitrary scales; the "absolute scale of temperature" resolves that issue, Sect. 3.3 below.

Based on the above definitions and general discussion, it is possible to summarise as follows. The state of a substance is in general terms a measure of its energy content and is quantifiable as $U(P,T)$ and by other functions of state yet to be introduced. Energy is interchangeable between its different forms and is interchangeable with work; it is conserved, that is, it cannot simply disappear or be "consumed"—the equivalence of mass and energy via nuclear reactions is not considered here, and mass also is regarded as being conserved. All of these statements are combined as an energy conservation equation:

$$dQ - dW = dU \tag{3.1}$$

which is known as the first law of thermodynamics. In this equation only U is a function of state, and the equation states that the net effect of dQ and dW, which are merely additions or subtractions of energy, is to change the state of the substance in the system. Some people replace dQ and dW in the equation by δQ and δW, respectively, as a reminder that Q and W are not functions of state, because many thermodynamic calculations rely on a set of differential relationships between the functions of state (Maxwell's equations—see e.g. Perrot [1998]). The absolute level of specific internal energy does not appear in the equation, only the change in level, dU. The algebraic signs of the heat and work terms are traditional for engineering use because the first major application of thermodynamics was to "heat engines", prime movers which converted the heat of combustion into work. Heat was always added to the system and work always left it (at least, that was the intention!). Thus, heat added to the system is positive and work done which leaves the system (i.e. work done by the system) is also positive.

Finally, it is essential to note that all of the above explanations have been built on the idea of a closed system—the cylinder with a piston in which a fixed mass of substance was locked up. The mathematics for this circumstance are the simplest, but there is no objection to allowing mass to enter and leave the system—the rules of continuity of mass simply have to be adhered to within the new boundary conditions. This is a more general case and vital for the present purposes, to analyse the natural and induced flows in resources and the flows in surface equipment and power stations. Calculation processes must be set up to deal with a continuous flow of geothermal fluid to generate power, which is defined as a rate of production of work.

3.2 The Steady Flow Energy Equation and the Definition of Specific Enthalpy

Many of the major engineered flows needing thermodynamic analysis are steady-state flows, invariant in time, and the Steady Flow Energy Equation (SFEE) is a common topic to be found in many thermodynamics texts. A power station

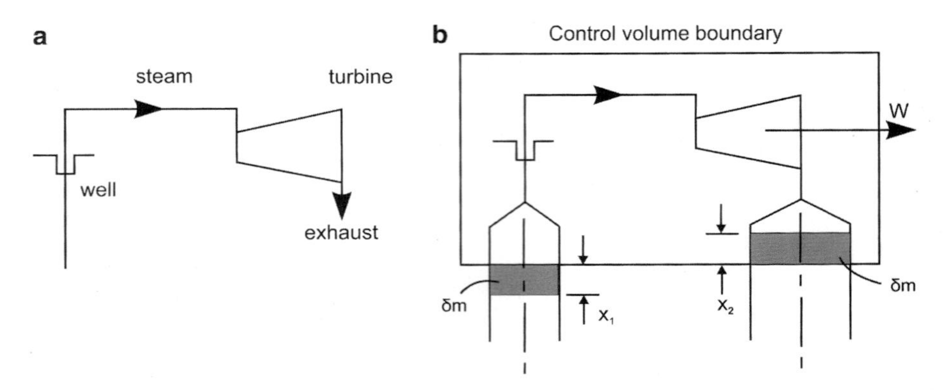

Fig. 3.2 (**a**) Diagram of a system comprising a well and turbine and (**b**) its representation as a control volume with work and increments of flow in the well and turbine exhaust (*shaded*) crossing the boundary

operating at a steady output is an example. A more general form of the energy equation incorporating transients is necessary to analyse unsteady conditions and also sometimes appears in reservoir engineering analyses. The SFEE is a restricted form of the general time-dependent equation, so the logical order would be to produce the general form first and then reduce it to the SFEE. However, the SFEE provides an easy introduction to specific enthalpy and the explanation of throttling (adiabatic pressure reduction) so will be examined first. Fluid density now needs to be introduced. It is defined as ρ (kg/m^3), the mass per unit volume of the substance, but it appears just as often in its inverse form, specific volume V (m^3/kg), where $V = 1/\rho$. Specific volume is a function of state, that is, it is determined if two of P,T, U or any of the other functions of state yet to be introduced are known.

3.2.1 The Steady Flow Energy Equation

The diagrammatic arrangement of a well supplying steam to a turbine is shown in Fig. 3.2a, and it is shown at a more suitable level of detail for the present purpose in Fig. 3.2b. What is to be developed is an equation, equivalent to Eq. (3.1), for a system with mass crossing its boundaries. The system boundary is defined, but it is a schematic boundary rather than an actual one. It is best referred to as a control volume, the space within which the relevant rules are to be applied.

The well in Fig. 3.2a discharges continuously and supplies a constant mass flow rate of steam to the turbine, which produces a steady work output and a steady flow of exhaust steam. Fluids are moving, but steadily, so there is no change of any variable with time at any point within the control volume. The general approach used in deriving the equation has been used by many authors, according to Rogers and Mayhew [1967]; a small mass of fluid, shown shaded in Fig. 3.2b, is followed as it enters the control volume from the well and a matching mass leaves in the turbine exhaust pipe. The small mass, δm, enters the control volume from the well.

It has specific internal energy U_I and energy in the form of kinetic energy, $\frac{1}{2} \rho_I u_I^2$, and potential energy gz_I relative to some datum elevation $z = 0$ which needs no further definition at present. It is pushed into the control volume in a time δt under a pressure difference, while an equal mass exits the control volume at the other end, with a different set of energy components, U_2, $\frac{1}{2} \rho_2 u_2^2$ and gz_2. The exact detail of the entering and leaving masses must be examined because work is involved. Suppose the control volume ends in the production casing of the well so that it exactly includes the shaded volume. The casing has an area of cross section A_I. The mass δm_I occupies a length of casing x_I and its volume $A_I x_I$ may also be written as $\delta m_I / \rho_I$. The work done on the control volume, which is negative according to the sign convention, is

$$\delta W_1 = P_1.A_1.x_1 = \frac{P_1}{\rho_1}.\delta m \tag{3.2}$$

At the exit, the same form of expression for δW is obtained:

$$\delta W_2 = P_2.A_2.x_2 = \frac{P_2}{\rho_2}.\delta m \tag{3.3}$$

but the work is now done by the system, pushing the mass δm out so it is positive. Work is also done by the system on the turbine, which appears as W and is also positive. The first law, Eq. (3.1), is now considered and the work and energy terms are introduced. The energy within the control volume as a whole, excluding the δm increments, is referred to simply as E. It is an important part of the concept that although the fluid is moving through the system and changing as it goes, at any point within the control volume all the variables describing the system are constant with time, so E does not change with time.

Thus, the form of Eq. (3.1):

$$dQ - dW = dU$$

can be adopted with the new terms so that

$$\left(dW + \delta m\left\{\frac{P_2}{\rho_2} - \frac{P_1}{\rho_1}\right\}\right) = \left(\delta m\left\{U_2 + \frac{u_2^2}{2} + gz_2\right\} - \delta m\left\{U_1 + \frac{u_1^2}{2} + gz_1\right\}\right) \tag{3.4}$$

The energy within the control volume, E, does not appear in the equation since it is unaffected—by definition the variables at any point in the control volume are invariant with time. No heat exchange has been built into the arrangement being represented (Fig. 3.2a), so there is no term equivalent to dQ. The signs of the heat and work exchange defined for Eq. (3.1) have been applied. There is continuous power output from the turbine, which draws attention to the fact that the units of power are J/s (Watts) and of work are J. The units of the equation are Joules.

Table 3.1 Examples of the properties of saturated water and steam at various temperatures

T_s (°C)	P_sV_f (kJ/kg)	h_f (kJ/kg)	P_sV_g (kJ/kg)	h_g (kJ/kg)
25.0	3.18	104.8	137.2	2546.5
175.0	1.0	741.2	193.3	2772.7
350.0	28.76	1670.9	145.48	2563.6

Reintroducing dQ to allow for heat exchange produces a more general equation, and replacing $1/\rho$ by the specific volume V and reorganising the terms result in

$$\frac{dQ}{\delta m}-\frac{dW}{\delta m}=\left(U_2+P_2V_2+\frac{u_2^2}{2}+gz_2\right)-\left(U_1+P_1V_1+\frac{u_1^2}{2}+gz_1\right) \tag{3.5}$$

The terms on the right-hand side are all specific (per kg), and those on the left are also per kg. The rate of work output is per kg of mass flowing through the system, so the introduction of a mass flow rate in kg/s will enable the rate of power output and heat loss or gain to be calculated. Although the PV term has the same units as PdV, the latter is the work dW done by a force acting on an area and moving a distance, as a piston in a cylinder, for example. The PV term in Eq. (3.5) relates to the energy within a volume of fluid as a result of it being at a pressure P. A change in PV might produce work, given suitable arrangements, but not necessarily, so the term must not be thought of as a work term.

3.2.2 Specific Enthalpy

Because U, P, and V are all functions of state, then so is the combined term $U + PV$, which is known generally as enthalpy, and in this equation is specific enthalpy (kJ/kg). It is customary to write the Steady Flow Energy Equation as

$$Q-W=(h_2-h_1)+\frac{1}{2}\left(u_2^2-u_1^2\right)+g(z_2-z_1) \tag{3.6}$$

where Q and W are now heat and work transfers per unit of mass flow (J/kg).

It is useful to know the order of magnitude of the terms in this equation and in the definition of h.

The saturation conditions for water are given in Table 3.1 for three values of T_s to cover the range of interest here, together with h and its components.

The first thing to notice in Table 3.1 is that the entries do not all increase with temperature. There is a maximum in the specific enthalpy of saturated steam at approximately 235 °C which has been speculated to be the cause of some phenomena observed in geothermal resources but which is not important for this discussion of the relative magnitude of PV as a fraction of h. For saturated water the P_sV_f term is 3 % or less of the specific enthalpy of saturated water—in other words the specific internal energy accounts for 97 % of the specific enthalpy. For saturated steam the P_sV_g term is 7 % or less of the specific enthalpy of saturated steam—the specific internal energy accounts for 93 % of the specific enthalpy.

3.2.3 The Kinetic and Potential Energy Terms in the SFEE

The order of magnitude of the terms in the SFEE should be examined for relative importance for each application. Taking the potential energy terms first, most engineering plant has a height of a few tens of metres. Taking 50 m as an example, the g_z term from top to bottom of the plant is 490.5 J/kg. Although the energy terms in Table 3.1 are in units of kJ/kg, Eq. (3.6) is in J/kg. Over 50 m then, the potential energy change is a mere 0.5 % of the smallest value of h_f in the table. However, geothermal wells are in a different league. For a well producing at a temperature of 350 °C from a depth of 4,000 m, the specific enthalpy at the wellhead would be 39 kJ/kg less than at the production zone. This is no doubt a small loss of energy in power station terms, but it is large enough to be accounted for.

The design velocity in water pipelines is usually of the order of 3 m/s, at which the kinetic energy term $u^2/2$ has a value of 0.0045 kJ/kg—a negligible energy consideration. The exit velocity of steam from a condensing steam turbine however can equate to a kinetic energy of 20 kJ/kg, and in large turbines the loss is enough to attract attention to the design to the exhaust ducting. The sonic velocity in steam, of order 425 m/s, provides a value of $u^2/2 = 91$ kJ/kg, or 3.6 % of the lowest tabulated h_g in Table 3.1.

3.2.4 Adiabatic Pressure Drop or Throttling

Sometimes steam is supplied through a pipe at a higher pressure than required—without a control valve, the flow through whatever is downstream would be too great. The control valve provides a local flow restriction. No work is done by the flow (there is no moving part to exert a force on anything and allow work to be performed) and the surface area of the valve is so small that the heat loss is negligible, even if the valve is uninsulated. There is no change of elevation. The SFEE, Eq. (3.6), thus reduces to

$$0 = (h_2 - h_1) + \frac{1}{2}\left(u_2^2 - u_1^2\right) \tag{3.7}$$

or if the kinetic energy terms are neglected,

$$h_2 = h_1 \tag{3.8}$$

The flow is said to be throttled, so throttling is thus a process in which there is little or no loss of energy. What is lost, however, is the opportunity to convert heat to work, so throttling of a flow from which electricity is to be produced must be avoided if possible. This is the next issue to be addressed in this chapter.

As a closing remark, the examination of the order of magnitude of the terms in the SFEE provides a clear message to those attempting to avoid global warming by using only mechanical energy such as wind power. The message is "thermal energy rules", which is why so much fossil fuel has been burned in the last 300 years.

3.3 Entropy and Absolute Temperature

There have been many different presentations of Carnot's work of 1824, e.g. Dugdale [1966] and Spielberg and Anderson [1987]. His ideas and the Carnot cycle remain difficult topics, but they must be understood if the concepts of entropy and absolute temperature are to be introduced.

Carnot set out to analyse heat engines, a new device at the time. To begin with it is sufficient to encapsulate his ideas in the comparison of heat flow with the flow of water downhill. The water flow can be interrupted with a machine and energy in the form of mechanical work can be extracted—it is not a necessity but an opportunity. The water continues to flow downhill after leaving the machine and would do so whether the machine was there or not. Likewise, heat flows from a high temperature to a lower one; it can be intercepted with a heat engine, a machine designed to convert heat into mechanical work, but it will flow regardless of whether the opportunity to convert some of it to work is taken or not. The classic diagram is shown as Fig. 3.3, in which heat flows from a source at temperature Θ_1 to a sink at temperature Θ_2 and is intercepted by a heat engine.

The new choice of symbol Θ for temperature is deliberate—T is reserved for °C, but Carnot's ideas require a less arbitrary definition of temperature which appears later in the discussion. Carnot thought of the heat engine from a practical point of view to the extent that he imagined a frictionless cylinder-and-piston reciprocating machine, containing a fixed mass of fluid. Any reduction in temperature of the heat, which can only flow from the higher temperature source to the lower-temperature

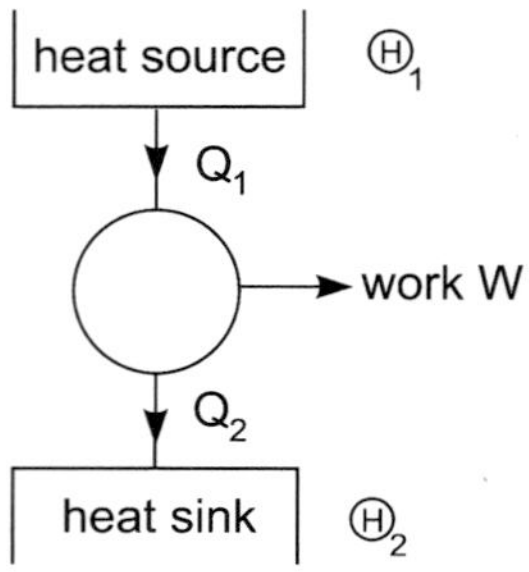

Fig. 3.3 Representation of a heat source supplying an engine which does work and rejects heat to a heat sink

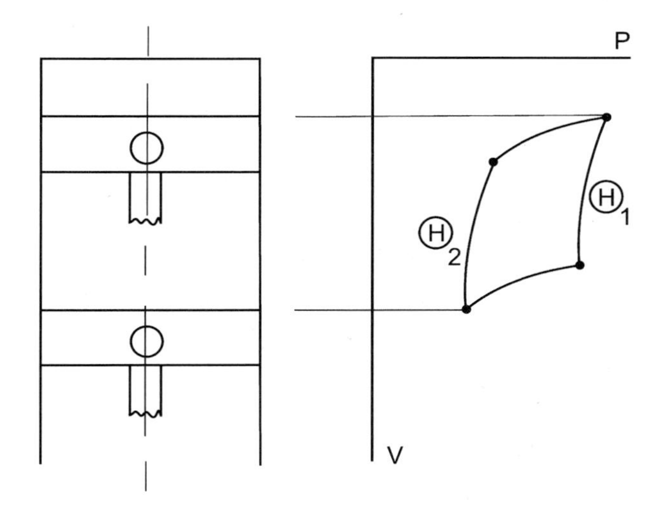

Fig. 3.4 A reciprocating engine following Carnot's cycle

source, he considered would be wasted unless it produced work as it went. Figure 3.4 shows the piston and cylinder, the axis of which lies parallel to the volume axis on the graph of volume against pressure of the working fluid mass, which he considered to be air—a helpful choice because anyone who has blown up a bicycle tyre knows that its temperature rises when it is compressed and can therefore be expected to fall if it expands.

The piston starts at the bottom of its stroke and as it rises the air is exposed to the low-temperature sink which keeps it at temperature Θ_2; how it is exposed is left to the imagination, perhaps the sink is portable and held up against the end of the cylinder. Heat must flow from the air to the sink as the air is being compressed. At some defined point, contact with the heat sink is removed, the cylinder is perfectly thermally insulated, and the air is compressed so that its temperature rises to Θ_1—it is not necessary to examine how the insulation can be applied or how the time to apply it could be decided upon. The piston has now reached the top of its stroke, at which time the air is exposed to the heat source at temperature Θ_1—the same temperature as the air. However, as the piston moves down and the air expands, the reduction in temperature that would normally take place is avoided by the presence of the heat source, which supplies heat to keep the air temperature at Θ_1. At a predetermined position the heat source is removed, the cylinder is insulated, the piston continues to move down, and the air expands, its temperature falling to Θ_2 exactly at the bottom of the stroke, the starting point.

The physics problems with this heat engine are obvious. All of the heat that flows from the source to the sink cannot possibly do so without passing through intermediate temperatures—heat flow without temperature reduction is not possible, heat only flows down a temperature gradient and must reach all the molecules of the working fluid. Yet the perfect engine requires that all heat flow must produce work. For the air to change temperature without transferring any heat to or from the piston and cylinder, the latter must have no thermal capacity, which is also impossible.

Finally, there is the practical problem of exchanging source, sink, and thermal insulation partway through the cycle. However, Carnot established the idea of a perfect heat engine. The net work produced is the integral of PdV around the cycle, which exactly equals the net heat flow. More heat is supplied from the source than is converted to work, so some is discharged to the sink. If this engine was reversed, supplying the amount of work produced would result in all of the heat being returned to the source; the engine is reversible—it does not just turn backwards, but reverses the changes in heat content of the source and sink exactly back to their starting values. Reversibility has come to mean the idealised exchange of heat or work in whatever situation. The work produced can thus be identified as $Q_I - Q_2$ in Fig. 3.3, and it is possible to define a thermal efficiency as

$$\eta_C = \frac{Q_1 - Q_2}{Q_1} = \frac{W}{Q_1} \tag{3.9}$$

where W is the work output. The thermal efficiency is the fraction or percentage of the heat supplied which is converted to work. The ability of the Carnot engine to convert heat to work depends on the temperatures of the source and sink, and it can be reasoned that the thermal efficiency can be expressed also as

$$\eta_C = \frac{\Theta_1 - \Theta_2}{\Theta_1} \tag{3.10}$$

The Fahrenheit and Celsius scales predated Carnot's work and were arbitrary, so bore no relation to the temperatures in this equation. Kelvin proposed a new temperature scale on the basis of Carnot's work. The units of the new "absolute" scale, named degrees Kelvin in his honour, were chosen to match the Celsius scale in the 0–100 °C range, which resulted in the absolute scale zero being a very low temperature of −273.15 °C. A temperature of 273.16 K became the triple point of water (the condition at which steam, ice, and liquid water can coexist). It follows from Eqs. (3.9) and (3.10) that

$$\frac{Q_1}{Q_2} = \frac{\Theta_1}{\Theta_2} \tag{3.11}$$

which can be rearranged as

$$\frac{Q_1}{\Theta_1} - \frac{Q_2}{\Theta_2} = 0 \tag{3.12}$$

This sets the scene for the introduction of the concept of entropy. An elegant set of proofs establishing that entropy is a function of state is given by Fermi [1936], [reprinted 1956], although there are many to choose from in thermodynamic texts. He first proves that in a cyclic change, as in Carnot's ideal engine, the integral around the cycle of all exchanges of heat is zero if they are carried out reversibly:

$$\oint \frac{dQ}{\Theta} = 0 \tag{3.13}$$

and more generally for all cycles, reversible or not,

$$\oint \frac{dQ}{\Theta} \leq 0 \tag{3.14}$$

He then observes that the quantity dQ/Θ varies around a cycle, and at two points labelled A and B (without needing to say where these are or how many stages the cycle has—Carnot's had only 4, as shown in Fig. 3.4), the integral

$$\int_A^B \frac{dQ}{\Theta}$$

has a value that is independent of how the fluid changed from A to B, provided that it did so reversibly. If the change was reversible, there would be no transients of temperature or motion, and the states would be equilibrium states. If the change took place in a series of stages similar to those in Carnot's engine, the order in which the additions or subtractions of heat and the performance of work does not affect the value of the integral. In this way he establishes that

$$ds = \frac{dQ}{\Theta} \tag{3.15}$$

where s is a function of state referred to as entropy. It is a specific property, with units of kJ/kgK. It has some important properties, most notably that for any change taking place in an isolated system, the final value of the entropy can never be less than the initial value—entropy remains constant in reversible changes but increases otherwise. This is what Eq. (3.12) shows. As a result of friction in a flow, heat is produced and the entropy increases. It is possible to decrease the entropy of a system only by performing work on it.

Carnot's principles are very relevant to geothermal engineering. High-grade heat can be released by burning fossil fuels. The question must be asked whether there is any benefit to be gained by using high-grade heat to produce power and materials to extract low-grade heat from geothermal resources, when only a small proportion of that low-grade heat can be converted to work (electricity). Low-grade heat is being produced in abundance by society, but it is distributed and often in very small amounts. Converting high-grade heat to power leaves some low-grade heat, and the lower the source temperature, the more heat is left in theory, and in practice even more heat is rejected. This issue is central to economic optimisation.

3.4 The Thermodynamic and Transport Properties of Water

3.4.1 Phase Change, Clapeyron's Equation and the Saturation Line

The important phase change here is that from liquid to gas. There is no fundamental distinction between gas and vapour. In the nineteenth century, it was observed that some gases could be condensed by compressing them only, whereas others had to be cooled as well, and this led to the two terms, vapour and gas. This difference in behaviour is due to the thermodynamic critical point parameters of the substance (P_c and T_c) relative to atmospheric conditions. The term vapour continues to be used by some authors to describe a gas when it is close to its saturation conditions. The flow of steam immediately downstream of the location of the boiling water producing it carries water droplets, due to heat loss and hence condensation. In this condition the steam may be referred to as "wet", and it is recognisable by its whiteness, which is not a property of the steam itself but a result of the water droplets internally reflecting light. Often a steam jet emerging from an orifice is totally transparent within a few centimetres of the exit, turning white downstream of that point. The transparent part is true steam, superheated, and free of the condensed water droplets in the white flow downstream; leaks of high-pressure steam in power stations are dangerous because they are invisible.

It is sometimes helpful to imagine liquid water with densely packed molecules ending at a surface above which the molecules are widely spaced and moving fast and randomly—they form a gas. The forces of intermolecular attraction at the surface are unbalanced, resulting in surface tension, which gradually decreases as P and T are increased towards the critical point. It can be imagined that the internal energy increases to saturation level at which the molecules disperse into their gaseous distribution, an idea useful in considering homogenous nucleation (see Chap. 7). Perrot [1998] explains that the saturation state is defined as the state at which the chemical potentials of liquid and gas are the same. Equilibrium points can be defined in terms of T and s, and the locus of points on the T–s diagram forms an envelope within (below) which the fluid is two phase and outside is either liquid or gas—Fig. 3.5.

There is no sharp distinction between liquid and gas except on or within the saturation envelope; put another way, fluid state can change from above the envelope on the left-hand side where it is liquid to above the envelope on the right-hand side where it is gas, without any surface between phases being observed. The envelope is the 1858 work of Clapeyron, who determined it as a P–V graph, since that was the age of reciprocating engines and reciprocating compressors for liquefying gases. He formulated an equation describing the saturation line as follows:

$$\left(\frac{dP}{dT}\right)_s = \frac{h_{fg}}{T_s.V_{fg}} \tag{3.16}$$

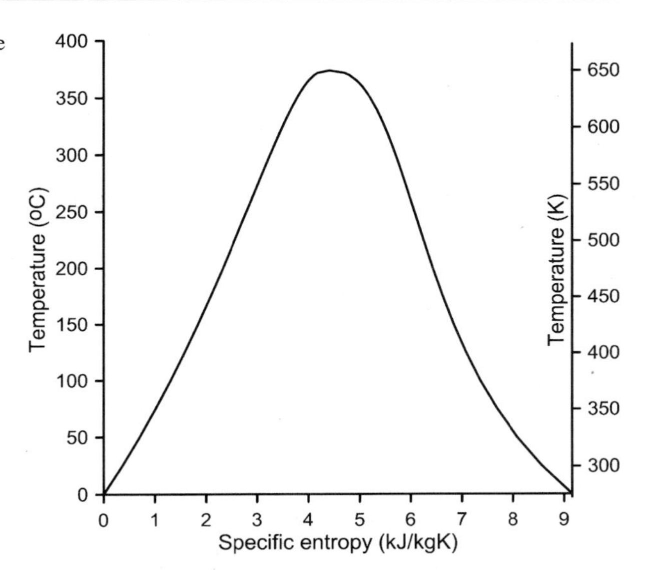

Fig. 3.5 Saturation envelope for water on *T–s* coordinates

It is not necessary to use the Clapeyron equation in this book, although it appears in connection with an equation in Chap. 7; it provided a means of defining the envelope and of actually measuring absolute temperature indirectly, by measuring fluid properties. The envelope can be plotted using steam tables. The essential point about the saturation line or condition for the present purposes is that P_s and T_s are a pair, so the three-parameter "rule" of functions of state reduces to two parameters at the saturation condition; it is only necessary to specify either P_s or T_s to fix all other functions of state at that value. In the form shown in Fig. 3.5, the saturation envelope is useful for plotting the thermodynamic changes taking place in a power station and comparing them with the ideal of reversible changes. The lowest pressure of practical interest here is condenser vacuum, so in Fig. 3.5 the temperature axis goes down only to the triple-point temperature.

3.4.2 Properties of Water

As the most common liquid on earth and the choice of working fluid from the earliest days of thermal power generation, it is at first sight surprising that the thermodynamic properties of water are still studied. The problem is that the precision with which the properties need to be known continues to increase, as well as the range of P and T over which they are needed, and their variation is complicated. For geothermal research the range extends from power station condenser pressures of 0.01 bars abs to ocean bed black smokers or deep subterranean aqueous solutions at thousands of bars abs and temperatures from 0 °C to well above critical (374 °C). For fossil-fuelled steam power engineering, very high precision is warranted because of the purity of the water used. The precision required is moderate

Table 3.2 A steamtable layout useful for geothermal engineering

P	T	h_f h_{fg} h_g	s_f s_{fg} s_g	ρ_f ρ_g	Cp_f Cp_g	λ_f λ_g	μ_f μ_g
Bars abs	°C	kJ/kg	kJ/kgK	kg/m^3	kJ/kgK	W/mK	kg/ms

for geothermal engineering because the fluids are not pure H_2O and the use of H_2O properties to describe geothermal water is already an approximation.

Traditionally, the thermodynamic properties required for power plant engineering calculations have been provided as "steam tables", in recognition that the variation of properties with P and T cannot be represented by simple formulae (their need predates hand calculators or computers other than the slide rule). Those steam tables focus on superheated steam. It is important that collaborating organisations use the same set of properties; otherwise a steam turbine manufacturer might contract to supply a turbine of a certain output when supplied with steam at specified rate and conditions, and the purchaser might wonder why his measurements suggested otherwise. Different sets of steam tables exist, with some significant differences in parameter values.

The International Association of the Properties of Water and Steam (IAPWS) [2012] updates the properties from time to time. Their publications are freely available for use. The needs for geothermal work differ from those for steam power engineering in general, because the main focus is on flashing and condensation processes, i.e. on saturation properties and rarely on superheat properties. A layout of property values which appears to be useful was produced by the author and Dr Sadiq Zarrouk for the University of Auckland Geothermal Institute based on the 1967 IFC formulation. The column headings and the units of the entries are shown in Table 3.2. The full table is in two parts, one with pressure in the first column, in regularly spaced increments, and the second with temperature in the first column.

In the specific enthalpy column, all three values are given: specific enthalpy of the saturated liquid, the difference between the specific enthalpies of saturated liquid and saturated steam and the specific enthalpy of saturated steam. The same applies to specific entropy. These help when calculating dryness fraction. Densities, specific heats, thermal conductivities, and viscosities are included although they are less often required.

A two-phase mixture can be defined in terms of its dryness fraction, X, which is the proportion by mass of the mixture which is steam; the proportion of water is $(1 - X)$. The specific properties, volume V, internal energy U, enthalpy h and entropy s, of the mixture are found by adding the saturated steam and water values in the appropriate proportions, thus for specific enthalpy,

$$h = (1 - X)h_f + Xh_g \tag{3.17}$$

This can be rearranged as

$$h = h_f + X(h_g - h_f) = h_f + Xh_{fg} \tag{3.18}$$

which is the reason for including h_{fg} in the tables.

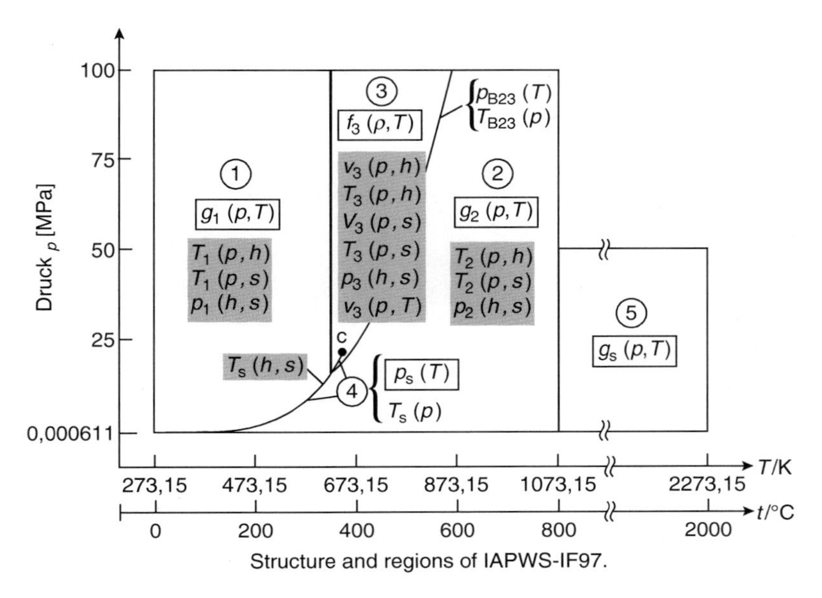

Fig. 3.6 Regions of the pressure–temperature plot into which the IF97 properties of water formulation are divided (Fig. 2.1 of Wagner and Kretzschmar [2008], published by Springer, second edition)

The mean density of the mixture must be found by following the additive rule above for specific volume:

$$V = (1 - X)V_f + XV_g \tag{3.19}$$

and then finding mean density as the reciprocal of the mixture specific volume. In this expression the specific volumes can be replaced by the reciprocal of the tabulated densities. The specific volume of water is often negligibly small compared to that of steam so the first term of Eq. (3.19) can sometimes be neglected, depending on the circumstances.

As regards the actual data, the 1967 Formulation for Industrial Use was replaced by a 1997 Formulation for Industrial Use (IAPWS-IF97), although the differences are negligible for geothermal use. The complete formulations divide the pressure–temperature space into regions, each with its own set of equations from which the properties can be calculated. The regions are shown in Fig. 3.6, reproduced from the publication by Wagner and Kretzschmar [2008].

For calculations where lower precision is acceptable in return for simpler numerical processing, IAPWS [1992] produced a set of equations describing the main property values along the saturation line, shown as region 4 in Fig. 3.6. These equations, which the IAPWS make available for unrestricted publication in all countries, are attached as Appendix A.

The compressibility of water is defined as

$$c = \frac{1}{\rho}\left(\frac{\partial \rho}{\partial P}\right)_T \tag{3.20}$$

and is needed for calculations of transient pressure changes in formations, Chap. 9. The compressibility of steam is less of a problem, as the properties of steam can be represented by $P/\rho = RT$ to acceptable accuracy, and the derivative substituted into Eq. (3.20). Values of water compressibility can be calculated from the expressions given for specific volume in IAPWS-IF97, region 1 in Fig. 3.6, but a simpler task for temperatures up to 100 °C and pressures to 1,000 bars abs would be to use the equations of Fine and Millero [1973] which are shown in Appendix B. Rogers and Mayhew [1967] also provide a data table showing the compressibility of water.

3.5 Hydrostatic Pressure and the Boiling Point for Depth Curve

It would not be wise to begin drilling a well in a known geothermal area without understanding the significance of hydrostatic pressure and what is known (in clumsy language) as "the boiling point for depth curve". The result would probably be an uncontrollable escape of any high-temperature water or steam that was intersected.

Hydrostatic pressure can be estimated roughly by assuming a density for water of 1,000 kg/m^3. (Although the word implies water, hydrostatic pressure is the name given to the pressure generated by a static column of any type of fluid.) The mass of a column of water 10 m tall and of cross-sectional area 1 m^2 would be 10^4 kg, and its weight would be g(10^4) or 9.81E4 kgm/s^2 (Newtons). The pressure exerted on the one square metre cross section would thus be 0.981 bars, a bar being 10^5 N/m^2, or approximately the equivalent of atmospheric pressure. In geothermal areas in particular, the temperature increases with depth, and since the density of water varies with temperature, it is more useful to describe the increase in pressure over an increment of depth dz as

$$dP = g.\rho(P, T).dz \tag{3.21}$$

from which the pressure at any depth can be calculated by integrating

$$P_z - P_0 = g\int_0^z \rho(P, T).dz \tag{3.22}$$

in which P_0 is the atmospheric pressure at the surface, where $z = 0$.

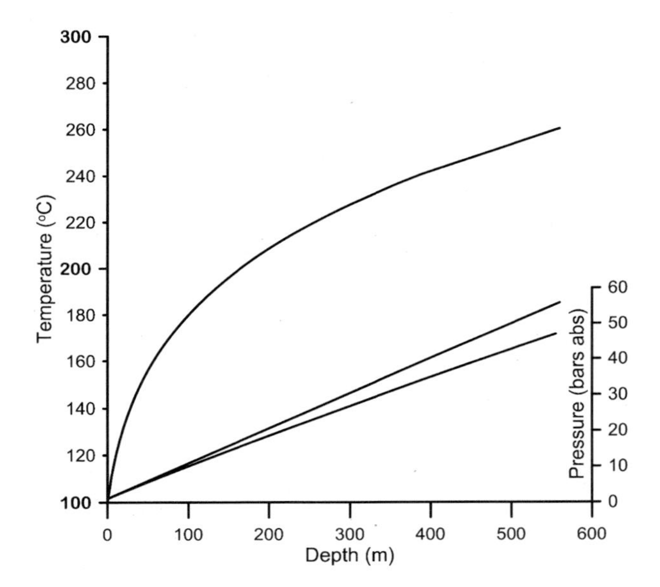

Fig. 3.7 Datum curves for hydrostatic pressure and the bpd curve

The variation of density with pressure and temperature is available as part of the IAPWS equations, so the variation of hydrostatic pressure with depth can be calculated for any given temperature distribution. The integration can be carried out using Simpson's rule or similar, for which the equation needs to be rearranged as

$$dz = f(P, T).dP = \frac{1}{g\rho(P, T)}.dP \tag{3.23}$$

A table of P_S as a function of z is given in Appendix C, calculated using the 1967 IFC Formulation for Industrial Use.

Two particular vertical pressure distributions are of interest, one with the well entirely cold, at surface temperature, and the other for the condition where the temperature variation with depth is such that the water is always just at its saturation pressure. These are shown in Fig. 3.7 as the two lower lines, using the right-hand axis. The upper of these two lines is straight, corresponding to the isothermal condition, which has a constant density for all depths; a temperature of 25 °C has been chosen, for which the density is 997 kg/m^3. The lower line is for the column being everywhere at saturation pressure, so the density decreases with depth because the temperature is increasing and the pressure falls progressively below the isothermal case with increasing depth. Provided the fluid is everywhere liquid, any measured vertical pressure distribution should fit between these limiting pressure cases if the water level is at the surface.

Having calculated the pressure distribution corresponding to saturation conditions at every depth, the matching temperature distribution can be found from a table (see Appendix C) or an equation giving P_s versus T_s, and the result is plotted in the figure, using the left-hand axis. This is the "boiling point for depth curve" (bpd curve). The curve is steep at first, and the gradient decreases with depth towards the critical point. When interpreting downhole temperature and pressure surveys, measurements are usually plotted on a graph like Fig. 3.7 which already has these distributions drawn in as datum curves.

What is at issue here is the form taken by a column of fluid in equilibrium. In the atmosphere, a compressible fluid, a stable distribution has specific entropy constant with height, but the physical reasoning for this involves air being able to mix freely to achieve this state. The air is at the same specific entropy at all levels in a stable atmosphere and changes pressure isentropically as it moves up or down. In an entirely cased well, orders of magnitude deeper than its diameter, fluid circulation is restricted—studies of the open thermosyphon clearly demonstrate this—see Sect. 6.1. Filled with isothermal cold water, the equilibrium pressure distribution is the linear distribution of Fig. 3.7. In the search for an equilibrium distribution, a well filled with water everywhere at its saturation conditions so that P and T are the saturation pair is tightly defined, but would it be in equilibrium? Under these conditions, Eq. (3.23) becomes

$$dz = f(P_s).dP = \frac{1}{g\rho(P_s)}.dP \tag{3.24}$$

The integration of this equation does not involve temperature, and the bpd temperature distribution is found separately. The well must be cased, as any exchange of fluid with the formations would have an influence. Despite the elegance of the idea, the bpd is not physically realistic as an equilibrium condition, as can be seen by examining thermal conduction. The temperature distribution gives rise to a conductive heat flux:

$$\dot{q}(z) = \lambda(P_s, T_s)\frac{dT}{dz} \tag{3.25}$$

The thermal conductivity, λ, at saturation conditions is a slowly varying function of temperature, and using the property data given by IAPWS [1998], the heat flux variation with z can be calculated. It is shown in Fig. 3.8 together with the bpd curve.

The heat flux increases towards the surface from about 400 m depth and below that increases for greater depths, more gradually. For the bpd distributions to represent an equilibrium state, the heat flux would have to be constant with z. This is in accordance with the general statement by Turner [1973] that Eq. (3.21) assumes no diffusion of heat or species in solution, the latter being the alternative to temperature as a means of producing density variations.

Despite these objections, some well measurements show temperature distributions which match the bpd curve, as will be demonstrated in Chap. 6. The reason is probably that the fluid in the well is two-phase but only just, with the dryness fraction almost zero. The steam would be present together with gas coming out of solution, as a few rising bubbles, sufficiently few that the liquid part of the column

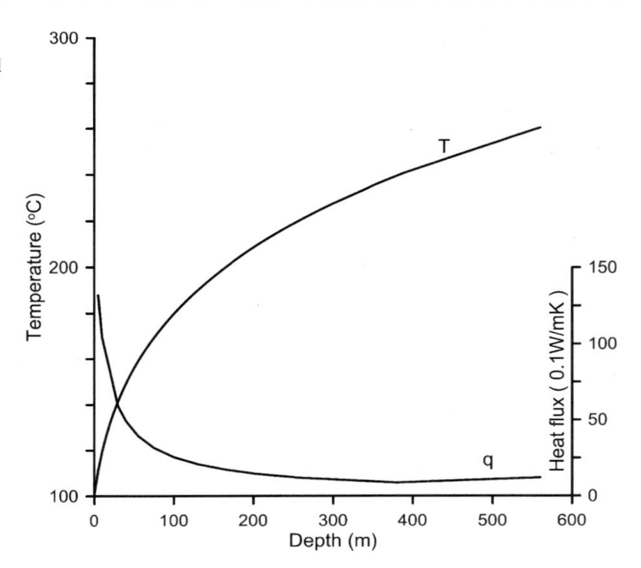

Fig. 3.8 Boiling point for depth (bpd) curve and related heat flux distribution

remained continuous so that the hydrostatic pressure is the same as without any bubbles at all, the conditions of the bpd curve. For this to be the case, the bubbles would need to be small, so as to provide negligible drag on the water column—upward lift would reduce the hydrostatic pressure gradient. If the volume occupied by the bubbles increased, the liquid column would become narrower, sinuous, and eventually discontinuous, it would be subject to lift, and the hydrostatic pressure distribution would be different. There would be a whole spectrum of possibilities, of which a narrow range will match the bpd curve. Within this range, the heat flux in the column with a bpd temperature distribution would always be upwards and greater than the values shown in Fig. 3.8.

The bpd curve is nevertheless an important datum against which to compare steady-state well measurements.

References

Dugdale JS (1966) Entropy and low temperature physics. Hutchinson University Library, London

Fermi E (1956) Thermodynamics. Dover (copyright 1936)

Fine RA, Millero FJ (1973) Compressibility of water as a function of pressure and temperature. J Chem Phys 59:10

IAPWS (International Association for the properties of Water and Steam) (1998) Revised release on the IAPWS Formulation 1985 for the thermal conductivity of ordinary water substance. London, England

IAPWS (International Association for the Properties of Water and Steam) (1992) Revised supplementary release on saturation properties of ordinary water substance. St. Petersburg, Russia (http://www.iapws.org)

Perrot P (1998) A to Z of thermodynamics. Oxford University Press, Oxford

Pippard AB (1961) The elements of classical thermodynamics. Cambridge University Press, Cambridge
Rogers GFC, Mayhew YR (1967) Engineering thermodynamics, work and heat transfer. Longmann, London
Spielberg N, Anderson BD (1987) Seven ideas that shook the Universe. Wiley, New York
Turner JS (1973) Buoyancy effects in fluids. Cambridge University Press, Cambridge
Wagner W, Kretzschmar H-J (2008) International Steam Tables – properties of water and steam based on the industrial formulation IAPWS- IF97. Springer, Heidelberg

The Equations Governing Heat and Single-Phase Fluid Flow and Their Simplification for Particular Applications

4

This chapter begins with an explanation of the fundamental equations governing the flow of fluids, which are expressions of the continuity of mass, momentum and heat. This leads to various simplifications to suit particular circumstances, such as the flow in circular pipes. Turbulent flow and the dimensionless forms of the equations, which are necessary for the design of experiments and the cataloguing of the results in a way that suits the requirements of equipment designers, are introduced. Natural convection and thermal conduction are briefly introduced. The basis of Darcy's law is discussed, and the governing equations are reduced to a form suitable for the analysis of the flow in permeable formations in geothermal resources.

4.1 Introduction

The study of the flow of heat and fluid is sometimes known as thermo-fluids. In geothermal engineering, the flow in pipes, separators and power station equipment must be analysed, and also the flow in the microscopic channels of permeable rocks. Several difficulties might arise for newcomers to the topic—equations appear to be simply "pulled out of the hat" like the magician's rabbit, they appear in units which are specific to a particular application, and they are written in variables which are specific to particular branches of science/engineering. All of the equations governing the flow of fluids and heat are related—after all, there are not many basic starting points, only Newton's laws of motion, the two laws of thermodynamics and the physical idea of continuity, that mass, momentum and energy cannot simply vanish. For ease of understanding, although not for brevity in writing, the equations have most meaning if they are presented in geometric terms, not in shorthand vector notation.

Gaining an intuitive understanding of fluid flow is probably easier than that of heat flow; water can be watched and experimented with easily, whereas heat is invisible. Probably for this reason, fluid flow was studied before thermodynamics or the flow of heat. Dryden et al. [1956] suggest that fluid friction was first studied by Mariotte (1620–1684) and that Guglielmini (1655–1710) put forward the

A. Watson, *Geothermal Engineering: Fundamentals and Applications*,
DOI 10.1007/978-1-4614-8569-8_4, © Springer Science+Business Media New York 2013

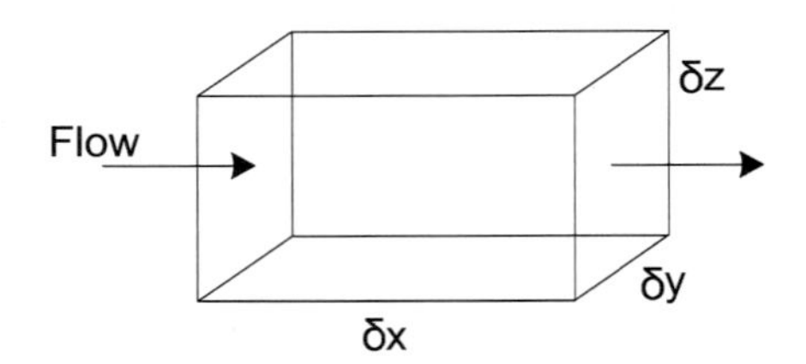

Fig. 4.1 A control volume, fixed in space

hypothesis that glaciers moved faster in the middle than at the edges because ice was a viscous fluid "like honey".

At any point in space within a flow which is to be analysed, five variables define the flow, namely, three velocity components, pressure and temperature, and their variation is to be described in three space dimensions and time. These variables appear in a set of equations which are formulated around three basic propositions, continuity of mass, momentum and energy; five equations are needed if five variables might have to be determined.

Continuity of mass is the easiest to deal with. Consider a control volume consisting of a rectangular framework (Fig. 4.1). The control volume is described in terms of rectangular Cartesian coordinates and has sides of length δx, δy and δz. The principle of conservation of mass simply states that the mass entering the control volume over any of the faces either leaves again or is stored within the control volume, a proposition that can be written as an equation.

The second and third propositions are the conservation of momentum and the conservation of energy, respectively. The latter is an energy balance—the energy that enters either stays in the control volume or leaves. It is more complicated than the equation of conservation of mass because energy is interchangeable with mechanical work, as recognised in the first law of thermodynamics Eq. (3.1), and the accounting is thus a bit more involved. Heat can enter or leave by thermal conduction, or as heat carried by moving fluid (convection), or it can be generated within the fluid. Real fluids are viscous, creating stresses in moving fluid which lead to frictional heat generation, the exchange of work to heat. Pressure changes result in compression or expansion and further exchange of work to heat. The result is that the energy equation has many more terms than the equation of continuity of mass.

Conservation of momentum is the most complicated of the three principles to apply, because momentum is conceptually more difficult. Momentum has direction as well as magnitude, so there are not one but three separate momentum equations, one for each of the three directions of motion. The physical idea of momentum transfer can be illustrated by considering a moving element of fluid passing a stationary one; viscosity causes the moving one to drag on the other, slowing it—in other words, some of its momentum is given to the stationary one which begins to move as a result. Momentum has been transferred. There is a shear stress between the two elements, and in general, a moving fluid exhibits a shear stress gradient and a velocity gradient in at least one direction; momentum is transferred down the shear stress gradient.

At least some of these equations must involve physical properties of the fluid—energy transfer by thermal conduction involves the thermal conductivity, mass flow rate involves density, and momentum and energy changes involve viscosity. An equation of state relates pressure, temperature and density, so it might be said that the complete system in thermo-fluids can be described by 6 equations, not 5. But that still leaves the transport properties to be provided, the properties which control the rate of transfer of heat and momentum, λ and μ, respectively, usually as functions of temperature, so the approach adopted here is to use 5 equations with the properties provided via relationships with P and T. (The symbol k is common for thermal conductivity but is reserved for permeability in this book.) For some of the problems facing geothermal engineers, the fluid can be assumed to have constant density, which simplifies the equations a good deal, and further simplifications can be introduced for particular problems; indeed, they must be because the equations have so many terms that they are impossible to deal with unless they are stripped down to bare essentials. Before computers made numerical solutions possible, the set of equations could often not be solved without making sweeping simplifications.

Two coordinate systems are useful for geothermal engineering, rectangular Cartesian geometry (x, y and z) and two-dimensional radially symmetric geometry (r and z), for example, to describe a flow in a pipe which is symmetrical about the axis so does not involve angular variations. Flow in a planar fracture could also be reduced to two dimensions. The only fluid referred to here is a Newtonian fluid, that is, a fluid for which shear stress τ is linearly related to velocity gradient, often taken as the definition of viscosity μ so that

$$\tau = \mu \frac{\partial u}{\partial x} \tag{4.1}$$

Strictly, this is the dynamic viscosity—there is another form of viscosity called the kinematic viscosity which is introduced later. "Viscosity" always means μ, and the other viscosity is always given its full name.

No attempt is made here to derive all of the equations formally, as the focus is on understanding them.

4.2 The Equations Governing the Flow of Fluid and Heat

4.2.1 The Continuity Equation

The continuity equation as it is called is based on the idea that the mass flow entering a control volume equals that which leaves less any that stays inside. If any stays inside, then the density must change accordingly. If the density of the fluid is ρ and u, v and w are the velocity components in directions x, y and z, respectively, the mass flow rate into the control volume in time δt can be summed over each face:

Mass flow entering the control volume (Fig. 4.1) in the x direction

$$= \rho u.\delta y.\delta z.\delta t \tag{4.2}$$

The mass flow rate leaving the control volume in the x direction must take account of any variation of ρu with x, so is written as

Mass flow rate leaving in the x direction

$$= \left(\rho u + \frac{\partial(\rho u)}{\partial x}\delta x\right)\delta y \delta z \delta t \tag{4.3}$$

The inflow minus the outflow in the x direction must contribute to the change in mass stored within the control volume, and this difference is

$$\rho u.\delta y \delta z \delta t - \left(\rho u + \frac{\partial(\rho u)}{\partial x}\delta x\right)\delta y \delta z \delta t = -\frac{\partial(\rho u)}{\partial x}.\delta x \delta y \delta z \delta t \tag{4.4}$$

Adding all three contributions gives the difference in mass stored within the control volume in a time interval δt as

$$-\left(\frac{\partial(\rho u)}{\partial x} + \frac{\partial(\rho v)}{\partial y} + \frac{\partial(\rho w)}{\partial z}\right)\delta x \delta y \delta z \delta t$$

Now the mass initially inside the control volume was density times volume, ρ $\delta x \delta y \delta z$, so the increase in this quantity in a time interval δt is

$$\frac{\partial \rho}{\partial t}.\delta x \delta y \delta z \delta t$$

Equating these gives

$$\frac{\partial \rho}{\partial t}.\delta x \delta y \delta z \delta t = -\left(\frac{\partial(\rho u)}{\partial x} + \frac{\partial(\rho v)}{\partial y} + \frac{\partial(\rho w)}{\partial z}\right)\delta x \delta y \delta z \delta t \tag{4.5}$$

or

$$\frac{\partial \rho}{\partial t} + \left(\frac{\partial(\rho u)}{\partial x} + \frac{\partial(\rho v)}{\partial y} + \frac{\partial(\rho w)}{\partial z}\right) = 0 \tag{4.6}$$

It can be seen directly that if the flow is steady (density and velocities do not change with time) and fluid has constant density, the equation reduces to

$$\frac{\partial u}{\partial x} + \frac{\partial v}{\partial y} + \frac{\partial w}{\partial z} = 0 \tag{4.7}$$

which is typical of the simplifications that can be made to help solve the final set, although there are two aspects of geothermal engineering which depend on the fluid density being variable, natural convection and transient pressure testing, so the density varying with time cannot be abandoned yet.

4.2.2 The Equations of Conservation of Momentum

The principle of conservation of momentum derives from Newton's second law of motion, that the net force in any particular direction acting on an element of fluid is equal to the rate of change of momentum of the element. Momentum and force are vectors, having size and direction, so there are three equations, one for each direction. The forces are gravity; pressure, which acts normal to each face of the element; and shear stresses due to the viscosity of the fluid. Gravity appears in only one direction, usually the z direction for geothermal engineering analyses, but could appear in the other equations if there were centrifugal or other forces present which depend on the mass of the element (referred to as body forces). Assuming a compressible Newtonian fluid with constant viscosity, the three momentum equations have the form

$$\rho\frac{Du}{Dt} = -\frac{\partial P}{\partial x} + \frac{\mu}{3}\frac{\partial}{\partial x}\left(\frac{\partial u}{\partial x} + \frac{\partial v}{\partial y} + \frac{\partial w}{\partial z}\right) + \mu\nabla^2 u \tag{4.8}$$

$$\rho\frac{Dv}{Dt} = -\frac{\partial P}{\partial y} + \frac{\mu}{3}\frac{\partial}{\partial y}\left(\frac{\partial u}{\partial x} + \frac{\partial v}{\partial y} + \frac{\partial w}{\partial z}\right) + \mu\nabla^2 v \tag{4.9}$$

$$\rho\frac{Dw}{Dt} = \rho g - \frac{\partial P}{\partial z} + \frac{\mu}{3}\frac{\partial}{\partial z}\left(\frac{\partial u}{\partial x} + \frac{\partial v}{\partial y} + \frac{\partial w}{\partial z}\right) + \mu\nabla^2 w \tag{4.10}$$

where

$$\nabla^2 = \frac{\partial^2}{\partial x^2} + \frac{\partial^2}{\partial y^2} + \frac{\partial^2}{\partial z^2}$$

These equations were first produced by Navier (1822) for incompressible fluids and later extended to compressible fluids by Stokes and others (Dryden et al. [1956]) and are known as the Navier–Stokes equations. The equations are written as per unit volume. The right-hand terms are all net forces on the element of fluid being considered in the derivation of the equations, and the left-hand term of each equation represents the rate of change of momentum of the element. Throughout the entire space occupied by the particular problem that is being tackled, the velocity component in the x direction of a moving element is changing with x, y, z and t in general, and the full change du can be expressed in terms of its components in each direction and in time:

$$du = \frac{\partial u}{\partial x}.\delta x + \frac{\partial u}{\partial y}.\delta y + \frac{\partial u}{\partial z}.\delta z + \frac{\partial u}{\partial t}.\delta t \tag{4.11}$$

which can be rearranged to

$$\frac{du}{\delta t} = \frac{Du}{Dt} = u\frac{\partial u}{\partial x} + v\frac{\partial u}{\partial y} + w\frac{\partial u}{\partial z} + \frac{\partial u}{\partial t} \tag{4.12}$$

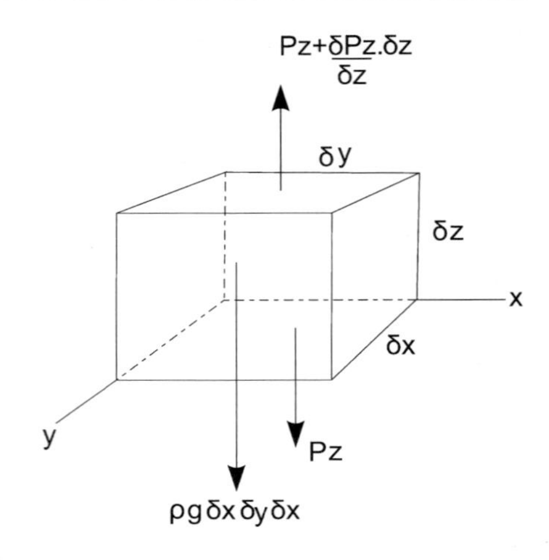

Fig. 4.2 Control volume showing *z*-axis pressure variation and gravity force

in which the traditional use of upper case *D* has been introduced. This is called the total derivative.

The occurrence of ρ outside the total derivative at first sight appears to indicate that fluid density has been assumed to be constant, but this is not so, and the density appears separately only because the derivation follows the motion of an element of given mass.

Considering the *z* direction, Fig. 4.2 shows the forces on the element due to pressure and gravity; the same algebraic pattern of increase with *z* is apparent, and if these two were the only forces on the element, the equation would read

$$\rho \frac{Dw}{Dt} = \rho g - \frac{\partial P}{\partial z} \tag{4.13}$$

and there would be two accompanying equations:

$$\rho \frac{Du}{Dt} = -\frac{\partial P}{\partial x} \tag{4.14}$$

$$\rho \frac{Dv}{Dt} = -\frac{\partial P}{\partial y} \tag{4.15}$$

This set could be derived from the Navier–Stokes equations (4.8)–(4.10) by setting $\mu = 0$, and they are known as Euler's equations. Despite real fluids having viscosity and exhibiting shear stresses, both features being ignored in forming Euler's equations, the equations are used in aerodynamics. The airflow past an object such as an aircraft wing or a power station cooling tower does not depend on the viscosity of the flow except very close to the surface of the object, and if this can be considered as a layer of almost uniform thickness which is very small compared to the dimensions of the object, then Euler's equations can be used to find the

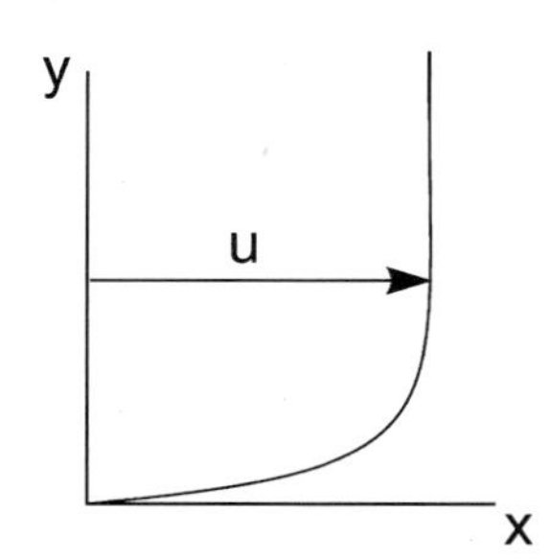

Fig. 4.3 Flow with a boundary layer, a high gradient narrow layer near the solid surface

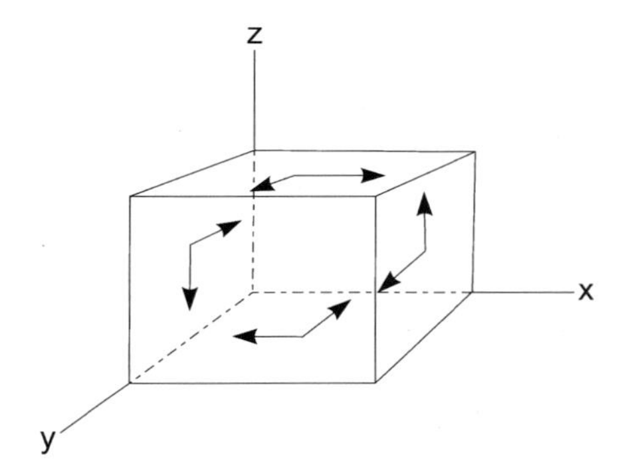

Fig. 4.4 Showing the shear stresses on opposite sides of the control volume (xy and zy planes only are shown, for clarity)

velocities in the flow outside this layer. Such a layer actually occurs in practice—the boundary layer, as shown in Fig. 4.3.

The steady-state form of Eq. (4.13), with Dw/Dt set to zero, has already been used in Chap. 3 to calculate the hydrostatic pressure gradient—(3.21), which was introduced intuitively.

Introducing viscosity results in many more terms in the equation, as suggested by Fig. 4.4 where the shear stresses are marked on only two pairs of faces.

Each face of the fluid element being tracked is subjected to two shear stresses, which are proportional to the viscosity, and one direct stress due to pressure. In a completely general approach, which is used for developing the equations for elastic solids for stress analysis, the direct stress is referred to as σ, which is given the positive sign if it is pulling the element outwards, and there is some benefit in using this terminology at present. There is a separate direct stress in each direction. This means that there are nine stress components, six shear stresses and three direct stresses, and the element has two faces in each of the three directions as it "sits" around this point. The stresses are shown in Table 4.1.

If the equations are assembled at this point, they are

$$\rho\frac{Du}{Dt} = \frac{\partial\sigma_x}{\partial x} + \frac{\partial\tau_{yx}}{\partial y} + \frac{\partial\tau_{zx}}{\partial z} \tag{4.16}$$

$$\rho\frac{Dv}{Dt} = \frac{\partial\tau_{xy}}{\partial x} + \frac{\partial\sigma_y}{\partial y} + \frac{\partial\tau_{zy}}{\partial z} \tag{4.17}$$

Table 4.1 Listing the stress components at a point

Direction	x plane	y plane	z plane
x	σ	τ_{yx}	τ_{zx}
y	τ_{xy}	σ	τ_{zy}
z	τ_{xz}	τ_{yz}	σ

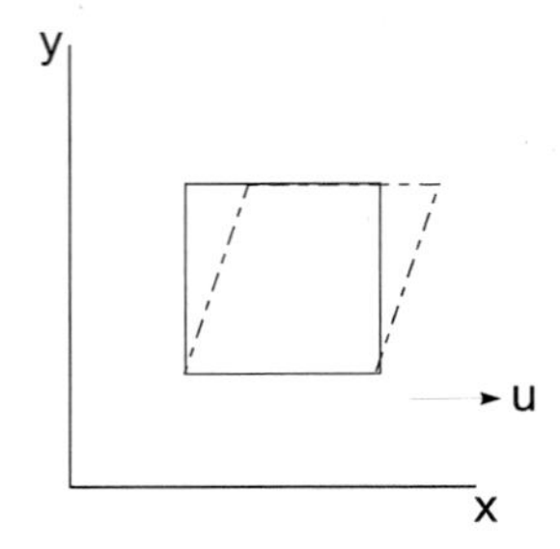

Fig. 4.5 Fluid distortion resulting from a velocity gradient

$$\rho\frac{Dw}{Dt} = \rho g + \frac{\partial \tau_{xz}}{\partial x} + \frac{\partial \tau_{yz}}{\partial y} + \frac{\partial \sigma_z}{\partial z} \tag{4.18}$$

The stress components in this set of equations distort the element, which is twisted and squashed. If the material was a solid, this distortion would be referred to as a strain, but, by definition, fluids cannot sustain direct forces but move instead, and the equivalent of strain in solids is rate of strain in fluids. The rate of strain is related to the velocity gradients. For example, a velocity gradient $\partial u/\partial y$ imposed on the diamond-shaped element in the sketch of Fig. 4.5 will distort it as shown.

If the analysis is followed through, it turns out that the normal rates of strain components are

$$\epsilon_x = 2\frac{\partial u}{\partial x}, \quad \epsilon_y = 2\frac{\partial v}{\partial y} \quad and\ \epsilon_z = 2\frac{\partial w}{\partial z} \tag{4.19}$$

and the shear rates of strain components are

$$\gamma_{xy} = \left(\frac{\partial v}{\partial x} + \frac{\partial u}{\partial y}\right), \quad \gamma_{yz} = \left(\frac{\partial w}{\partial y} + \frac{\partial v}{\partial z}\right) \quad and\ \gamma_{xz} = \left(\frac{\partial w}{\partial x} + \frac{\partial u}{\partial z}\right) \tag{4.20}$$

The physical law relating shear stress to rate of strain is not Eq. (4.1) but

$$\tau_{xy} = \tau_{yx} = \mu\gamma_{xy}, \quad \tau_{yz} = \tau_{xz} = \mu\gamma_{yz} \ and \ \tau_{zx} = \tau_{xz} = \mu\gamma_{zx} \tag{4.21}$$

For one-dimensional flow, this reduces to Eq. (4.1), which is thus revealed to be a simplified definition of viscosity.

It is not necessary to follow the complete development of the Navier–Stokes equations—if required, it can be found in Lamb [1906] or other books—and for the moment they can be taken as Eqs. (4.8)–(4.10), forms which will be reduced to suit the problem being examined.

4.2.3 The Equation of Continuity of Energy

The approach is different for this equation, in that an element the same shape as the control volume is tracked as it moves with the flow. The basic form of the equation can only be the first law of thermodynamics Eq. (3.1) since that is the fundamental statement of conservation of energy. The element being tracked receives heat as a result of thermal conduction, or from any internal generation (e.g. by the decay of radioactive isotopes), or as a result of frictional heating since the fluid is viscous and the element is subjected to forces. A rearranged Eq. (3.1) and the equation of continuity of energy written for comparison are

$$dU = -dQ + dW \tag{4.22}$$

$$\rho \frac{DU}{Dt} = \left(\frac{\partial \left(\frac{\lambda \partial T}{\partial x} \right)}{\partial x} + \frac{\partial \left(\frac{\lambda \partial T}{\partial y} \right)}{\partial y} + \frac{\partial \left(\frac{\lambda \partial T}{\partial z} \right)}{\partial z} \right) + \dot{H} + \mu \varnothing + d\dot{W} \tag{4.23}$$

where λ = thermal conductivity (W/mK).

The energy conservation equation is a "rate" equation, whereas the first law equation is not. There is no "shaft work" (work extracted or input from a machine), but work is done by the direct and shear stresses on the element. The term on the left-hand side of Eq. (4.23) is the net change in energy content of the element as a result of the sources of heat and work on the right-hand side; it has the units of kJ/m^3, because as with the control volume, the area of surfaces presented by the element being followed must be considered—energy flows are defined in fluxes, i.e. flows per unit area not per unit mass. Comparing the two equations is a reminder of the convention in thermodynamics that the fluid does work and absorbs heat, and these energy contributions are positive.

In dealing with the momentum equations, the momentum balance was first produced in terms of stresses; then as a separate stage, the transport properties of the fluid were introduced, to convert the stresses to velocity gradients. In that case, a Newtonian fluid was chosen, but non-Newtonian fluids could have been considered in principle. In developing the energy equation from the beginning, the heat balance is considered in terms of heat fluxes, and the transport properties then introduced—Newton's definition of thermal conductivity in this case, which is

$$\dot{q} = \lambda \frac{\partial T}{\partial x} \tag{4.24}$$

The result is the first term on the right-hand side of Eq. (4.23). The net heat flow into the element, using Fig. 4.1, is the difference between that entering and that leaving, the latter being defined in a general form as for the continuity equation:

$$\text{Net flux} = \lambda\frac{\partial T}{\partial x}\delta y\delta z - \left(\lambda\frac{\partial T}{\partial x} + \frac{\partial\left(\lambda\frac{\partial T}{\partial x}\right)}{\partial x}\delta x\right).\delta y\delta z$$

$$= \frac{\partial\left(\lambda\frac{\partial T}{\partial x}\right)}{\partial x}\delta x\delta y\delta z \tag{4.25}$$

Adding up the differences gives the full term shown in Eq. (4.23) which is again per unit volume.

The heat generation term $\dot{H}$ is simply the heat generated per unit volume. The term $\mu\varnothing$ is a rate of heat generation resulting from fluid friction, hence the appearance of μ, and $\varnothing$ is a collection of velocity gradient terms arising from the shear stresses, a part of the work term in Eq. (4.23) known as the dissipation function. In the majority of fluid mechanics problems, $\varnothing$ is negligible; however, the dissipation function is a significant source of heat in high-speed flows. Since the major surfaces exposed to the flow of a high-speed aircraft are orientated in the line of the flow, the x direction, and the flow on a surface takes the form of a boundary layer with high-velocity gradients near the surface, as in Fig. 4.3 above, the term often appears as

$$\mu\left(\frac{\partial u}{\partial x}\right)^2$$

Finally, part of the term $d\dot{W}$ is the work done by the direct stresses on each face of the element, the direct stresses being the pressure.

It will be recalled from Sect. 3.2.2 that specific enthalpy is defined as $h = U + PV$. Equation (4.23) was written in terms of specific internal energy, the fundamental parameter, but the energy equation is more often needed in the form in which T is the dependent variable, if only because physical properties are usually provided as functions of T. U can be written as C_vT, where C_v is the specific heat at constant volume, but the specific heat at constant pressure, C_p, is more often available, and specific enthalpy is $C_p\,T$. Making use of the definition of h and differentiating gives

$$\frac{Dh}{Dt} = \frac{DU}{Dt} + \frac{1}{\rho}\frac{DP}{Dt} - \frac{P}{\rho^2}\frac{D\rho}{Dt} \tag{4.26}$$

This provides enough information to rewrite the energy equation in terms of specific enthalpy rather than specific internal energy, but it requires the last two terms on the right-hand side to be incorporated into the equation or some basis for neglecting them to be found. Exact forms of the energy equation in terms of total derivatives of U, h and specific entropy, s, have been the subject of several technical notes in the published literature. There is no alternative but to examine the terms for the particular problem being studied, and Bird et al. [2007] provide the best starting point. The most common form in mechanical engineering fluid mechanics and heat transfer is that for an incompressible fluid, with $C_p = C_v$, constant λ and μ and no internal heat generation or dissipation function; thus,

$$\rho C_p \frac{DT}{Dt} = \lambda\left(\frac{\partial^2 T}{\partial x^2} + \frac{\partial^2 T}{\partial y^2} + \frac{\partial^2 T}{\partial z^2}\right) \tag{4.27}$$

4.2.4 Bernoulli's Equation

Bernoulli's equation is one of those "rabbits out of the hat" referred to in Sect. 4.1. It is a mechanical energy continuity equation, so it is encapsulated within the energy equation above, but since it is essentially an equation for a one-dimensional isothermal flow, it is easiest to produce it by simplifying the momentum equations. Bernoulli's equation is

$$\frac{P_2}{\rho} + \frac{u_2^2}{2} + gz_2 = \frac{P_1}{\rho} + \frac{u_1^2}{2} + gz_1 \tag{4.28}$$

from which it can be seen that it describes the relationship between three energy terms at two locations represented by suffixes 1 and 2, along the flow path taken by a steady fluid flow (Fig. 4.6).

The flow path is referred to as a stream tube because fluid flows along it as if along a variable diameter pipe, in any general direction. No fluid crosses its "walls" which are not solid containment and thus provide no shear stress. Only in this sense is the equation one-dimensional, but the potential energy terms incorporate the vertical z-axis so the equation must acknowledge the coordinate system. If the stream tube is a vertical well through which there is a steady isothermal flow, Eqs. (4.13)–(4.15) reduce to Eq. (4.13) only, as there are no lateral flows. Thus,

$$\rho \frac{Dw}{Dt} = \rho g - \frac{\partial P}{\partial z} \tag{4.13}$$

becomes

$$\rho w \frac{dw}{dz} = \rho\left(\frac{d}{dz}\left(\frac{w^2}{2}\right)\right) = \rho g - \frac{dP}{dz} \tag{4.29}$$

and integrating between two levels in the well,

$$\int_1^2 \frac{d}{dz}\left(\frac{w^2}{2}\right).dz = \int_1^2 g.dz - \frac{1}{\rho}\int_1^2 \frac{dP}{dz}.dz \tag{4.30}$$

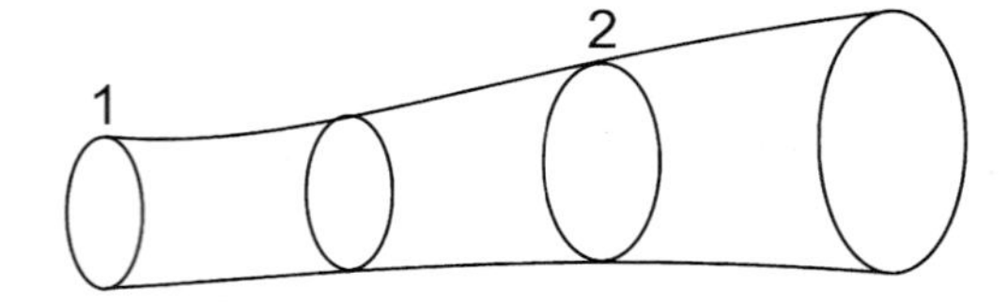

Fig. 4.6 A stream tube along which the total energy remains constant

The result is Eq. (4.28) above, Bernoulli's equation. It looks like the Steady Flow Energy Equation, (3.6), as indeed it should—it is the same, with the specific internal energy terms, work and heat supply terms absent and only one of the two terms making up specific enthalpy is present, PV written as P/ρ. It is a mechanical energy continuity equation.

4.3 Flow in Pipes and Engineered Equipment

Focussing on pipe flow is a convenient way of introducing dimensionless variables, which are valuable in helping to plan experiments and organise the results of parametric surveys and mathematical solutions. In order of difficulty, the ranking of flows is laminar single phase, turbulent single phase and then two-phase; laminar flow equations can be solved analytically (in principle), but empirical relationships are necessary to make any progress in solving turbulent flow equations and even more are required for two-phase flow equations.

A single-phase constant property fluid will flow uniformly through a smooth pipe at low flow rates; the flow is said to be laminar (from the Latin *lamina*, meaning in thin layers or uniformly stratified). If the flow is gradually increased, it becomes turbulent of its own accord—turbulence may be triggered by an external disturbance, but it is caused and maintained by the shear stress in the fluid which is due to viscosity. As it flows through a constant diameter smooth pipe in laminar flow, a particle of fluid follows a smooth, predictable path, but in turbulent flow, it follows a very irregular and transient path, even if the total flow rate is constant—turbulent flow is discussed later. There are solutions to the governing equations for laminar flow without any empiricism, but no solutions for turbulent flow that do not involve empirical information.

4.3.1 The Equations for Steady Laminar Flow in a Pipe

A steady laminar pipe flow of an incompressible fluid with constant viscosity and thermal conductivity is fully described by the following four equations, one each for continuity of mass and energy and two for momentum:

$$\frac{\partial u}{\partial x} + \frac{1}{r}\frac{\partial}{\partial r}(rv) = 0 \tag{4.31}$$

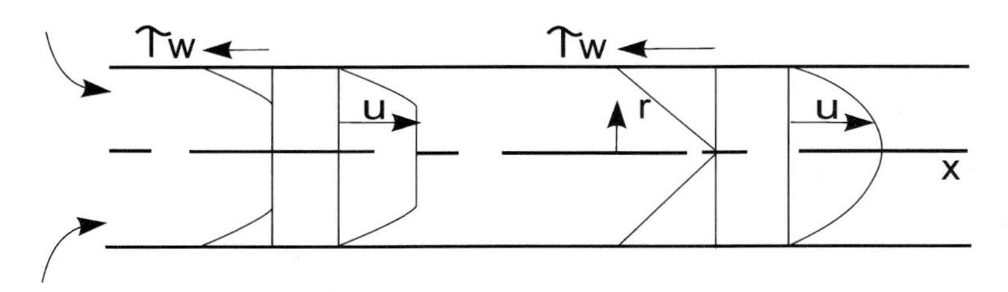

Fig. 4.7 Axial velocity and shear stress distributions in laminar flow, shortly after entry in a pipe and when the flow is fully deve'oped

$$u\frac{\partial u}{\partial x}+v\frac{\partial u}{\partial r}=-\frac{1}{\rho}\frac{\partial P}{\partial x}+\frac{\mu}{\rho}\left(\frac{\partial^2 u}{\partial x^2}+\frac{1}{r}\frac{\partial}{\partial r}\left(r\frac{\partial u}{\partial r}\right)\right) \tag{4.32}$$

$$u\frac{\partial v}{\partial x}+v\frac{\partial v}{\partial r}=-\frac{1}{\rho}\frac{\partial P}{\partial r}+\frac{\mu}{\rho}\left(\frac{\partial^2 v}{\partial x^2}+\frac{1}{r}\left(\frac{\partial}{\partial r}r\frac{\partial v}{\partial r}\right)\right) \tag{4.33}$$

$$u\frac{\partial T}{\partial x}+v\frac{\partial T}{\partial r}=\frac{\lambda}{\rho C_p}\left(\frac{\partial^2 T}{\partial x^2}+\frac{1}{r}\left(\frac{\partial}{\partial r}r\frac{\partial T}{\partial r}\right)\right) \tag{4.34}$$

No internal heat source is present in the fluid, and the dissipation function is neglected. There are only two momentum equations because circumferential variations, described by the angle, are absent; the flow is symmetrical about the pipe axis. The pipe axis in these equations is in the x direction—the pipe is horizontal to avoid the introduction of gravity.

Considering only the development of the flow pattern, the velocity distribution changes continuously after entering the pipe, eventually becoming "fully developed" some distance downstream, a distance usually called the entry length, as shown in Fig. 4.7. Even though the mass flow rate entering is constant, the parameters describing an individual element of fluid change in time as it flows downstream, but at any given point in the pipe, the parameters are steady. Figure 4.7 shows the shear stress and axial velocity variations with radius at two locations downstream of the entry, which is at the left. The downstream distributions (right-hand side) are the fully developed ones; those nearest the entry are only partly developed—there is still a core of fluid which is unaffected by the viscous drag on the pipe wall. Momentum can be considered to be transferred radially in a pipe flow, along the shear stress gradient, just as heat is transferred along a temperature gradient. The shear stress distribution matching the fully developed velocity distribution is linear with radius; in the entry region, there is no shear stress in the core of the flow.

To find the velocity distributions in this region, all four equations above must be solved. Boundary conditions on the pipe wall must be prescribed for temperature or temperature gradient (heat flux). Solutions can be found in many engineering textbooks on heat transfer, for example, Kays [1966], but they are not of direct

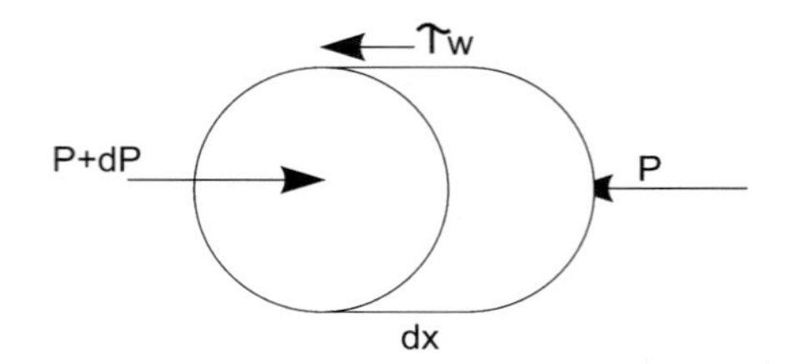

Fig. 4.8 A simple force balance relating pressure gradient to wall shear stress in fully developed pipe flow

interest here, except to show that the set of equations can actually be solved given simple enough conditions of flow. The development of the flow and the shear stress distribution is important however as an illustration of momentum transfer.

When the flow is fully developed, the axial pressure gradient will remain constant and will be related to the shear stress at the pipe wall; the relationship can be found from a simple force balance over a section of pipe length dx, as shown in Fig. 4.8.

The pipe radius is a, and the pressure difference over the length dx is dP; thus,

$$2\pi a.dx.\tau_w = \pi a^2.dP \text{ or } \tau_w = \frac{a}{2}\frac{dP}{dx} \tag{4.35}$$

The definition of shear stress Eq. (4.1) allows the pressure gradient to be related to the velocity gradient at the wall:

$$\tau_w = \mu\left(\frac{\partial u}{\partial r}\right)_{r=a} \tag{4.36}$$

This forms a boundary condition for the solution of the momentum equations. In fact, in the fully developed condition, the radial velocity v is zero and the flow is invariant with time and x, so only one momentum equation remains:

$$0 = -\frac{1}{\rho}\frac{\partial P}{\partial x} + \mu\left(\frac{1}{r}\frac{\partial}{\partial r}\left(r\frac{\partial u}{\partial r}\right)\right) \tag{4.37}$$

which can be integrated using the boundary condition above plus the symmetry which exists about the pipe axis:

$$\frac{du}{dr} = 0 \text{ at } r = 0 \tag{4.38}$$

giving the velocity distribution as

$$u = \frac{a^2}{4\mu}\left(-\frac{dP}{dx}\right)\left(1 - \left(\frac{r}{a}\right)^2\right) \tag{4.39}$$

This solution was obtained by Poiseuille in 1838 and is said to form the basis of Darcy's law, as discussed later. A similar solution can be carried out for the flow between parallel plates, which is representative of the laminar flow in rock fractures.

4.3.2 Turbulent Flow

In 1883, Osborne Reynolds demonstrated, by introducing a stream of dye into water flowing through a glass pipe, that as the flow rate increased the flow changed from laminar to turbulent. The fluid motion in a turbulent flow has a local random variation, so adjacent fluid elements transfer momentum and heat not only as a result of viscosity and thermal conductivity but also by fast fluid elements moving into a zone of slower flow and hot fluid elements moving into a zone of cooler flow. If the set of governing equations is to be solved to find the velocity and temperature gradients, the viscosity and thermal conductivity need to be the effective values in the turbulent flow and not simply the fluid property values. This is the approach adopted for solving turbulent pipe flow problems, and various empirical mathematical expressions for the turbulent diffusivity of momentum and heat have been proposed. The characteristics of turbulent flows must be found experimentally, which requires a shift in focus from solving the equations towards how to plan the experiments and catalogue the results.

The transition from laminar to turbulent flow was found to take place when the property group $\mathrm{Re} = \rho\bar{u}d/\mu$ has a value of 2,100; this group became known as the Reynolds number, and its origins are best demonstrated in the next section, in which the governing equations are converted to dimensionless form.

Consider the axial pressure gradient along a pipe through which a fluid is being pumped; this is an important parameter for engineers since it dictates the power consumption of the pump. If the pressure gradient is measured for a certain diameter of pipe and the pipe diameter is doubled, what will the pressure gradient be then? If a range of fluid properties, pipe diameters and mass flow rates are to be used, what are the rules about scaling up the measurements made with a single set of parameters? Converting the governing equations to dimensionless form can show how to select the parameters for the experiments. This is a necessary step towards finding the scaling rules—it reveals the parameters by which to catalogue the experimental results—but the actual rules are found from the experimental measurements.

4.3.3 Making the Governing Equations Dimensionless to Reveal the Controlling Parameters

The governing equations for the steady flow of an incompressible fluid with constant λ and μ in a smooth pipe are Eqs. (4.31)–(4.34). They can be written in a form suitable for any sized pipe by changing the space variables. If the radius of the pipe is a, then any radius r can be expressed as r/a, but it is more usual to use the diameter d, so that the ratios, indicated by suffix D, (meaning dimensionless) are

$$r_D = r/d \tag{4.40}$$

and similarly the length along the pipe, x, can be expressed as a number of diameters; thus,

$$x_D = \frac{x}{d} \tag{4.41}$$

These are referred to as dimensionless variables. The entry length over which the shear stress distribution becomes established is measured in terms of pipe diameters, in other words, as x_D; it is typically 20 diameters for a turbulent pipe flow. Turning to the velocities, it seems intuitively reasonable to compare all velocities with the mean velocity in the pipe and to form the dimensionless velocities as

$$u_D = \frac{u}{\bar{u}} \text{ and } v_D = \frac{v}{\bar{u}} \tag{4.42}$$

The best way of making the other variables dimensionless can be seen after inspecting the equations with only x and r converted, so beginning with Eq. (4.32) and introducing r_D and x_D leaves (after some rearrangement)

$$u_D \frac{\partial u_D}{\partial x_D} + v_D \frac{\partial u_D}{\partial r_D} = -\frac{1}{\rho \bar{u}^2}\left(\frac{\partial P}{\partial x_D}\right) + \frac{\mu}{\rho \bar{u} d}\left(\frac{\partial^2 u_D}{\partial x^2} + \frac{1}{r_D}\frac{\partial}{\partial r_D}\left(r_D \frac{\partial v_D}{\partial r_D}\right)\right) \tag{4.43}$$

The Reynolds number $\rho \bar{u} d/\mu$ is evident. The term $\rho \bar{u}^2$ is familiar as a kinetic energy term and has the dimensions of pressure. It can be chosen as the way to make pressure dimensionless; thus,

$$P_D = \frac{P}{\rho \bar{u}^2} \tag{4.44}$$

which can be introduced to Eq. (4.42) to leave the first term on the right-hand side as simply $\left(\frac{\partial P_D}{\partial x_D}\right)$.

Equation (4.34), the energy equation, has T to the power of one in several terms, and it can be made dimensionless with respect to any constant temperature or temperature difference—the latter is more usual, say ΔT_{ref} (a reference temperature difference chosen later) so that dimensionless temperature T_D replaces $T/\Delta T_{ref}$. Introducing this and the substitutions already defined, the final set of equations becomes

$$\frac{\partial u_D}{\partial x_D} + \frac{1}{r_D}\frac{\partial}{\partial r_D}(r_D v_D) = 0 \tag{4.45}$$

$$u_D\frac{\partial u_D}{\partial x_D} + v_D\frac{\partial u_D}{\partial r_D} = -\left(\frac{\partial P_D}{\partial x_D}\right) + \frac{1}{\mathrm{Re}}\left(\frac{\partial^2 u_D}{\partial x_D^2} + \frac{1}{r_D}\frac{\partial}{\partial r_D}\left(r_D\frac{\partial u_D}{\partial r_D}\right)\right) \tag{4.46}$$

$$u_D\frac{\partial v_D}{\partial x_D} + v_D\frac{\partial v_D}{\partial r_D} = -\left(\frac{\partial P_D}{\partial r_D}\right) + \frac{1}{\mathrm{Re}}\left(\frac{\partial^2 v_D}{\partial x_D^2} + \frac{1}{r_D}\frac{\partial}{\partial r_D}\left(r_D\frac{\partial v_D}{\partial r_D}\right)\right) \tag{4.47}$$

$$u_D\frac{\partial T_D}{\partial x_D} + v_D\frac{\partial T_D}{\partial r_D} = \frac{1}{\mathrm{Re.\,Pr}}\left(\frac{\partial^2 T_D}{\partial x_D^2} + \frac{1}{r_D}\frac{\partial}{\partial r_D}\left(r_D\frac{\partial T_D}{\partial r_D}\right)\right) \tag{4.48}$$

The term 1/Re.Pr needs some explanation. Equation (4.34) has the group $\lambda/\rho C_p$ preceding the right-hand term of second derivatives of temperature (a thermal diffusion term), and this group has now become 1/Re.Pr after the substitutions. The Prandtl number (named after Ludwig Prandtl) is the ratio of thermal diffusivity, κ, and kinematic viscosity, ν:

$$\mathrm{Pr} = \frac{\nu}{\kappa} \tag{4.49}$$

Where

$$\nu = \frac{\mu}{\rho} \text{ and } \kappa = \frac{\lambda}{\rho C_p}$$

The Prandtl number for water varies between 2 and 11 depending on temperature and for heavy oils is of the order of thousands.

All of the variables in the above set of equations are dimensionless, by which is meant that the actual size of the pipe, the flow rate and the fluid properties are invisible when looking at the equations. The set can represent a capillary tube in the laboratory or a geothermal steam pipe. For a given set of boundary conditions, say a fixed pipe wall temperature for all x, a catalogue of solutions could be obtained (in principle at least) for a wide range of Reynolds numbers and Prandtl numbers, the solutions being the velocity and temperature distributions in the pipe, the axial rate of rise of mean temperature and the axial pressure gradient. For any given pipe and fluid, the Reynolds and Prandtl numbers for the flow could be calculated and the appropriate solution selected from the catalogue. The pumping power and length of pipe required to transfer the required amount of heat could then be calculated.

The dimensionless groups were derived above from the governing equations because it offers the clearest demonstration of their origins. However many problems arise for which the equations are unknown, and in this case, dimensional analysis to arrive at the controlling groups can be carried out using Buckingham's Pi theorem, an explanation of which can be found in many engineering fluid mechanics and heat transfer texts. Shemenda [1994], referred to in Chap. 2, provides a good explanation and example of this approach in planning a laboratory experiment to investigate tectonic plate subduction.

4.3.4 Presenting the Experimental Results for Use in Design

Designers of pipework need to know how much pumping power will be required to produce the required flow rate through a pipe of given size. If the pipework is part of a heat exchanger, then the temperature difference between pipe and fluid necessary to produce a required heat flux from wall to fluid and the length of pipe required to raise the mean temperature of the flow to a certain value will be the questions. This information is contained in two new dimensionless parameters which do not appear in the treatment of the governing equations. They are friction factor and Nusselt number.

The shear stress at the pipe wall (or any other wall exposed to the flow) was introduced earlier as Eq. (4.36):

$$\tau_w = \mu \left(\frac{\partial u}{\partial r} \right)_{r=a} \tag{4.36}$$

This applies to both laminar and turbulent flows, because it is found that the random turbulent motion damps out immediately adjacent to the wall, in the limit. The shear stress at the wall in some circumstances can be found by measuring the velocity gradient, but more usually, it is deduced from a measurement of axial pressure gradient in the pipe, using the force balance, similar to Eq. (4.35) but based on pipe diameter rather than radius:

$$\tau_w . \pi d = \frac{\pi d^2}{4} \cdot \frac{dP}{dx} \tag{4.50}$$

With the idea that the shear stress is proportional to the kinetic energy of the flow, a new parameter called the friction factor can be defined:

$$f = \frac{\tau_w}{\frac{1}{2}\rho \bar{u}^2} \tag{4.51}$$

Given a value for f, the designer can calculate the wall shear stress, having found $\bar{u}$ from the mass flow rate and pipe diameter. Pressure gradient can thus be calculated, and for a given length of pipe over which the pressure drop is ΔP,

$$\text{Pumping power} = \frac{\dot{m}\Delta P}{\rho} \tag{4.52}$$

Friction factor has been measured experimentally and is plotted as a function of Reynolds number for isothermal flow. The pipe must be categorised in terms of the smoothness of its wall. Unfortunately, lack of international collaboration in the past has resulted in two different definitions of friction factor in use. That used in this book is the Fanning friction factor generally used in Europe and by most US academics, but there is another, the Darcy–Weisbach or Moody friction factor which is exactly four times the Fanning friction factor and is used in some sectors of US industry and no doubt elsewhere. This is a source of errors in engineering design as the symbol and name are sometimes used without definition.

The Fanning friction factor is correlated with Reynolds number as follows for isothermal flows in smooth pipes, the turbulent flow correlation being based on experiments:

Laminar flow (theoretical solution):

$$f = 16/\mathrm{Re} \tag{4.53}$$

Turbulent flow:

$$f = 0.079Re^{-0.25} \quad for\ 5,000 < Re < 30,000 \tag{4.54}$$

$$f = 0.046Re^{-0.2} \quad for\ 30,000 < Re < 1,000,000 \tag{4.55}$$

Not all ducts of interest are circular pipes—for example, the slotted liner in a well forms an annular perforated wall in a nominally circular hole. The equivalent diameter concept is an old established approach for translating the correlations of friction factor $f = function\ (Re)$ made with circular pipes to noncircular ducts. An equivalent diameter is defined as

$$d_e = \frac{4A}{P_{wet}} \tag{4.56}$$

where A is the cross-sectional area and P_{wet} the perimeter of the duct which provides shear stress on the flow (the "wetted" perimeter). For a circular pipe, the equivalent diameter turns out to be the actual diameter. This approach works well for many ducts, including annuli, unless the cross-sectional shape distorts the flow too much; it would not work for the ducts with a cross section like those of Fig. 4.9.

If heat is being transferred to the fluid through the pipe wall, there exist a temperature distribution in the fluid and a temperature gradient at the wall just as there was a shear stress gradient. The fluid is stationary exactly at the wall, so heat conduction is the only means of heat transfer. The heat flux at the wall, $\dot{q}_w$, could in principle be found by measuring the radial distribution of temperature and determining the slope of the temperature gradient at the wall, and by definition,

$$\dot{q}_w = \lambda \left(\frac{dT}{dr} \right)_w \tag{4.57}$$

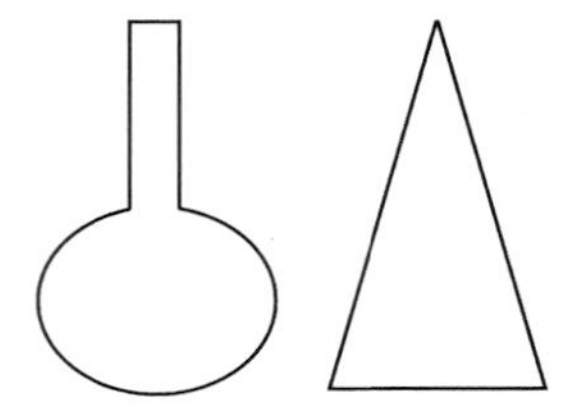

Fig. 4.9 Shapes for which the equivalent diameter concept would not work

Instead, the rate of heat transfer is usually found by measuring the rise in mean temperature over a length of the pipe in question. With a uniform heat flux distribution and a mean entry temperature of T_o, the mean fluid temperature, a distance L downstream of the entry must be given by the heat balance:

$$\dot{m}C_p\left(\overline{T} - T_0\right) = \dot{q}_w P_w L \tag{4.58}$$

where C_p is the fluid specific heat and $\dot{m}$ the mass flow rate. It is postulated that the heat flux is proportional to the temperature difference with a constant of proportionality $\hbar$ known as the heat transfer coefficient:

$$\dot{q}_w = \hbar\left(T_w - \overline{T}\right) \tag{4.59}$$

where T_w is the wall temperature and $\overline{T}$ the average temperature of the flow. If the wall temperature is measured at the distance L, Eqs. (4.58) and (4.59) allow the heat transfer coefficient to be found. (Equation (4.59) was proposed by Newton (1642–1725) and is sometimes referred to as Newton's law of cooling.) Because of the limitations of the assumption, it is often necessary to consider that the heat transfer coefficient *is* $\hbar(T)$.

The heat transfer coefficient is the parameter used in design, but it is an experimentally determined coefficient, so it must be catalogued in some manner. This is achieved by comparing it with the heat flux that would result from the same temperature difference if the fluid was not moving and the heat was transferred only by conduction. The ratio is the Nusselt number, Nu, defined for a pipe flow as

$$Nu = \frac{\hbar d}{\lambda} \tag{4.60}$$

A certain arbitrariness in the choice of linear dimension will be noticed. There is no independent linear dimension associated with the temperature difference, so the pipe diameter is used. The same problem arises with Re, and it is essential to check the definition of any dimensionless parameter. Nusselt numbers are given as correlations and, as expected from Eq. (4.48), have the form

$$Nu = function(\mathrm{Re}, \mathrm{Pr}) \tag{4.61}$$

4.4 Natural Convection

The purpose of this section is to illustrate natural convection as a branch of the sections above and also to further illustrate the approach to making equations dimensionless as a means of identifying the controlling dimensionless groups. Natural convection in single-phase fluids, as opposed to permeable media, plays very little part in geothermal engineering because there are no volumes of fluid left unaffected by forced flows. There are two very minor exceptions, natural draft cooling towers (of which there is only one in a geothermal power station, in New Zealand) and a suggestion that natural circulation can occur in closed wells.

Natural convection takes place in the non-Newtonian fluid of the mantle, in the magma adjacent to subducting crust and in the permeable medium of the solid crust near a geothermal resource before any fluid motion is induced by the discharge of wells or the injection of cooled water. That in the mantle and magma is complicated by the fact that the geometry is spherical and the Earth is rotating, introducing a Coriolis force. The governing equations that would form a suitable starting point for analysing this difficult problem are given by Bird et al. [2007]. Some of the mantle and magma contain an internal heat source from the decay of radioactive isotopes, and possibly heat sources arising from chemical phase change following temperature variations.

Natural convection is best introduced in two-dimensional rectangular coordinates, in a semi-infinite open body of fluid. Consider the flow induced around a vertical flat plate held at a uniform temperature in a fluid of infinite extent and initially at temperature T_∞ (Fig. 4.10).

The plate is initially at the same temperature as the fluid in which it is immersed; the fluid is stationary, horizontal pressure gradients are zero and the vertical pressure gradient is hydrostatic. At time $t = 0$, the plate temperature becomes T_w everywhere, and heat conducts from it into the fluid, the density of which decreases with temperature. The hydrostatic pressure distribution adjacent to the plate, in the heated fluid, now differs from that further away from the plate, and the heated fluid rises. The vertical pressure gradient varies continuously with distance away from the plate, until it becomes that of the unheated fluid, which is the initial hydrostatic gradient. The warmed fluid is said to rise as a result of an upward buoyancy force, but that statement implies that the rest of the fluid plays no part when in fact any element of fluid which has had its temperature increased experiences a couple due to the difference between the vertical pressure gradient at its location and the hydrostatic gradient in the unheated, stationary fluid. In setting up an experiment of this type, a very large vessel is required, as the plate must be in an effectively semi-infinite space. This will be discussed again later, because the vertical pressure gradient within the rising plume of the entire geothermal resource can be estimated from well measurements.

Axially symmetric radial geometry about a vertical z-axis is more applicable to internal convection in vertical tubes and resource plumes above a deep plutonic heat source. The governing equations are the continuity, Navier–Stokes and energy equations for a fluid with constant viscosity and thermal conductivity, but the

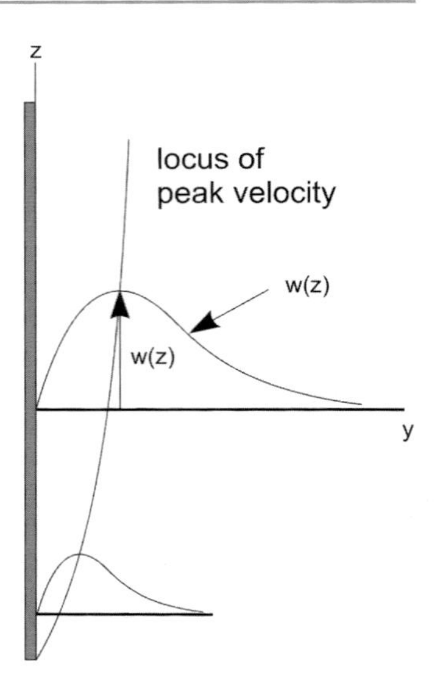

Fig. 4.10 Showing the development of the velocity distribution on a vertical flat isothermal hot plate in a fluid

variation of density with temperature must be permitted as it is the cause of the fluid motion. The usual approximation is due to Boussinesq (1903) and involves the assumption that density is constant except in the gravitational term where it is allowed to vary with temperature according to an expansion coefficient β (°C^{-1}). With this assumption, the z direction (vertical) momentum equation becomes [see for comparison Eqs. (4.32) or (4.33)]

$$\frac{\partial w}{\partial t} + w\frac{\partial w}{\partial z} + v\frac{\partial w}{\partial r} = g\beta(T - T_0) - \frac{1}{\rho}\frac{\partial P}{\partial z} + \frac{\mu}{\rho}\left(\frac{\partial^2 w}{\partial z^2} + \frac{1}{r}\frac{\partial}{\partial r}\left(r\frac{\partial w}{\partial r}\right)\right) \tag{4.62}$$

This equation is one of the complete set of continuity momentum and energy, but it is used in isolation here to illustrate the way natural convection equations are made dimensionless. Unlike in forced flow, there is no reference velocity as defined in Eq. (4.42). Instead velocities can be made dimensionless by nominating a reference velocity as $\mathrm{v}_{\mathrm{ref}}$ and then later adopting the most suitable form for it; thus,

$$w_D = \frac{w}{\mathrm{v}_{ref}} \text{ and } v_D = \frac{v}{\mathrm{v}_{ref}} \tag{4.63}$$

Making these substitutions and reducing the dimensionless equation to its simplest form, the remaining dimensional parameters will be found to form a group representing the buoyancy term. This group is called Grashof number, Gr:

$$Gr = \frac{g\beta(T - T_\infty)L^3}{\nu^2} \tag{4.64}$$

Once the energy equation is included, it is found that the group controlling the buoyancy term is Gr.Pr which is given the name Rayleigh number. The Rayleigh number signifies the magnitude of the natural convective driving forces and is used in both porous media and fluid flows. It appears in correlations of Nusselt number, from which heat transfer rates can be calculated.

4.5 Thermal Conduction in Solids

Thermal conduction takes place in both solids and fluids, but in solids, it forms a class of heat transfer investigations that is amenable to mathematical solution, if by rather difficult methods in many cases, e.g. see Carslaw and Jaeger [1946 reprinted 2000]. In engineered systems such as heat exchangers, heat transfer by conduction within the fluids is of very minor significance compared to convection, but it can be very important in its effect on the temperature of solids (e.g. pipe flanges and structural components). The thermal conductivity of rock is small, so the centre temperature of large masses containing decaying radioactive isotopes can reach high temperatures, which is the appeal of enhanced geothermal systems technology.

Because there is no motion, the energy equation alone is required for solution, and it reduces from Eq. (4.27) to

$$\frac{\partial T}{\partial t} = \kappa \nabla^2 T + \frac{\dot{H}}{\rho C_P} \tag{4.65}$$

in which $\nabla^2 = \frac{\partial^2}{\partial z^2} + \frac{1}{r}\frac{\partial}{\partial r}\left(r\frac{\partial}{\partial r}\right)$ for axisymmetric radial coordinates and $\nabla^2 = \frac{\partial^2}{\partial x^2} + \frac{\partial^2}{\partial y^2} + \frac{\partial^2}{\partial z^2}$ for rectangular Cartesian coordinates.

If Eq. (4.65) was made dimensionless, two groups would appear, one for the heat source and the other known as the Fourier number. The heat source group is of lesser interest here, so suppose it is zero and the equation is reduced to one linear dimension, for example, to describe heat conduction along a bar or through a plate or slab. Then Eq. (4.65) reduces to

$$\frac{\partial T}{\partial t} = \kappa \frac{\partial^2 T}{\partial x^2} \tag{4.66}$$

Choosing characteristic time, length and temperature (or temperature difference) as reference values, t_{ref}, x_{ref} and ΔT_{ref}, respectively, and introducing dimensionless variables:

$x_D = \frac{x}{L}$, $t_D = \frac{t}{t_{ref}}$ and $T_D = \frac{T}{\Delta T_{ref}}$, then the equation becomes

$$\frac{\partial T_D}{\partial t_D} = \left(\frac{\kappa t_{ref}}{x_{ref}^2}\right)\frac{\partial^2 T_D}{\partial x_D^2} \tag{4.67}$$

in which the dimensionless group of characteristic parameters is known as the Fourier number. The group x_{ref}^2/κ (s) is a characteristic time and could be introduced to bring the equation to its simplest form. When hot fluid is placed in sudden contact with a conducting material, e.g. steam admitted to a cold well casing or pipe, it is occasionally helpful to know how long it takes the pipe wall to reach the steam temperature throughout its thickness. Carslaw and Jaeger [2000] provide the solution, but for an order of magnitude estimate, the characteristic time just mentioned can be used. For a steel pipe, $\kappa = 15\text{E-}6\text{m}^2/\text{s}$, so a 10 mm wall thickness responds in about 7 s. As the heat flux conducts into the material, some of the heat is required to raise the temperature of the material locally, because it can only conduct further if the temperature is high enough to establish the necessary gradient. Thus the rate of penetration depends not only on the thermal conductivity but on the specific heat and density, i.e. on the thermal diffusivity.

The solutions for a solid isothermal circular disc of infinite extent penetrated by a hole, the perimeter of which is raised to a high temperature or has an applied heat flux, are similar to certain solutions of flow in permeable formations and will be introduced in Chap. 9. Thermal conduction problems were amongst the first to be solved by finite difference methods, form an easy introduction to the latter and are discussed in Chap. 13 in relation to reservoir simulation.

A well-documented problem of internal heat generation is that of a long circular cylinder with uniform internal heat generation and an isothermal outer surface at temperature T_0 because this arrangement resembles the fuel rod of a nuclear reactor (in fact, a nuclear fuel rod has a non-uniform heat generation rate both radially and axially, but the solution is nevertheless used for first estimates). Internal heat generation resulting from exothermal chemical reaction is important in laying concrete because the setting reaction is exothermic, and for large dams, an estimate of centre temperature reached must be made in order to define the volume of liquid cement which can be added in a single pour, since allowing the temperature to rise reduces the strength of the final product. Neglecting axial conduction, Eq. (4.65) in steady-state form reduces to

$$0 = \frac{\kappa}{r}\frac{\partial}{\partial r}\left(r\frac{\partial T}{\partial r}\right) + \frac{\dot{H}}{\rho C_P} \tag{4.68}$$

and with symmetry about $r = 0$ and the outer surface temperature held at zero, this integrates to give a parabolic temperature distribution with a centre-to-surface temperature difference of

$$\Delta T = \frac{\dot{H}a^2}{4\lambda} \tag{4.69}$$

Carslaw and Jaeger [2000] provide solutions for the centre temperature of a slab and a sphere, both with internal heat generation at rate $\dot{H}$, with zero initial temperature and surface temperature held at zero throughout. The slab thickness is 2 L and the sphere has a radius a. Transient solutions are provided, leading to the steady state for which the centre-to-surface temperature differences are, respectively, for slab and sphere:

$$\Delta T = \frac{\dot{H}L^2}{2\lambda} \text{ and } \Delta T = \frac{\dot{H}a^2}{6\lambda} \tag{4.70}$$

In the case of the sphere, the equilibrium temperature difference is 95 % established after a time 0.4 a^2/κ and for the slab approximately 2 L^2/κ. For rock, $\kappa = 1.2\text{E-}6\text{m}^2/\text{s}$ is a reasonably representative value, so response times are of order of magnitude 25,000 years for linear dimensions of 1 km and 2.5 million years for 10 km.

Finally, it will be noticed that all of the centre-to-surface equilibrium temperature distributions are of the same algebraic form, so this conductive temperature difference could be used as the reference temperature difference for appropriate problems.

4.6 Flow in Permeable (Porous) Media

The materials normally dealt with by earth scientists, groundwater, geotechnical and geothermal engineers, namely, rocks, are more often porous than totally solid, but that does not mean that they are permeable. Yet the literature on the flow in permeable materials has become known as "flow in porous media". Any porous material mentioned here is assumed to be permeable also.

4.6.1 Darcy's Law

There is a very wide range of porous and permeable materials, all with the common factor of small-scale flow passages within them having some degree of randomness. The application of the equations developed above is easy with well-defined flow paths and with regular boundaries and tidy, easily prescribed boundary conditions but far from easy with porous media.

In a footnote by Bear and Cheng [2010] (ascribed to Narasimhan), it is suggested that Darcy (1856) introduced his "law" after considering the solutions of the Navier–Stokes and continuity equations for the steady laminar flow of incompressible viscous fluid through a capillary tube, which had earlier been investigated by Poiseuille (1838). The solution is Eq. (4.39) above which, slightly rearranged, is

$$u = -\frac{a^2}{4\mu}\left(1 - \left(\frac{r}{a}\right)^2\right)\left(\frac{dP}{dx}\right) \tag{4.71}$$

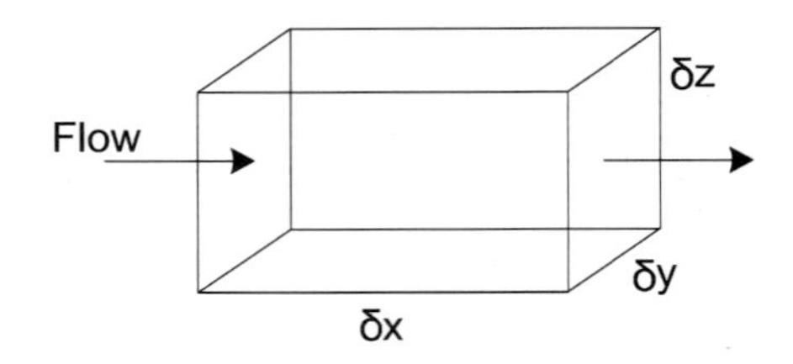

Fig. 4.11 Control volume for the derivation of the continuity equation for flow through a permeable medium

Being confronted with the problem of random passages through a permeable porous material instead of Poiseuille's capillary tubes, it is suggested that Darcy might have written:

$$q_v = -\frac{k}{\mu}\left(\frac{dP}{dx}\right) \tag{4.72}$$

where k is the permeability, an experimentally derived property which represents the equivalent of the capillary geometry for the passages of the material. This apparently simple experimental law has found extensive application to studies of the flow of groundwater, oil and gas through natural materials and to flows of other fluids through manufactured porous material.

4.6.2 Reducing the Set of Continuity and Momentum Equations to a Single Equation

The equations governing the flow through a permeable material in Cartesian coordinates are expected to be reductions of Eq. (4.6) and Eqs. (4.8)–(4.10), but Darcy's law is of itself the solution to the set of Eqs. (4.8)–(4.10). Combining the law with the continuity equation provides an equation governing the flow in the permeable, porous media. The formal approach begins with the continuity equation, derived similarly to Eq. (4.6) using Fig. 4.11.

The volume within the element available for storage of fluid is $\phi\, \delta y \delta x \delta z$, the pore volume, since ϕ is the porosity. The only area available for flow in the $\delta y \delta z$ planes at entry to and exit from the control volume is that of the intersected pores, but an overall view is taken by defining a volumetric flux q_v (m^3/s per m^2 or m/s) over the whole area $\delta y \delta z$, which is simply a velocity u (m/s) in a different form, as if the fluid has the whole cross-sectional area to flow through. The actual velocity of the fluid entering the open pores is much greater than this. Thus the mass flow rates entering and leaving the control volume in a time δt can be written as

$$\text{Mass entering} = (\rho q_v) dydz \tag{4.73}$$

$$\text{Mass leaving} = \left(\rho q_v + \left(\frac{\partial(\rho q_v)}{\partial x}\right).\delta x\right)\delta y \delta z \delta t \tag{4.74}$$

The volume available for storage is the pore volume, and the mass stored in this volume must equal the difference between that entering and leaving:

$$\phi.\delta\rho = -\frac{\partial(\rho q_v)}{\partial x}\delta t \tag{4.75}$$

Or

$$\phi\left(\frac{\partial\rho}{\partial t}\right) = -\frac{\partial(\rho q_v)}{\partial x} \tag{4.76}$$

giving the full equation as

$$\phi\frac{\partial\rho}{\partial t} + \left(\frac{\partial(\rho q_{vx})}{\partial x} + \frac{\partial(\rho q_{vy})}{\partial y} + \frac{\partial(\rho q_{vz})}{\partial z}\right) = 0 \tag{4.77}$$

which is the same as Eq. (4.6) with the introduction of porosity.

Instead of deriving momentum equation equivalents to Eqs. (4.8)–(4.10), Darcy's law can be merged with the continuity equation, (4.77), to give

$$\phi\mu\frac{\partial\rho}{\partial t} = \frac{\partial}{\partial x}\left(k\rho\frac{\partial P}{\partial x}\right) + \frac{\partial}{\partial y}\left(k\rho\frac{\partial P}{\partial y}\right) + \frac{\partial}{\partial z}\left(k\rho\frac{\partial P}{\partial z}\right) \tag{4.78}$$

if dynamic viscosity is constant. This equation may be simplified further according to the particular type of problem being analysed.

4.6.3 If the Fluid Is Incompressible

For constant k and ρ, Eq. (4.78) reduces to

$$\nabla^2 P = 0 \tag{4.79}$$

which is known as Laplace's equation and has many applications (e.g. thermal conduction, aerodynamics). There can be no pressure variations in an incompressible fluid, i.e. if density is constant.

4.6.4 If the Fluid Has Small Constant Compressibility c

The aim is to simplify Eq. (4.78). Compressibility is defined as

$$c = \frac{1}{\rho}\left(\frac{\partial\rho}{\partial P}\right)_T \tag{4.80}$$

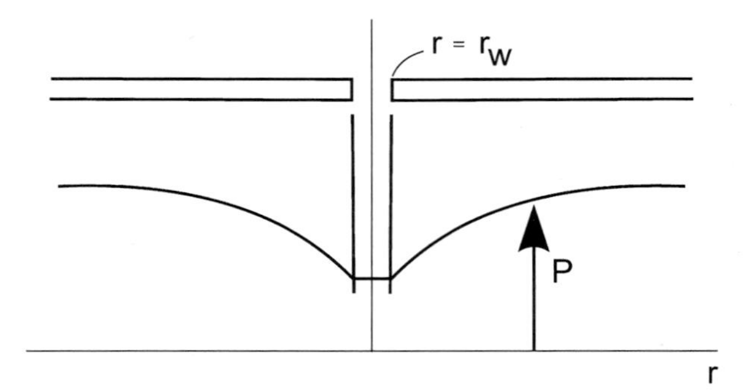

Fig. 4.12 The pressure distribution around a discharging well some time after the start of discharge—the cone of depression

from which it can be seen that if the compressibility is small so also is $(\partial\rho/\partial P)_T$ and thus $\rho c \approx$ constant. The left-hand side of Eq. (4.78) can now be improved upon by noting that

$$\frac{\partial \rho}{\partial t} = \frac{\partial \rho}{\partial P} \cdot \frac{\partial P}{\partial t} = \rho c \frac{\partial P}{\partial t} \tag{4.81}$$

Adopting the first term on the right of Eq. (4.78) only, for simplicity, making the substitution and also assuming k is constant gives

$$\phi \mu c \rho \frac{\partial P}{\partial t} = k \frac{\partial}{\partial x}\left(\rho \frac{\partial P}{\partial x}\right) \tag{4.82}$$

Taking ρc into the right-hand bracket on the grounds that it is almost constant and with c and k already defined as constants, the final equation is

$$\frac{\partial P}{\partial t} = \left(\frac{k}{\phi \mu c}\right) \nabla^2 P. \tag{4.83}$$

This is the Fourier equation and, in radial geometry, it is the basis of tests to measure formation properties. Various solutions are described in Chap. 9, but in the meantime, when applied to the case of a horizontal permeable formation of uniform thickness and infinite extent penetrated by a discharging well, its solution shows that the radial pressure distribution at any time declines with increasing steepness towards the well as shown in Fig. 4.12. The cylindrical area of the formation open to the well is called the sandface, and the decline in pressure around the well is often called the "cone of depression", both terms coming from groundwater hydrology. For a given formation, the cone of depression is a function of time and flow rate being extracted, the pressure decline varies non-linearly with radius, and the pressure reduction gradually extends outwards.

References

Bear J, Cheng AH-D (2010) Theory and applications of transport in porous media, vol 23. Springer, Heidelberg
Bird RB, Stewart WE, Lightfoot EN (2007) Transport phenomena. Wiley, New York
Carslaw HS, Jaeger JC (2000) Conduction of heat in solids. Oxford Science, Oxford, originally 1946
Dryden HL, Murnaghan FP, Bateman H (1956) Hydrodynamics. Dover, Kent
Kays WM (1966) Convective heat and mass transfer. McGraw-Hill, New York
Lamb H (1906) Hydrodynamics. Cambridge (Dover, 1945)
Shemenda AI (1994) Subduction: insights from physical modeling. Kluwer Academic Publishers, Dordrecht

5 Geothermal Drilling and Well Design

This chapter begins by describing the construction of geothermal wells. The drilling process and equipment follow, but this is under continuous development, and the techniques described have been chosen to illustrate the issues rather than the latest practice. A section on well design begins with the basics of stress analysis and theories of failure and then describes casing selection and particular modes of failure to be designed against, in both the construction of the well and its later operation. Wellhead equipment is described. The units in this chapter are mixed, some SI and some imperial, because the materials required for drilling are manufactured in imperial units.

5.1 Introduction

Geothermal wells must be designed and constructed in an engineering sense—they are not simply holes in the ground but are made up of lengths of steel tubing, referred to as casing, cemented into the ground with a valve on the top. Their function is to connect the resource to the surface in such a way that discharge from the well is controllable at all times, and well design follows tight specifications. Once it begins, drilling incurs a high cost over a short period of time and calls for careful management and an experienced drilling team. Difficult decisions often have to be made on the spot if problems arise. Although well design, drilling equipment and the procedures to be followed during drilling and completion of a well are all soundly based on engineering principles, the practice is almost an art rather than a science. The materials and conditions that the drill will encounter have a reasonably high degree of uncertainty, the components are often highly stressed and the whole activity is "heavyweight"—there is the potential for things to go badly wrong without a careful and experienced manager (drilling superintendent).

High-temperature geothermal wells range in depth from a few hundred metres to about 4,000 m, wells at Kakkonda, Japan, being the prime examples with

A. Watson, *Geothermal Engineering: Fundamentals and Applications*,
DOI 10.1007/978-1-4614-8569-8_5, © Springer Science+Business Media New York 2013

temperatures of over 350 °C. Petroleum wells may be much deeper, and the deepest well drilled of any type has a depth of 12,000 m. Formation temperature is a major limitation for geothermal wells due to its effect on the strength of the drilling equipment. Wells are drilled to quite specific targets set by geologists, either faults, formations or an intrusion or pluton, and often they are not straight and vertical but curved. It is possible to drill them geometrically vertical to within a few metres, but sometimes wells are deliberately deviated by hundreds of metres to reach targets which cannot be reached from immediately above. Drilling is an expensive activity, and deviated wells are more expensive than vertical ones to reach the same target depth, so the final choice of well track is the result of optimisation of several factors, including pipeline routes and separator locations for delivery and processing of the discharge or delivery of separated water in the case of injection wells.

A common diameter for the inner casing of a well to supply a power station is 9 5/8″ (0.244 m), but even if it had a diameter of 20″, as some modern wells have, even a 500 m deep well would be a very slender structure. This affects both its performance as a duct for fluid flow and the care required during construction. Slender tubes bend easily, and like bridges, wells are at most risk of failure during their construction. The drilling process follows a detailed plan and takes place at a prepared site; it is a source of environmental impact, and site requirements will be better understood if well construction and the actual drilling process are explained first.

5.2 Well Construction

Figure 5.1 is the design for a 2,000 m deep production well.

The upper 800 m is made up of three concentric casings, referred to as casing strings when all the separate 10 m lengths from which they are formed are screwed together, namely:

- A 90 m length of casing 20″ in diameter cemented into a hole 92 m deep and 26″ diameter, called the surface casing.
- A 250 m length of 13 3/8″ casing cemented into a hole 252 m deep and 17 ½″ diameter. This is called the anchor casing because when the well is finished, it will carry the wellhead valve.
- An 800 m length of 9 5/8″ casing cemented into a hole 802 m deep and 12 ¼″ diameter. This is called the production casing because the produced fluid will flow through it.

The production casing seals off the well from the formations through which it is drilled, preventing cross-contamination of produced fluids and shallow groundwater. Sometimes a short length, say 30 m, of "conductor" casing is placed first, where an unsupported hole in the upper levels of the ground would collapse before 90 m depth had been reached. In some circumstances, more casings might be needed to drill to 2,000 m, for example, an intermediate casing between anchor and production; this is dependent on the vertical temperature profile, as explained later.

The lower part of the well is an 8 1/2″ diameter hole into which is placed a 7 5/8″ diameter steel tube perforated by a large number of axially orientated rectangular

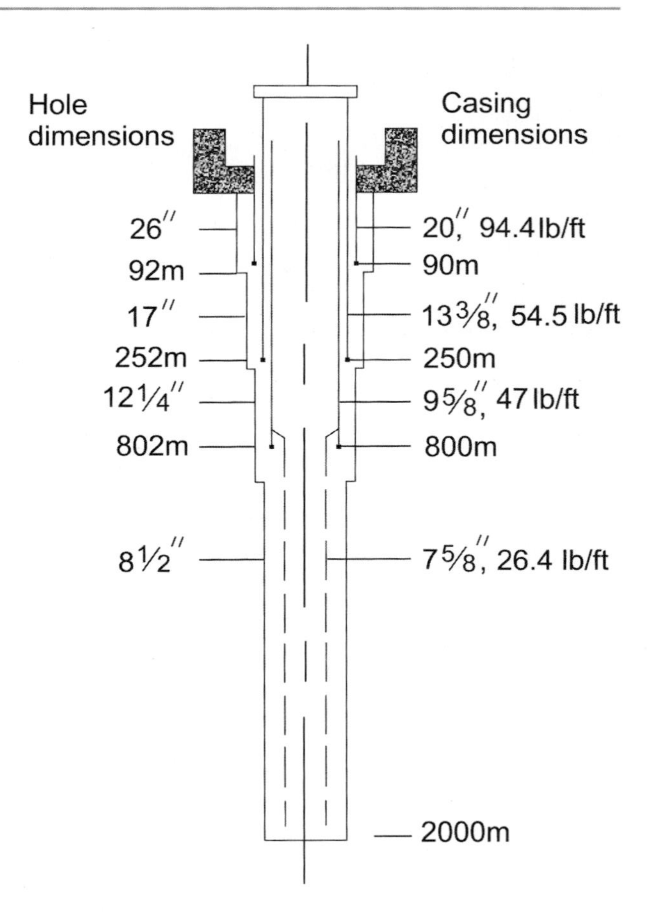

Fig. 5.1 An example of a geothermal well

slots which, in contrast to the production casing, are there to allow fluids to flow easily from the penetrated formations into the well and up to the wellhead. Although not specified on the diagram, the slotted liner either stands on the bottom of the hole so that it protrudes a few metres into the production casing or is attached to the bottom of the production casing by a special fitting and hangs from it, ending just above the bottom of the hole. The drawing specifies the weight/ft of all the casings, which are made up of standard 10 m lengths screwed together. For example, the slotted liner has a weight of 26.4 lb/ft, giving it a total weight of 104 tonnes. This is important information, since lifting equipment with this capacity must be used to put it in place. If it is suspended from the production casing when finally in position, it will produce a tensile force that the production casing must be capable of withstanding. These are all aspects of the well design process, as are the numbers of casings and the depths of each.

The casings are cemented into the ground, and where they pass through an already cemented casing, the annular gap is filled with cement. Once the hole is drilled for the first (shallowest) length of casing, the casing is gradually assembled in the hole in 10 m lengths. The piece which will be lowest in the well is first fitted with a "shoe", to protect the bottom edge, and a non-return valve. It is lowered into the well, held while the next piece is screwed to it, the pair are lowered and held and the process repeated until the required length is assembled and in place, hanging in the hole. A volume of cement slightly greater than the volume of the annulus formed by the outer surface of the tube and the surrounding hole is pumped into the tube, and a rubber piston inserted after it. Water is pumped into the tube behind the piston, driving it downwards and causing the cement to flow upwards to fill the annulus, eventually overflowing at the surface. Once the overflow has been seen, the annulus is assumed to be full of cement, which is allowed time to set. After the cement has set, a flange is welded to the top of the casing, a drilling wellhead fitted and a collection of blowout preventers and valves, described later. The top of this casing and all successive ones is a few metres higher than it will be eventually, to carry the drilling fluids to a high level; when it has been cemented and drilling is taking place through a smaller diameter cemented casing, it can be cut down to its final height above surface. To return to the process, drilling resumes after the first casing is set in place, extending the well to 252 m below the surface with a hole of 17 1/2″ diameter. The rubber piston, non-return valve and any cement remaining inside the casing are drilled through.

The last cemented casing stands a few metres above ground level so that the returning drilling fluid is lifted to a level from which it can flow to screens and coolers. In New Zealand, geothermal wellheads stand in concrete cellars, which act as containment for drilling fluid spills during drilling and make operational well-head maintenance easier, although this is not a universal procedure. In their final form, the well casings terminate at cellar floor level, the extra length required for drilling wellheads being cut off as construction proceeds. The anchor casing is an exception, as it is left standing above the cellar floor and carries the casing head flange (CHF) to which is attached the permanent main shut-off valve (the "master valve") and the wellhead assembly. Any pressure within the well produces forces tending to lift the anchor casing out of the ground when the valves are shut, hence its name. Once the well is completed, the CHF becomes the datum for depth measurements in the well. Throughout the life of a well, the CHF moves relative to the ground around it due to thermal expansion according to the temperature distribution in the well. The movement is of order 10–100 mm, which, although important enough to be taken into account in designing the pipeline attached to a wellhead, is insignificant for well depth measurements.

While a production casing diameter of 9 5/8″ as described in this example is perhaps the most commonly used for power production, larger or smaller diameters are sometimes specified. For a shallow formation producing low-pressure steam, the extra cost of a larger diameter casing might be justified by extra electricity generation as a result of smaller pressure drop in the production casing. Smaller diameter and hence lighter weight casings, typically 4″, might be the only way of drilling a well in a remote location, if access is restricted to only a small truck mounted rig, and this is also the approach used to drill geothermal wells for supplying hot water to domestic and light commercial properties in many places.

5.3 The Drilling Process

5.3.1 Drilling a Hole

Holes are drilled into homogenous materials such as metal by a cutting action; a workshop drill bit for metal has two edges that cut as the bit rotates and two spiral slots along its length to guide the cuttings out of the hole. Without this escape route for the cuttings, the bit cannot penetrate. Rock is rarely homogenous and is brittle, i.e. strong in compression and weak in tension. It cannot be cut by the type of drill bit just described. Instead a rock bit makes progress by crushing, which is achieved by fitting the end of the bit with toothed conical rollers that follow a horizontal circular path as the bit is rotated—for pictures and a history, see ASME [2012]. The drill bit fits on the end of a steel tube typically 10 m long, which has threaded ends so that more tubes can be added to allow the drill to reach greater depths. Consistent with the crushing action, the steel tubes immediately adjacent to the bit are heavier than normal and are called drill collars; they are selected to give the proper weight on the bit for the rock type being drilled. The normal drill pipe is added in 10 m lengths as the well progresses.

The rock cuttings are in the form of fine gravel and sand and are flushed to the surface by the flow of a dense, viscous drilling fluid which is pumped down to the bit through the drill string, the name given to the connected steel tubes carrying the bit. Cuttings must be cleared from the hole if the drill is to make progress, preferably back to the surface, although as will be seen later, it is possible for them to be pushed into the formation. The drilling fluid emerges from the bit through nozzles as a set of powerful jets which pick up the cuttings and carry them back to the surface in the annular gap between the drill string and the hole. The cuttings sometimes fail to lift completely and become trapped, preventing the drill bit from turning; in the extreme, the drill string may break by twisting off. Retrieving broken equipment from a well is known as "fishing", a frustrating activity by its very nature and because of the expense of having the rig and crew busy but not making progress. Massei and Bianchi [1995] have reported on non-destructive testing of drill string components to avoid failures in service. The drilling fluid is usually a mixture of water and bentonite clay with minor additives, which has the appearance of a black slurry or thin mud and by virtue of its density and viscosity is capable of lifting the cuttings. At the surface, the cuttings are sieved out, and the cleaned mud is continuously recirculated—it must also be cooled when the well is penetrating hot formations. The cuttings are collected and inspected by the rig geologist, who is able to map the stratigraphy of the well, i.e. the formations through which the bit passes, and the depths of their interfaces.

5.3.2 The Drilling Rig

Geothermal wells are drilled using a rotary drilling rig of the type developed for petroleum drilling; a rotary rig is one in which the bit is rotated about its axis. The first oil wells were drilled by a "percussion" method in which the bit was

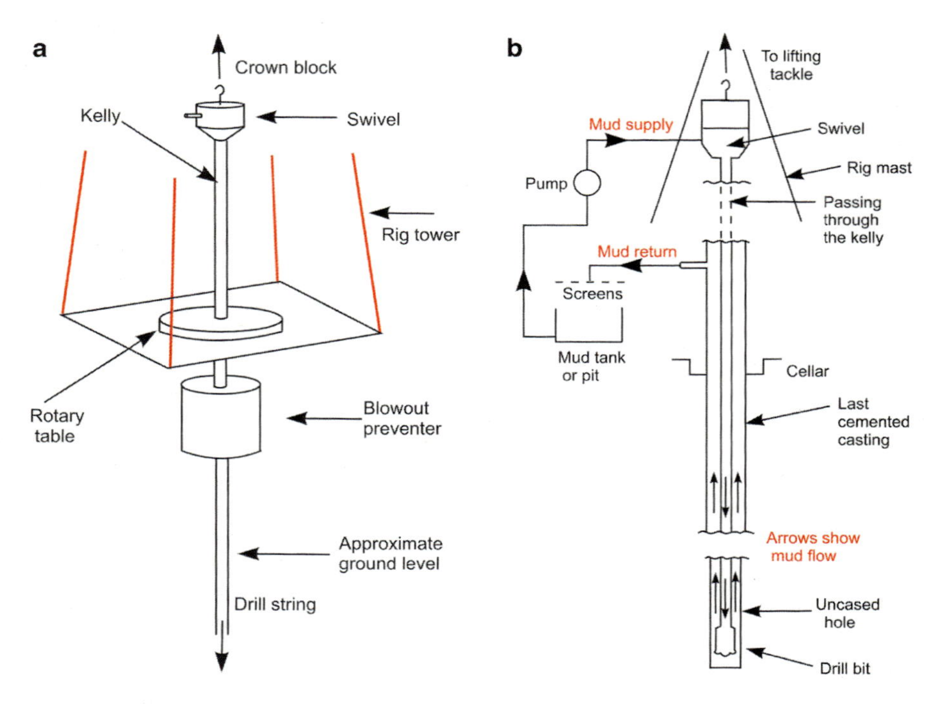

Fig. 5.2 Showing the main components of a drilling rig

suspended by a flexible cable and repeatedly lifted and dropped on the bottom of the hole, so that it chiselled its way downwards by virtue of its large mass and pointed end. Such rigs were more sophisticated than might be thought at first sight, for example, the cable being of twisted steel strands imparted some rotation to the bit, and this was studied and used to advantage to achieve faster penetration.

The main features of the drilling rig and mud supply system are shown in Fig. 5.2. Part a of the sketches shows details of the rig floor and Part b an overall view. The swivel can be used as a reference point.

Drilling rigs come in a variety of sizes, from small ones permanently mounted on a truck to large ones that comprise several truck loads and must be assembled on site. The drill bit must be rotated, and the traditional surface drive is explained here; more modern alternatives are discussed later. Drilling rigs are recognisable as a tall derrick or tower structure often of open framework, and many of the components have traditional names originating in early oil drilling. The tower is essentially a crane, which must be capable of lifting the entire length of the drill string when the bit has reached the designed depth and also the full lengths of each casing string used in the construction of the well. To remove the bit from the hole, the entire drill string must be lifted 10 m at a time, held while the top length is unscrewed and removed, lifted again, and so on, and the drill tower is the means of achieving this. The weight it can carry dictates the depth and diameter of well that can be drilled, for example, a rig capable of drilling a well with an open hole diameter of 8 ½″, and a depth of 3,000 m may need to provide a pull of 150 tonnes and be capable of a power output of up to 1,000 hp. At the top of the tower is the "crown block", a multiple pulley block coupled by a wire rope to the "travelling" block to which the hook is fitted, the number of pulleys providing the mechanical advantage for lifting. A mechanical advantage of 10:1 means that the tension in the wire rope is 1/10th of the hook load, an advantage, but the travelling block must rise 10 times the distance through which the weight is lifted, a disadvantage. Hook loads may be up to 400 tonnes, and the motive power is usually provided by a diesel engine or electric motor. Lengths of drill pipe or casing stand vertically in racks alongside the tower, ready for use.

The activity of adding or removing a length of drill pipe or casing takes place on the drill floor, a platform a few metres above ground level. The entire drill string is very slender so is weak in torsion, and it twists so that the bottom lags behind the top as it is rotated. It must transmit torque to the bit, but it must also be free to move axially while transmitting this torque as the well deepens, and drilling mud must be continuously pumped through it to lift the cuttings back to the surface. This poses a mechanical problem at the rig, to which the traditional solution is a horizontal circular disc known as the rotary table supported on the rig floor by bearings around its perimeter. It can be rotated about its axis by a diesel or electric motor. In the centre of the rotary table is a square or hexagonal hole. To the top of the drill string, which is slightly longer than the depth of the hole, is attached a special length of pipe known as the "kelly", which is square or hexagonal in cross section and fits in a matching bushed hole in the centre of the rotary table. Rotating the rotary table thus rotates the kelly and hence the entire drill string. The kelly is about 13 m long, deliberately longer than each drill pipe, and moves downwards as the drilling proceeds. When the upper end of the kelly approaches the rotary table, the rotation is halted, and the drill string is lifted until the bottom of the kelly is clear above the top of the rotary table, in preparation for adding another length of drill pipe. Wedges (slips) are then jammed into the space between the drill string and the bush, thus supporting the drill string to prevent it falling back into the hole; the kelly is unscrewed, another 10 m length of drill pipe added to the top of the drill string and the kelly reconnected. The wedges are specially shaped to fit this square or hexagonal hole and have teeth so that the weight of the drill string causes them to grip it tightly. Once all joints have been tightened up, the drill string is lifted, the wedges are removed, the drill string is lowered, and bit rotation begins again. The detailed engineering of the rotary table, bush and Kelly is aimed at reducing wear and tear caused by the necessary repeated screwing and unscrewing and the use of wedges.

Two alternative ways of rotating the bit are available, both using dedicated motors which align axially and effectively become part of the drill string. A mud motor is a positive displacement type which works in the same way as a monopump. In the latter, a single spiral rotor fits within an apparently matching tube; however, a cavity exists between the two and moves along the axis as the rotor turns, carrying any fluid in it from one end of the pump to the other. As a motor, the fluid is pressurised, enters the motor at one end and drives the rotor as it moves to the low-pressure exit end. In the form used for drilling, the fluid is drilling mud, and the motor is fitted into the drill string near to the bit. The other method is called a top

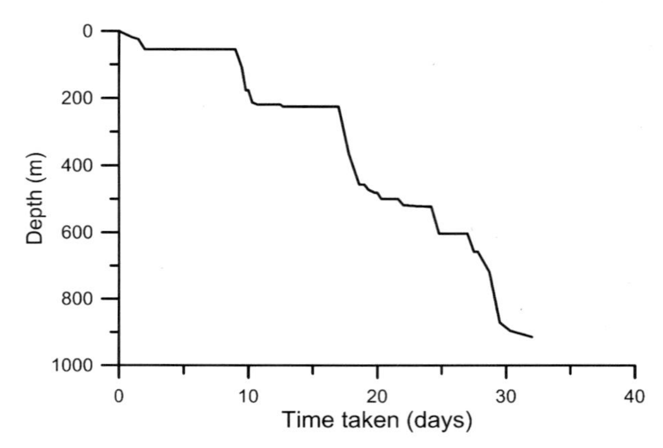

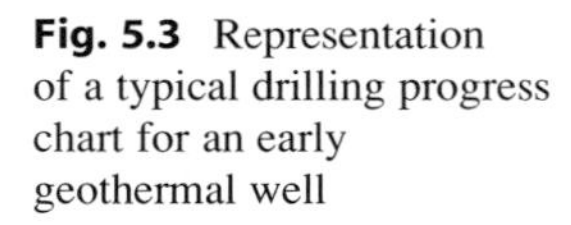
Fig. 5.3 Representation of a typical drilling progress chart for an early geothermal well

drive and consists of a motor, electric or hydraulic, which is carried near the crown block of the rig and drives the drill string, which is suspended from it. This type has advantages in allowing more than one length of drill pipe to be added or removed from the string at every lift.

A rotating drill string rubbing against the side of the casing through which it was operating would damage the casing and could wear a hole through it. This would make an exchange of fluids between the well and the surroundings more likely; at the very least, this could cause contamination, but it might also provide a route for hot geothermal water to reach the surface, where it would break out as steam. Such breakouts (blowouts) can only be controlled from inside the well, which requires access to the wellhead, and this might be impossible. Rubbing of the drill string on the casing can be avoided by keeping the drill string in tension, which is achieved by making the lower end heavy. Apart from the drill collars already described, the lower few lengths of drill pipe may be stiffer than the rest, to reduce twisting and keep the bit in line with the hole.

Figure 5.3 shows a plot of depth versus drilling time for a geothermal well.

The rate of penetration of the bit is a principal factor affecting the cost of the completed well, but the speed and efficiency of a drilling team in "running" the drill string into the hole and pulling it out are important also. Two periods of several days are identifiable in Fig. 5.3, during which the hole stays at the same depth, and these are when the surface and anchor casing (to 250 m) are cemented and perhaps other problems arose. Otherwise, progress is made in irregular steps. Bits may need to be changed, requiring the entire length of the drill string to be removed from the hole, unscrewed piece by piece and then reassembled with the new bit. When it is necessary to lift the entire drill string out of the hole or replace it, the kelly is not needed since no rotational torque is applied. A different lifting device called the elevator replaces the kelly and swivel between the travelling block and the top of the drill string. Rather than being a screwed connection to the drill string, the elevator simply clamps around the small diameter part of the pipe beneath the joint—the screwed fittings built into each end of each piece of drill pipe have a larger diameter than the remainder.

In practice, the graph of Fig. 5.3 is plotted alongside a stratigraphic column drawn up by the rig geologist and a detailed log of penetration rates and rig activities such as drilling fluid used, periods of cementing and reasons for making no progress. This data is used to improve the drilling of the next well in the same area. The geological stratigraphy found is essential information about the resource, used in the interpretation of well measurements and in reservoir modelling.

Commercial examples of most types of drilling equipment can be found on manufacturer's websites.

5.3.3 Drilling Fluids

It is necessary to introduce fluid into the well while it is being drilled, to pick up the cuttings and lift them to the surface, to cool the drill string and bit and to reduce wear and power requirements by acting as a lubricant. The options are drilling mud, water, aerated mud and air alone, and this list is roughly in order of popularity. A totally reliable water supply is essential whatever drilling fluid is used, because in the case of problems, the well may need to be "quenched" by injecting cold water to prevent discharge. The original mud used for oil well drilling in the USA around the turn of the century was just that, mud stirred up in a shallow pit dug into soils with high clay content. Today it is supplied to a required chemical and physical specification; the mud particle size ranges from 0.5 to 2 μm, and its main constituent is bentonite. It is mixed with water at the site and held in an open pond adjacent to the rig. It is pumped from the pond through a flexible hose and "swivel" fitting at the top of the kelly into the hollow drill string and down to the bit at the bottom of the hole, as indicated by Fig. 5.2. The swivel allows the mud delivery pipe to be stationary while supplying mud to a rotating shaft—it is essentially a box fitted with glands through which the shaft passes. A high delivery pressure is required and positive displacement reciprocating pumps are commonly used. On reaching the bit, the mud emerges from nozzles in the bit body between the rollers, as explained above. The nozzles eventually erode and must be replaced.

The mud with the cuttings carried with it rises up the annular space between the hole and the outside of the drill string and emerges from the casing at wellhead. It then passes back to the pit via a series of vibrating wire screens known as shale shakers, which remove the larger cuttings. After the screens, cutting particles up to about 200 μm diameter may remain in suspension, although the greatest proportion of solids in the mud, over 50 %, will be the particles of mud itself. The screens are usually followed with a settling tank in which the mud stream moves so slowly that the larger particles remaining sink to the bottom. Centrifugal separators may then be used to remove as much as possible of the remaining small cuttings until only 20–30 % of the solids in suspension are cuttings, of less than 30 μm. Some mud is lost during its passage up the well, being forced under pressure into the permeable rock formations through which the hole is drilled. The pores of the rock may be smaller than the mud particles, and a layer of mud may seal off the permeable zones. This may be advantageous if the formation is weak and liable to fall into the hole, but it may be disadvantageous in the long term by reducing permeability and hence the flow of geothermal fluids from the formation into the well.

By the time it reaches the surface, the mud will have increased in temperature since it was pumped from the pit, depending on the reservoir temperatures reached and the mechanical work done by the bit. The mud cools and lubricates the moving parts of the bit, and some of the chemical additives are selected for their lubricating properties; the mud is cooled at the surface during the filtration process.

An important function of the drilling mud is the prevention of blowouts, by filling the well with a high-density fluid so that the pressure at any formation can be exceeded by the hydrostatic pressure in the well. Otherwise, the removal of cuttings from the hole is the highest priority function that the mud performs, and cake formation, cooling and lubrication have lower priorities. All of these require that circulation be maintained, that is, the mud pumped into the well returns to the surface. In some geothermal resources, circulation is difficult to retain using mud, because of low pressure in the formations perhaps, and there is a need to reduce the hydrostatic pressure due to the drilling fluid. In this case, air or aerated mud (foam) is used, which has the added benefit of making faster progress in deepening the well, although at the expense of higher cost. Ultimately, if circulation cannot be retained, then the drilling fluid used is water, which carries the cuttings into the formation but is otherwise benign. Drilling without circulation is referred to as drilling blind because there is no information for the rig geologist to know what type of formation is being drilled through. Before resorting to that, it may be possible to recover circulation by injecting material to plug the loss zone until after the well is completed, when the temporary blockage can be removed by allowing the well to discharge fully open, or by acid injection to chemically remove it.

5.3.4 Drilling Site Requirements

The drilling site must be flat and is typically 50 by 100 m or larger to allow the transport vehicles easy access and placement of the rig. Fuel oils, drilling mud and materials and staff temporary accommodation must be provided. Since the site is likely to have to be cleared of vegetation, storm water drainage is important. Areas of cleared forest in volcanic soils are prone to slips if they are steep; silt run-off into natural waterways damages flora and fauna and must be prevented. One or more pits to contain drilling mud and contaminated waste water are required. If the ground is already contaminated with geothermal chemical species, then these may be unlined, so that water can percolate away leaving solid materials which can be disposed of elsewhere if necessary at the end of drilling or simply buried. The site will also have to accommodate shale shakers, mud cooling and mixing equipment and a silencer for the eventual well discharge, although the drilling rig and equipment will have departed by then. A reliable permanent water supply is essential, and the New Zealand Code of Practice for deep geothermal wells [1991] indicates that a 2,000 m deep well such as that shown in Fig. 5.1 requires a water supply of 2,000–2,500 l/min (of the order of 3,000 tonnes/day). Temporary pumps with pipelines several kilometres long are not unusual.

Depending on the shallow geology, the site may need to be consolidated by grouting, that is, drilling a pattern of holes around the well site to a depth of a few metres and pumping in cement to provide more compressive strength.

5.3.5 Blowout Prevention

Imagine a partly completed well. The anchor casing has been installed and cemented, and the hole for the production casing is being drilled. Drilling mud or water is being pumped down the centre of the drill string, and it may be returning through the annulus between the drill string and the anchor casing, or it may not be returning at all in which case the annulus will contain some stagnant fluid. Regardless, if the bit penetrates a geological formation or feature such as a fault and suddenly exposes these flow paths to a pressure much higher than previously experienced, fluid will tend to be pushed up and ejected at the wellhead, reducing the hydrostatic pressure at the bottom of the hole and increasing the discharge. The simple geometry of the casing and drill string provides an opportunity to mechanically stop this flow. A valve could be fitted on the passage through the drill string, and a valve with gates that fit around the drill string could be fitted on the anchor casing. In oil and gas drilling, the risk from blowout is very high because the fluids are inflammable, and a valve is fitted with gates to cut the drill string as a last resort. These valves are called blowout preventers, and sets are available for each casing size to be used for the well, fitted in turn as the casings are set, as the last line of defence.

5.4 Well Design

The design process is to imagine how the well might fail, select dimensions and casing depths to avoid risk both in construction and in use, calculate the forces involved and choose a casing size that will withstand those forces. When subjected to various forces, a particular component of a structure experiences a stress distribution. Failure of materials is often thought of as resulting from a single stress which exceeds a certain level, but in general, failure is more often the result of several stresses in different directions, which combine to cause failure. Failure is a general term that requires particular definition for each case—in some instances, failure means breaking a component into two pieces, but in others, allowing the stress to reach a level at which the material stretches or deflects might constitute a failure. The design process is usually one of trial and error—the forces are estimated, material grade and thickness are chosen and the stresses are calculated and compared to allowable stress limits. The design process had to be examined early in the development of geothermal drilling. Clearly, it was understood in Italy in 1912 when Larderello was first brought into use and, for liquid-dominated resources, was reported on by Dench [1970].

5.4.1 The Properties, Strength and Failure Criteria of Steel

If a bar of steel of diameter d is pulled in the direction of the axis of the bar with a force F, it extends linearly in proportion to the force applied. This is called elastic behaviour. The axial stress in the bar is

$$\sigma_a = F/\left(\pi d^2/4\right) \tag{5.1}$$

Strain is defined as the proportional extension of the bar. If the original length of the bar is L and it extends δL under the pull, the strain ε is

$$\varepsilon = \delta L/L \tag{5.2}$$

A graph of σ_a versus ε is linear but only up to a certain stress called the yield stress, after which the bar begins to stretch in a non-linear manner (disproportionately to the force applied). The behaviour is elastic only up to the yield stress, that is, it returns to its original length when the stress is removed. The linear elastic part is of slope σ_a/ε, known as Young's modulus, E, which has the value 2×10^{11} Pa for steel. Carbon steel such as is used for casing reduces in strength as the operating temperature increases. The strength reduction factor can be applied to whatever allowable design stress is chosen at room temperature.

Once the yield stress is exceeded, the steel will not return to its original length but becomes permanently deformed (extended); at stresses above yield, the steel is said to be plastic, and the next important occurrence is failure, when the ultimate tensile stress is reached and the steel breaks. For some applications, stresses must not be allowed to exceed the yield stress. In other applications, failure means fracturing, so the ultimate tensile stress may be used as the upper limit. Steel is capable of safely carrying stresses much greater than the yield, and in some circumstances, where weight of the component or cost is sufficiently important, the designed stress may be closer to the ultimate. Whatever the circumstances, some margin of safety is necessary to allow for material property variations and uncertainty in the method of calculating stress, and it is usual to adopt a safe working stress lower than the stress defined as the failure criterion. In common steel structures where weight and cost are not the controlling factors, a safe working stress of 2/3 of yield is often used, that is, components are designed so that the calculated stress is no greater than the safe working stress. In some aspects of casing design, the yield stress is regarded as the safe working stress.

Steel is also elastic in compression, up to the yield stress; however, the yield and ultimate stresses are always measured by subjecting the material to a tensile test. The yield stress is difficult to measure exactly, so the stress required to produce a 0.5 % strain (permanent strain of 0.005) may be defined as the yield stress, sometimes called the 0.5 % proof stress. The American Petroleum Institute publishes a list of standards for drilling, in which the parameters discussed here are defined.

The majority of real engineering components, including casing, are subjected to forces in all three directions, and the material fails because of the combined stresses. In casing, these stresses may be produced by a combination of tension or compression, bending (which induces tension and compression) and torsion. If the yield stress is to be used as the failure criterion, then the question is what combination of stresses on a piece of casing will cause the material to yield?

There are several theories that provide answers, of which two frequently used ones in general mechanical engineering are the maximum principle stress difference theory and the von Mises theory. Leaving aside the joints, a casing is a cylinder with a wall that is thin compared to the diameter. An internally pressurised cylinder with closed ends is subjected to axial stress due to the pressure acting on the ends and a hoop stress due to the wall trying to expand under the internal pressure. These are both tensile stresses, defined as follows:

$$\text{axial stress} = \sigma_a = P.A/\pi D.t = PD/4t \tag{5.3}$$

where P is the pressure difference between inside and outside, D is cylinder diameter, A is cross-sectional area $= \pi D^2/4$, t is wall thickness and πDt is an approximation to the cross-sectional area of the wall, acceptable since the wall is thin.

$$\text{hoop stress} = \sigma_h = PD/2t \tag{5.4}$$

The hoop stress is twice the axial, which is why steel pipelines or gas bottles burst by splitting along the axis rather than at the ends.

A cylinder is best described by cylindrical coordinates, x, r and θ, and the hoop stress acts in the θ direction and the axial in the x direction. Because the wall is thin, the stress in the radial direction is zero. Stresses in these three directions are called the principal stresses and the axes the principal axes. If the x-axis was replaced by one at an angle to the cylinder axis, the hoop and axial stresses would combine and result in a tensile stress plus a shear stress along this new axis. Principal axes are the axes along which the stresses resolve without any shear stress. Whilst the principle axes are intuitively obvious for thin cylinders, for general three-dimensional stress problems described by rectangular Cartesian coordinates, they must be deduced.

Adopting σ_f as the failure stress and assuming x, y, z Cartesian coordinates and principle stresses σ_1, σ_2 and σ_3, the maximum stress difference theory gives

$$\sigma_{max} - \sigma_{min} = \sigma_f \tag{5.5}$$

where σ_{max} and σ_{min} are the maximum and minimum of σ_1, σ_2 and σ_3. For the thin cylinder, the principle stresses can be allocated as

$$\sigma_1 = \sigma_h, \sigma_2 = \sigma_a \text{ and } \sigma_3 = 0 \tag{5.6}$$

so the maximum stress difference is

$$\sigma_1 - \sigma_3 = \sigma_f \tag{5.7}$$

By this theory, the material fails when the hoop stress equals the failure stress measured in an axial tensile test, $\sigma_h = \sigma_f$.

The von Mises theory is that the material fails when the following combination of principle stresses reaches the stress decided as the failure stress:

$$\sqrt{\left[1/2\left\{(\sigma_1 - \sigma_2)^2 + (\sigma_2 - \sigma_3)^2 + (\sigma_3 - \sigma_1)^2\right\}\right]} = \sigma_f \tag{5.8}$$

Substituting from Eq. (5.6) again gives

$$\sqrt{\left[1/2(\sigma_h - \sigma_a)^2 + {\sigma_a}^2 + {\sigma_h}^2\right]} = \sigma_f \tag{5.9}$$

In practice, the forces on the casing can only be estimated, since the real situation is usually much more complicated than envisaged, and API Bulletin 5C3 [1994] gives formulae for various failure loads on casing and API 5C2 provides tabulated values of the failure loads for each casing type. (Note that details of all API Bulletins can be found on the API website.) These are the result of theory adjusted by experimentation and experience and should be adhered to, with an understanding of the theoretical approach used as an aid to thinking.

5.4.2 Failure Due to Instability: Buckling and Collapse

It is possible for the material to fail under loads that should produce stresses much less than the yield stress, if the shape of the component allows any initial deformation to magnify without increase in load. This will lead to high bending stresses and failure unless the load is removed quickly. The best-known example of this is the buckling of a strut. If the ends of a strut of length L are pin jointed, that is, the ends stay in line but can rotate as required, there is a level of axial force that will cause any slight initial departure from perfect straightness to grow. If the load is maintained, the ends of the strut will move towards each other to produce a large sideways deflection until the stress caused by bending makes the material fail—schoolboys break rulers this way. The force that will induce this condition is known as the Euler buckling load and is $\pi^2 EI/L^2$ for the strut just described, where I is the second moment of area, a parameter with the dimensions of m^4 dependent on the cross-sectional shape. This type of buckling is relevant in handling a long length of built-up casing, which may bow if it comes under axial compression under its own weight and will touch the wall of the hole in which it is placed (casing or open hole), leading to poor cementing and less strength in the combined structure than anticipated—see Leaver [1982].

The collapse of a cylinder under external pressure is another form of instability relevant to drilling. A circular cylinder under external pressure is unstable if the pressure is high enough and the walls thin enough, because local departures from the true circle increase under the applied pressure and the cylinder may collapse before the stresses reach yield.

5.4.3 The Properties and Failure of Rock

Rock is an inadequate description of the materials through which a well might be drilled. In a geothermal resource with volcanic origins, various rock types are encountered; some are brittle while others will have been altered by the flow of geothermal fluids and be plastic and easily deformed, a form of clay. The dense brittle rocks (rhyolite, andesite, diorite and greywacke) are strong in compression but weak in tension, and fractures arise from stress concentrations due to imperfections in the microstructure of the rock, minor voids, for example. Geothermal areas have been subjected to tectonic stresses which induced faults and hence a wide range of fracture sizes, so the rock is in a failed state from the outset. Existing fractures may extend, and new ones form during drilling as a result of pressurising the hole.

Analogous to hydrostatic pressure, lithostatic pressure is the pressure that a fluid with the density of rock would experience at depth. Rock is not a fluid, so it does not transfer pressure equally in all directions—one particular stratum may carry the formations above it leaving those below at a lower level of stress. Tectonic stresses may act horizontally. In addition, permeable rocks or those with open connected fractures will exhibit pore pressure related to the hydrostatic pressure of water in connected formations. In general, the lithostatic pressure will ensure that any horizontal fractures remain under a compressive load tending to keep them closed, although surface roughness may leave gaps through which fluid flow is possible.

5.4.4 The Properties of Casing Steel and Casing Specification

The properties of casing steel must now be considered. Casing is manufactured to standards and specifications set by the American Petroleum Institute (API), and the complete specification of the 9 5/8″ casing shown in the well of Fig. 5.1, for example, might be Grade K-55, 47 lb/ft, R3, buttress casing. API Specification 5A classifies types of steel for casings and sets out strength requirements and dimensions. Grades J, K and L signify a metallurgical type commonly used for geothermal wells; the notation “K-55” refers to a minimum tensile strength requirement of 55,000 lb/ft^2, and the remainder of the details refer to the thread type. API Bulletin 5C2 defines performance properties of casing (as well as tubing and drill pipe) and shows that there are various wall thicknesses with an outside diameter of 9 5/8″ and that these are specified in terms of nominal weight per foot, thicker casing being heavier. The weights per foot for 9 5/8″ for J, K and L grades range from 36.0 to 53.5 lbs/ft. Kurata et al. [1995] reviewed the suitability of casing types for high-temperature geothermal resources. Knowing the weight per foot allows the total weight of each casing string to be calculated; the casing has to be lifted and lowered into the hole by the drilling rig, which must have suitable lifting capacity. The ends of each casing length are threaded so that they can be coupled, and the coupling design and performance are specified. In short, the materials available to construct the well are all tightly defined, and the designer’s task is to find out what

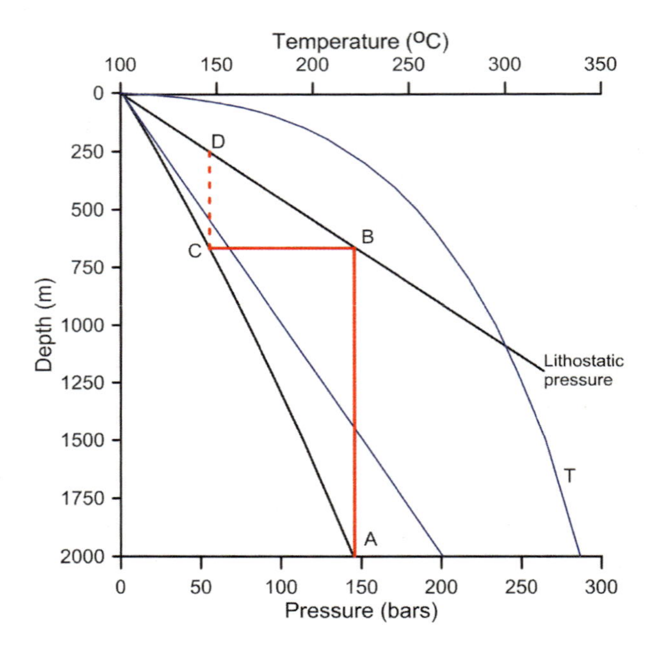

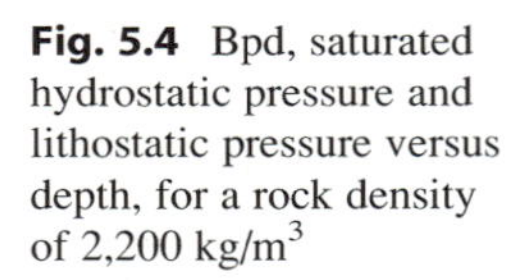
Fig. 5.4 Bpd, saturated hydrostatic pressure and lithostatic pressure versus depth, for a rock density of 2,200 kg/m^3

duty the casing has to perform and to specify which particular type is to be used for each string. Different casing thicknesses (weights) are sometimes necessary in different parts of the same string.

5.4.5 Selection of Casing Depths

The primary issue in deciding at what depth each casing should be cemented is protection against blowout—how deep can open hole be drilled? High-pressure fluid intercepted at depth might be able to communicate with a fracture in the open hole which will provide a route back to the surface; the consequences are extremely bad if this happens, as it amounts to an uncontrolled release of resource fluid which can destroy the rig and continue discharging for years, producing a crater.

The physics of this issue relates to the bpd curve and its matching hydrostatic pressure distribution and the lithostatic pressure distribution in the area. These are plotted on Fig. 5.4.

The hydrostatic lines and bpd curve will be recognised as Fig. 3.7 with the axes reorientated. In addition, a pressure–depth line is shown for lithostatic pressure, calculated using a rock density of 2,200 kg/m^3. Assume that the plan is to drill a well to a depth of 2,000 m. If the ground was saturated with water at saturation temperature from the surface downwards, the temperature and pressure distributions would be the bpd curve, marked T, and the corresponding hydrostatic pressure curve on which point A lies. A well drilled to 2,000 m and filled with steam at the bottom temperature could have a pressure versus depth graph shown by the line A–B, continuing up to the wellhead at the surface. This is because the hydrostatic pressure of a column of steam is negligible compared to that of a column of water. If the production casing of the well ended at depth B (about 650 m) and there was a fracture in the rock at that point, in principle, the pressure at that fracture would be enough to lift the entire weight of the rock above—the lithostatic pressure—the fracture could be forced open and steam could find its way to the surface. This reveals how the casing depths are selected. Drilling to 2,000 m requires the protection of a cemented casing to a depth B plus a margin of safety. The next shallower casing depth can be selected by continuing the process; drilling to depth B could encounter saturated steam conditions at point C, which could supply high-pressure steam up to the depth D, the dotted line on the graph. A formalised approach is given in the New Zealand Code of Practice [1991].

Near the surface, the method is unreliable, and experience with adjacent wells is used to decide on the depth of conductor casing. The NZ Code of Practice for Deep Drilling [1991] states that most blowouts occur from depths of 150 m or less and recommends exploratory drilling of small diameter wells if there is any uncertainty.

The completed well will collect all fluid from below the bottom of the production casing, referred to as the production casing shoe. Fluids from formations above this depth are said to be “cased out”. The production casing shoe depth is determined at the time that the well is designed, but it is possible to adjust the actual depth slightly according to the temperature found while drilling, and this might be advantageous if it would block off (case-out) sources of cold water that would simply reduce the discharge enthalpy of the finished well. However, it is difficult to determine the formation temperature during drilling because the formations are cooled by the drilling fluids. Two approaches have been adopted, the technique known as a static formation temperature test (SFTT) and by measuring the circulating mud temperatures. The SFTT is a form of the transient pressure testing of Chap. 9—see, for example, Brennand [1984] and a recent paper which provides a review, Bassam et al. [2010]. The measurement of mud temperatures was examined by Takahashi et al. [1997]. This issue is particularly important during exploration drilling, when perhaps only a single well has been scheduled and information about deeper formations can be totally obscured by failing to case out a cold formation or, even worse, by casing out the hottest formation in an outflow.

5.5 Summary of Modes of Well Failure

5.5.1 Drilling Operations That Could Cause Failure

The following operations subject the casing to forces which could cause it to fail:

- Lowering the casing into the hole—the tensile stress in the uppermost length (nearest the hook) could exceed the yield due to the weight of casing below it. The uppermost lengths may need to be stronger than the rest.
- Lowering the casing into a hole which is full of mud—the displaced mud exerts the hydrostatic pressure appropriate to its depth. Radial pressures inside and

outside the casing are balanced so long as the inside is open to the mud, but the pressure on the end (however small a cross section that might appear to be) acts upwards putting the casing in compression as a strut. The casing string is very slender and must be stiff enough not to buckle.
- In deviated wells, the curvature of the bend could raise combined stresses beyond the yield stress or be at a radius of curvature that the drill collars or stiffer pipe sections cannot take up. The curvature must be appropriate.
- Pumping cement could produce high internal pressure or high external pressure on the casing if the cement is not continuous throughout the casing and annulus.
- In an effort to reduce the risk of blowout, mud density could be increased to such a level that its hydrostatic pressure is sufficient to fracture the rock in the open hole. This is to be avoided. Cementing could also fracture the rock.

5.5.2 Possible Modes of Failure of a Completed Well

Histories of well failures and deterioration have been reported, e.g. Bixley and Hattersly [1983], Zarrouk [2004] and Southon [2005]. Among the most commonly recognised are the following:
- Bursting due to the expansion of a pocket of water trapped in the cement between the production and anchor casing. Sometimes the cement annulus between the hole and the casing or between the casings is discontinuous, and there is a pocket of water left in the space. This water is cold when it forms in the wet cement, and it becomes trapped when the cement hardens. It expands thermally when the well is first discharged and may produce a high enough pressure to deform or burst the production casing inwards, leaving a flow restriction or jagged edges on which instrument lines could catch. It can be repaired by cementing a patch in the form of a length of smaller diameter casing, but this reduces the cross-sectional area and hence the well discharge.
- Failure of the uppermost length of anchor casing that carries high internal pressure at high temperature when the well is shut, exacerbated by corrosion at ground level (discussed in Sect. 5.6).
- Failure by internal corrosion if very acidic fluids with dissolved magmatic gases are intersected by the well.
- Failure of one or more casings as a result of compaction of one of the strata open to the well. If a formation compacts and the well casings remain firmly cemented in all other formations, shallow and deep, then the casings at the compacting formation come under a compressive load and may fail, depending on the amount of compaction.

5.5.3 Inspection Methods

However unreasonably, given the usually corrosive environment in which they operate and the severe temperature variations to which they are exposed, geothermal wells are expected to have an indefinite life. However, the New Zealand

Standard [1991] gives a prescription for permanent abandonment when the time comes. Various inspection techniques have been developed to measure the condition of the production casing. Stevens [2000] describes an instrument capable of being lowered into a hot, recently discharged well to give an immediate surface output of the measurements. An eddy current principle is used to measure wall thickness, an electronic calliper to measure diameter and a surface roughness indicator. Lejano et al. [2010] adopted the approach of measuring the internal diameter of a sample of 10 wells in the Leyte field, Philippines, using a 60-finger mechanical calliper tool. The remaining wall thickness was deduced from the measurements, which were detailed enough to show surface roughness and pitting. Thus a bursting pressure could be established for any particular well measured which provided a "not to be exceeded" pressure for the guidance of future operations on that well.

5.6 The Completed Wellhead

The completed well must be capable of being safely closed against the pressure exerted by the reservoir fluids. The wellhead pressure varies with time after closure of the wellhead valve; the forces on the wellhead are carried by the anchor casing and the wellhead components. The anchor casing above ground level is unlike the buried part because it is an unsupported pipe (the buried part is supported by an outer casing and cement), and it is open to corrosion by being exposed to the atmosphere and hence oxygen and water.

Dealing with the latter point first, the geothermal gas H_2S is heavier than air and is likely to collect in the well cellar, where it will dissolve in rainwater. It attacks the anchor casing at ground level, and to avoid this, venting of the cement annuli of the outer casings at ground level is encouraged. The cement is removed for a few centimetres depth and replaced by loose gravel (permeable), with perhaps a cover and short vent pipe, and the heat of the well evaporates any water. Painting the outside of the casing that is exposed to the atmosphere will help reduce corrosion damage. The cellar usually has a large drain to a lower level, so that CO_2 and H_2S can flow away, being denser than air.

The anchor casing above ground is a simple pipe, and its wall thickness can be selected using the Standard ANSI/ASME B31.1-1980 Power Piping, which gives a formula for thickness:

$$t = P.D/[2(\sigma_f + 0.4P)] \tag{5.10}$$

where t is wall thickness, P is pressure, D is the outside diameter of the casing and σ_f is the safe working stress (regarded as the failure criterion). To this minimum thickness must be added a corrosion allowance; typical corrosion allowances in steel piping and process equipment are 3 mm. An allowance for strength reduction due to high operating temperature is also necessary, as given in the standard.

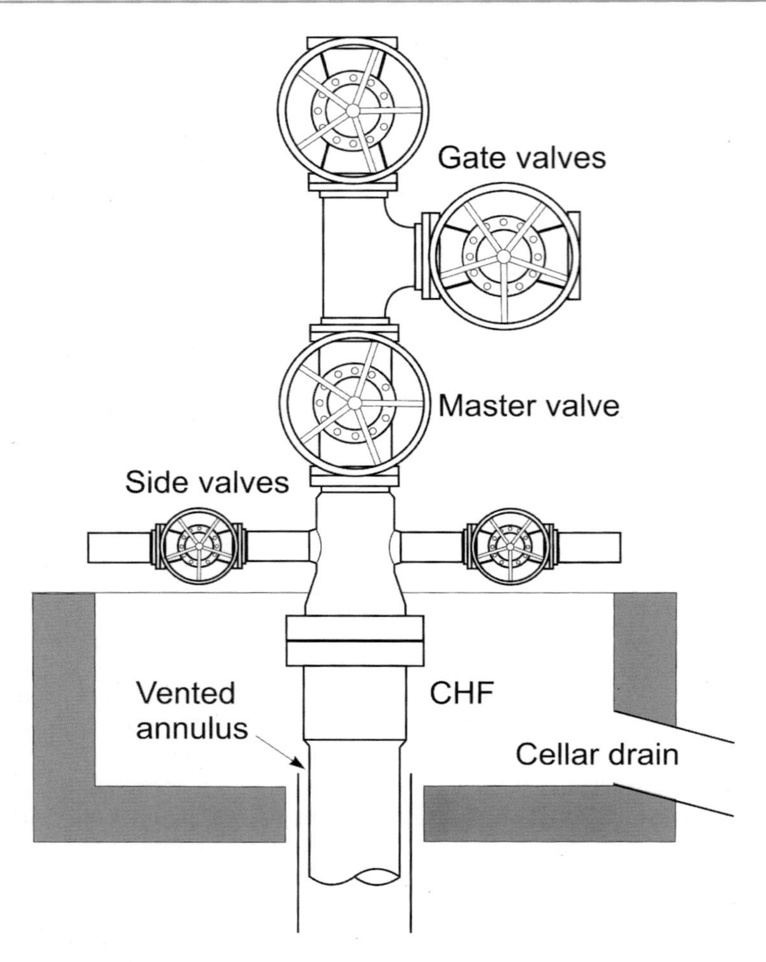

Fig. 5.5 A typical geothermal wellhead

A typical arrangement of valves on the wellhead is shown in Fig. 5.5. Immediately above the casing head flange is a reducer (referred to as an expansion spool) to allow 10″ valves to be fitted. The production casing is cut off to allow this reducer to be fitted, and a sufficient clearance must be left to ensure that the production casing cannot apply a load to the anchor casing by expanding and making contact with the interior of the reducer—Fig. 5.1 diagrammatically shows the production casing cut short within the wellhead. The master valve is attached to the reducer and then a T-section to allow 2 control valves. All of these wellhead valves are gate valves, a type in which the flow passage is directly through the valve body, without any restriction. To stop the flow, a plate is driven across the valve normal to the flow direction. To control the flow, i.e. restrict the flow, the plate is partially closed. This results in abrasion damage to the plate which results in it forming a "leaky" seal when closed, so the rule is that the master valve is either open or shut, and its gate is never exposed to the flow, so as to preserve its sealing ability. Master valves often have a more elaborate plate mechanism than other gate valves, to ensure a perfect seal. Any flow control required is achieved with the valves above the master valve or other means (see Sect. 8.1.1). When opening the well, the control valves are left closed until the master valve is fully opened, and the control valves are closed first when discharge has to be stopped. The expansion spool carries two side valves which are used for chemical sampling or similar duty.

References

American Petroleum Institute. http://www.api.org

API (1994) Bulletin 5C3 - Bulletin on formulas and calculations for casing, tubing, drill pipe, and line pipe properties, 6th edn. American Petroleum Institute, Washington, DC

ASME. http://files.asme.org/MEMagazine/Web/20779.pdf

Bassam A, Santoyo E, Andaverde J, Hernandez JA, Espinoza-Ojeda OM (2010) Estimation of static formation temperatures in geothermal wells by using an artificial neural network approach. Comput Geosci 36(9):1191–1199

Bixley PF, Hattersly SD (1983) Long term casing performance of Wairakei production wells. In: Proceedings of the 5th New Zealand geothermal workshop

Brennand AW (1984) A new method for the analysis of static formation temperature tests. In: Proceeding of the 6th New Zealand geothermal workshop, University of Auckland, Auckland

Dench ND (1970) Casing string design for geothermal wells. Geothermics. Special Issue 2, Part 2: 1485–96

Kurata Y, Sanada N, Nanjo H, Ikeuchi J (1995) Casing pipe materials for deep geothermal wells. Geothermal Resources Council Trans 19:105–109

Leaver JD (1982) Failure mode analysis for casing and liners in geothermal production wells. In: Proceedings of the 4th New Zealand geothermal workshop

Lejano DMZ, Colina RN, Yglopez DM, Andrino RP, Malate RCM, Sta-Anna FXM (2010) Casing inspection caliper campaign in the Leyte geothermal field, Philippines. Stanford geothermal workshop

Massei S, Bianchi C (1995) Failure control of drill string components: non-destructive inspection. World geothermal congress

New Zealand Standard (1991) Code of practice for deep geothermal wells. NZS 2403:1991, Standards Association of New Zealand

Southon JNA (2005) Geothermal well design, construction and failures. In: Proceeding of the world geothermal conference

Stevens L (2000) Monitoring of casing integrity in geothermal wells. In: Proceeding of the world geothermal conference

Takahashi W, Osato K, Takasugi S, White SP (1997) Estimation of the formation temperature from the inlet and outlet mud temperatures while drilling. In: Proceedings of the 22nd workshop on geothermal reservoir engineering, University of Stanford

Zarrouk SJ (2004) External casing corrosion in New Zealand's geothermal fields. In: Proceedings of the 26th New Zealand geothermal workshop

6 Well Measurements from Completion Tests to the First Discharge

Once the drilling and construction of a well is finished, various tests are usually carried out before any attempt is made to discharge it. It would be helpful to know the undisturbed vertical temperature distribution in the resource, which could be related to the stratigraphy found from drilling. But the temperature distribution has been disturbed by the use of drilling fluids, and some time must elapse before it recovers. Even then, it is unlikely to recover to its undisturbed state because of the effect of the well. This chapter focuses on liquid-dominated resources, which are the general case, with water levels close to the surface. Internal flows in the finished well are discussed, followed by a description of the relatively standard set of tests done when construction of the well is complete, but before the rig is removed. The interpretation of temperature and pressure surveys carried out over the few weeks that the well is usually left to heat up are then discussed, with examples, and finally the problem of initiating a discharge is addressed.

6.1 Internal Flow in Wells

Johnston and Adams [1916] published a paper in the journal Economic Geology describing experiments using mercury-in-glass thermometers and electrical resistance thermometers to measure the temperature distribution in a 1,600 m deep well. Their aim was to see whether the presence of coal or oil reserves could be correlated with local increases in temperature. Just as interesting as their methods and results is a statement in their closing discussion to the effect that it is useless to have an accurate method of temperature measurement unless the temperature registered by the thermometer is that of the adjacent rock. Knowing the vertical temperature distribution in the formations of a geothermal resource is essential for estimating the amount of energy available. Wells are the only means of access, but the temperature distribution in the well very rarely matches that in the formation because the fluid in the well is often moving. Wainwright [1970] reported the methods used for downhole surveys and Grant [1979] addressed the issue of interpretation.

A. Watson, *Geothermal Engineering: Fundamentals and Applications*,
DOI 10.1007/978-1-4614-8569-8_6,

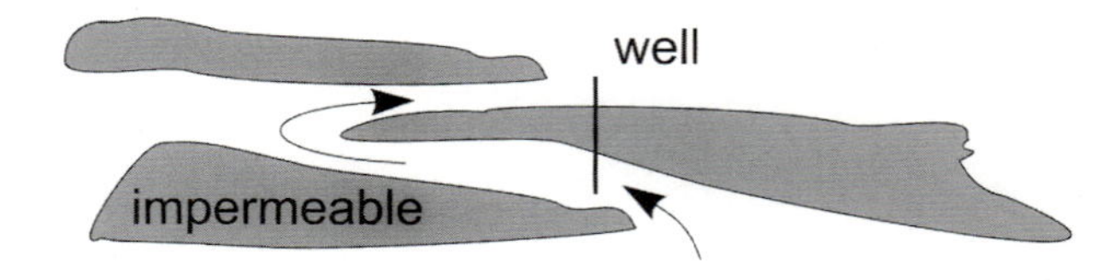

Fig. 6.1 Sketch of formations and circuitous flow path

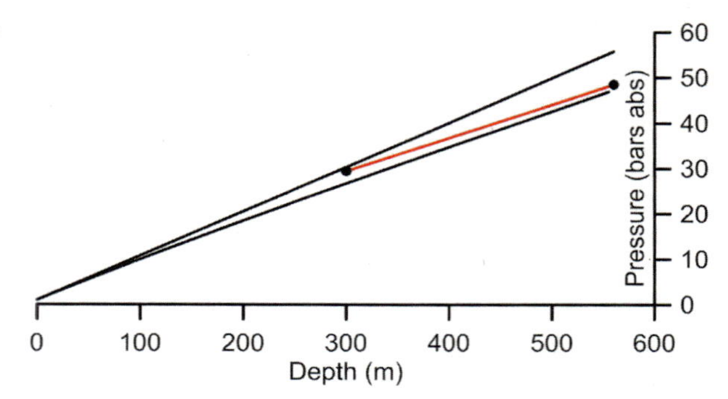

Fig. 6.2 Showing a pressure difference between two formations sufficient to cause a downflow in a well open to both

The problem with geothermal wells is that they are drilled into regions where the temperatures are high enough to change the density of liquid water and often sufficient to produce steam zones. In volcanic locations the strata can be very irregular, both in thickness and permeability, for example, impermeable lenses of fine sediments or impermeable lava flows, alternating with very permeable material. Large-scale (resource-sized) circulation patterns above the heat source may be identifiable, but on a small scale the route taken by the convecting fluid is often circuitous; the sketch of Fig. 6.1 illustrates the point.

Although a well has the relative dimensions of a human hair, the uncased lower section, typically 8 ½″ diameter, represents a major short circuit in the hydrological flows of the locality, for example, if it was open where marked on Fig. 6.1. The permeable zones are left white in the figure and the impermeable ones are grey. Even though there is a permeable route upwards, it may be long and sinuous and will incur a frictional pressure drop. If the part of the well shown is the slotted liner, a significant pressure difference can exist over it, greater than hydrostatic, and as a result very large internal flows can occur, usually with cold water entering at shallow levels and flowing down to hotter, deeper levels. Figure 6.2 shows this graphically. The two pressure distributions are part of the bpd datum set described in Sect. 3.5 and both are equilibrium distributions. In the upper curve, the entire well is filled with water at 25 °C, and in the lower curve with water at boiling point-previous concerns about the bpd distribution do not affect the reasoning here. If the well is open to a formation at 300 m containing water at a temperature just higher than 25 °C, and all formations are impermeable down to 550 m, where a second permeable formation exists, this time filled with water at just below saturation temperature, the pressure gradient between the two formations will be the line joining the two dots in the figure.

This is not an equilibrium distribution but has a steeper slope, and water will flow down the well. Fluid can also enter the well from a deep formation and flow upwards to leave to a shallower one. This example shows that the well becomes an integral part of the resource rather than a sealed tube filled with water. Figure 6.2 is a fictitious example, but the behaviour is common.

Grant et al. [1983] reported that convection cells had been observed in New Zealand wells, and cite a particular case identified by a flowmeter, of a 100 m long cell in the slotted liner. The suggestion is that the temperature gradient in the well alone can promote internal convection, without any flow entering or leaving the formations. For this to take place, the well would have to be effectively a sealed tube. Internal natural convective circulation flows in closed cylinders have been measured for various applications, but in the laboratory where it is easy to make sure that the cylinder is indeed closed, in contrast to a well. Murgatroyd and Watson [1970] measured velocity and temperature distributions, fluid rising in the centre and falling down the tube inner surface. In their case the fluid had a uniform internal heat source $\dot{H}$ and the vertical cylinder was cooled on the outside, so the entire cylinder, 50 mm diameter and 50 diameters long, was occupied by a single cell; fluid circulation was unavoidable as a result of the internal heat source. A configuration which has attracted more attention is the open thermosyphon, a vertical cylinder closed at the bottom but opening into a large-volume reservoir at the top, with cylinder wall held at uniform temperature. Lighthill [1953] produced theoretical predictions which were later proven to be valid and found that a convection cell was established, with downward flow in the core of the tube and rising flow in an annular column adjacent to the hot wall—the fluid was cooling the tube. But the cell did not fill the tube—the water in the bottom remained stagnant above a certain length/diameter ratio, dependent on the Rayleigh number. Geothermal resources exhibit steady temperature gradients in the undisturbed state, especially permeable resources, and a cylinder of length/diameter ratio of greater than 450, as in the case cited, would offer considerable internal friction between the rising and falling fluid. This is a problem that can be tackled numerically; Zarrouk [1999] reproduced many of the classical cellular convection experimental results in porous media flows using the numerical code Phoenix, and the same method could be used. As an example of the use of dimensionless numbers, consider planning an experiment to replicate the well mentioned by Grant et al. [1983] above. Recall that from Sect. 4.4, the set of equations governing natural convection of a constant property fluid in a vertical tube has only one parameter when written in dimensionless form, the Rayleigh number. This means that for two similar but not identical systems, the solution of the equations in dimensionless variables will be identical if the Rayleigh number is the same. The Rayleigh number is Gr.Pr,

where $Gr = g\beta(T - T_{\infty})a^3/\nu^2$ and $\mathrm{Pr} = \nu/\kappa$ from Eqs. (4.64) and (4.50), respectively. A laboratory experiment would require the same length/diameter ratio as the real case, and a 100 m length of well could only be represented by a small diameter tube. The temperature difference would need to be scaled up to keep the Rayleigh numbers the same.

6.2 A Typical Measurement Programme for a New Well

Determining the vertical temperature distribution in the drilled area is important but difficult because of the effects of drilling and of the well itself. It is best attempted before the well has been discharged or has received fluid other than during drilling and the short injection test carried out when construction of the well is completed. The latter tests, called completion tests, are usually carried out as follows:

- After construction of the well is finished, wash out any remaining mud.
- Lower a sinker bar to measure the depth and make sure the well is clear so that instruments will not get caught on any projections. A sinker bar is a cylinder or birdcage-like frame made of copper or lead, which will fairly quickly corrode away if it gets stuck or falls to the bottom of the well. The bar is lowered on a wire arranged to run through a measurement system which can be set to zero when the bar is at wellhead (casing head flange) and which will accurately record the depth of the bar. The wire is housed on a drum driven by a winch, and referred to as wireline equipment.
- Pump cold water into the well at a low flow rate while carrying out a temperature and pressure survey. Repeat at twice the flow rate. If a flowmeter (spinner) is available, it is preferable to use it for a simultaneous survey.
- Place a pressure recorder at the main loss zone (at the depth which has indicated by either drilling or water loss surveys that the formation there will accept the most water, a matter of judgement) and pump water in at various rates, leaving time for the pressure to become steady or adopt a steady rate of change—a matter of experience. The pressure will increase with each increase in flow rate, and the rate of increase can provide estimates of formation permeability.
- After the last flow rate, stop injecting water but leave the instrument in place. The pressure will have increased because of the injection, and the measured decay in pressure can provide an estimate of formation permeability.

The last two bullet points are simple forms of transient pressure test, which will be explained in Chap. 9. The water injection test is designed to give the injectivity, I (kg/s per MPa abs), which is defined by the equation

$$\dot{m} = I(P_w - P_\infty) \tag{6.1}$$

where P_w is the pressure in the well opposite to the formation, otherwise known as the sandface pressure and P_∞ is the undisturbed formation pressure a long way from the well. The volumetric flow rate (and hence mass flow rate) can be kept constant in these tests by using the positive displacement mud pumps of the rig.

Grant et al. [1982] (see also Grant and Bixley [2011]) provide a correlation of well discharge capacity against injectivity, based on this old established test, which enables the power output to be estimated shortly after construction of the well is finished. The pressure difference between the sandface and the undisturbed formation pressure should, on theoretical grounds, be a logarithmic function of mass flow rate and not the linear function of Eq. (6.1), as will be discussed in Chap. 9. However experience shows that measurements do follow the linear relationship, at least up to pressures high enough to open fractures or increase the effect of secondary zones.

After the completion tests are finished, the well is left closed for a period of typically 4 weeks during which temperature and pressure distributions called heat-up surveys are measured. Heat-up surveys are not always necessary, of course; for example, the new well may be an infill well in an area already drilled and understood. The frequency of measurements in New Zealand is often prescribed as 1, 2, 4, 7, 21 and 28 days after completion testing. The well will then be discharged while output measurements are made (see Chap. 8).

The temperature, pressure and local flow rate can be measured with a single instrument which either records the data within the instrument or transmits a signal back to the surface. The latter is preferable, since it removes any judgement about how long to maintain any step in the measurement procedure, but there are temperature limitations for electrical cable to the surface. The Kuster Company [2012] has a long record of producing instrumentation for geothermal and petroleum well measurements, and the instruments are described on their website, and other companies also produce instrumentation and wireline equipment. If the instruments being used provide signals at the surface, then the measured T, P and flow can be related directly to the signal from the wireline equipment and a plot of P and T versus depth can be produced as the instrument is moving. The instrument has a thermal capacity so some time must elapse before the temperature sensor reaches the temperature of its surroundings; a steady slow rate of descent may suffice, or the instrument may be lowered in stages, halting at each stage until the measurement is steady. For older-type instruments which record internally, the depth of the instrument at any time must be recorded separately at the surface; the instrument records the change in P, T and flow as a function of time. When the instrument is returned to the surface, T versus time is translated into T versus depth, manually. Lowering in stages is essential for this instrumentation, so that the graph of T versus time looks like a series of steps, and each step length must be longer in time than the uncertainty in the measurement of time—tricks to mark the halt at each depth are often employed, such as lifting the instrument a metre or so then lowering it again at the beginning of every stage measurement, which leaves a "pip" on the record, a continuous line progressing at a constant rate.

6.3 The Interpretation of Temperature and Pressure Surveys

Wells show wide variations in behaviour when not discharging, and it is tempting to try to explain how every particular variation in pressure and temperature distribution has come about. The information to hand is often sparse, however, and the speculations may be impossible to verify, so may not form concrete evidence of resource characteristics. Some of the phenomena observed are very interesting in terms of physical processes taking place; see, for example, Clotworthy [1988].

The interpretation of the P and T versus z (depth) distributions is carried out by plotting them on graphs which already have the datum hydrostatic and bpd curves plotted.

If the well is not discharging, the fluid occupying it is free to exchange heat with the formations through which it passes, and it may tend to take on the surrounding temperature distribution. The open part of the well can exchange fluid as well as heat, as already mentioned. If the well was to intersect only one permeable formation of small thickness, then it would act as a manometer, the pressure in the well at the formation would be the same as throughout the formation, and the water level in the well would be at a depth consistent with the temperature distribution, since density is a function of temperature. Wells often have multiple loss zones but several simpler cases are also common and will be considered below.

If the well can be left open without discharging and the water level is some distance below wellhead, the pressure from the surface to the water level will simply be atmospheric pressure. The density of air at 1 bar abs pressure and 25 °C is 1.293 kg/m^3 so the hydrostatic pressure change over a depth of 10 m is $\rho gh = 127$ Pa or 0.00127 bars, which does not show on the graph—water is 800 times more dense, giving a change of 1 bar in 10 m. (Note that the pressure due to a stationary column of gas is also called "hydrostatic".) The datum curves are related to the water level, not the wellhead, so must be moved to begin at the water level. For wells such as this, an instrument to detect the water level can be made in the form of a weight lowered on a pair of wires forming the open ends of a circuit of battery and light, so that the circuit is completed and the light comes on when the weight dips into the water.

The interpretation will be illustrated by a series of examples in which the survey results are plotted in the usual way, with depth as the horizontal axis, with temperature as the vertical axis at the left-hand side and pressure at the right and with the datum curves of bpd, cold hydrostatic pressure and saturated water hydrostatic pressure superimposed. A geological log of the well and drilling reports should always be examined at the same time as the well surveys, but this chapter focuses on interpretation of the surveys themselves, so no reference to geological logs is made.

The plotting and datum curve adjustments may be made manually using graphs and tracing paper, or electronically, the principle is the same—for all of the examples shown here, points have been read from the original drawing-board-sized hand plots and replotted electronically or read from data tables. The wellhead pressure gauge reading in the case of a shut well with gas or steam over a liquid level in the well is the pressure above the surface of the water; the pressure distribution begins at this value but the origin of the datum curves lies on the extrapolation of the hydrostatic pressure curve back to the z-axis (which, although it is the vertical axis, is horizontal in this plot). For a well filled with cold water, this extrapolation is straightforward, as shown in Fig. 6.3.

In Fig. 6.3 (which has been made up for this explanation), the supposed measured pressure distribution is identified by the triangular points. The well is shut, and the water level is some distance below the wellhead, which has a positive wellhead pressure; the water level might have been depressed by injecting compressed air. If the well was opened, the water level would rise to the level marked as

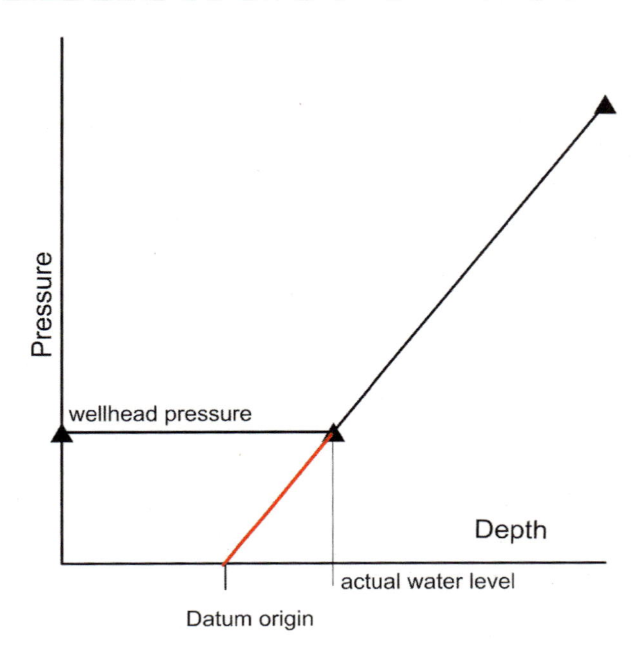

Fig. 6.3 Graph of pressure versus depth for a closed well filled with cold water, the level of which has been depressed by a wellhead pressure

"datum origin"; this is the depth at which the datum curves of Fig. 3.7 (bpd and the two pressure gradients) would have their origin and can be called the effective water level for the present purposes. Finding the water level is not always so obvious, as the following example shows.

6.3.1 Finding the Water Level

Searching for the water level was not the reason the surveys shown in Fig. 6.4 were made on Bore 67 at Wairakei, 2 years apart, in 1974 and 1976, but they provide a good example. The well was 670.6 m deep from casing head flange (CHF) and had slotted liner from 472 m to the bottom. It is not clear when the well had originally been completed. The graph states that it had been "on bleed" for 5 days before the 1974 surveys and that the wellhead pressure was 11.6/11.14 bars abs, indicating that the pressure had changed during the survey. The instruments will have been run fastened together with the wire passing through a gland (stuffing box) because of the wellhead pressure. The well had been on bleed for 6 days before the 1976 surveys and the wellhead pressures recorded were 9.4 and 10.2 bars abs. The term "on bleed" means that a side valve on the wellhead was partially opened to allow a very small discharge of steam, with the intention of keeping the upper parts of the casing hot to allow the well to be brought into production quickly with minimum thermal stressing. The flow rate is set by eye.

The first data plot has 10 points, and the second only 4, placed in the liquid-filled part, suggesting that the purpose was to see how the reservoir pressure had changed

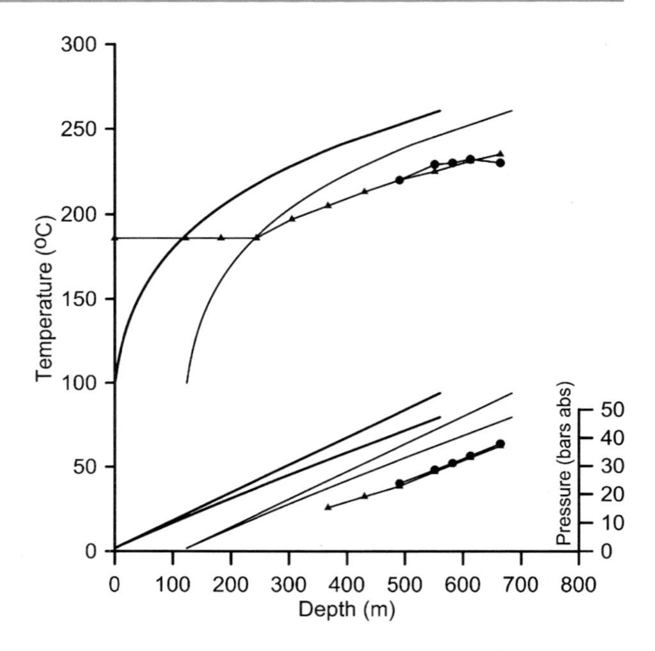

Fig. 6.4 Survey data of a well "on bleed"—first interpretation *(measurement data reproduced by permission of Century Drilling and Energy Services (NZ) Ltd)*

over the 2-year period; the operators knew exactly what they wanted to measure. The figure shows two sets of datum curves, one of them the standard from CHF and the other adjusted so that the end of the linear part of the temperature plot just lies on the bpd curve. This looks like the location of the effective water level, because the temperature change looks definite—a column of saturated steam to wellhead lies over boiling water. The measured value of 186 °C is T_s for a pressure of $P_s = 11.49$ bar abs, which lies within the reported wellhead pressure (11.6/11.14 bars abs). The rest of the temperature distribution falls increasingly below the bpd curve with depth—all very satisfactory until the pressure measurements are compared with the newly placed datum set, and it is found that they do not match. Poor calibration might spring to mind. But if the datum set is adjusted so that the pressure measurements lie along the saturated liquid hydrostatic line, the effective water level appears to be 220 m and the temperature measurements intersect the bpd curve at a higher temperature, about 220 °C, Fig. 6.5.

The image of steam sitting over a defined water level must be changed. The bleed is producing a high enough flow rate that there is no clear water level, but instead there is steam only to a depth of about 220 m, as shown by the linear temperature distribution at the value corresponding to the wellhead pressure, after which the well is filled with a bubbly two-phase mixture, becoming wetter and more dense with depth until about 500 m, where the temperature distribution crosses the bpd curve, below which the pressure distribution follows the liquid line as the well is filled with liquid water at that depth. The pressure gradient is declining up the well from the top of the liquid, within the two-phase region, because the mixture

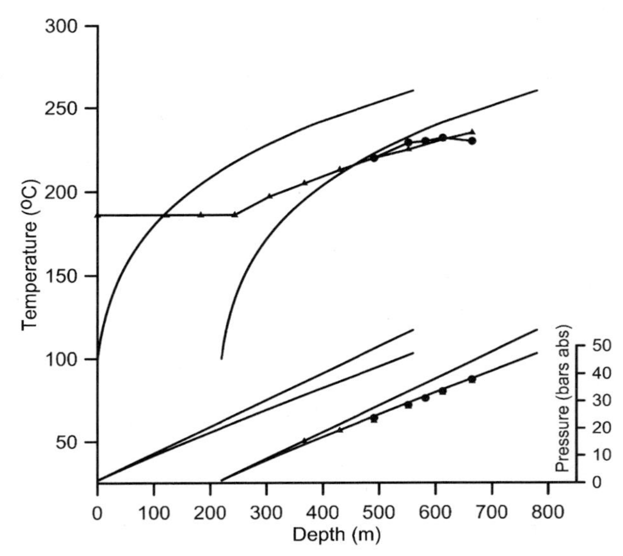

Fig. 6.5 Correct interpretation of the survey data of Fig. 6.4 *(measurement data reproduced by permission of Century Drilling and Energy Services (NZ) Ltd)*

becomes drier with reduction in pressure (spacially, not in time, the well is in a steady state).

This example has been explained to illustrate the train of thought. In reality, a plot of saturation temperature distribution corresponding to the measured pressures would have led to a quicker conclusion.

6.3.2 A Well with a Downflow

Figure 6.6 is an example from a Wairakei well, chosen to demonstrate evidence of a downflow. The well was drilled to a depth of 643 m and cased to 275 m and had been on bleed for 2½ years at the time of these measurements in 1978. The wellhead pressure then was 4.0 bars abs, for which the saturation temperature is 143.6 °C, which agrees with the shallowest temperature survey data point. The measurements are relatively sparse, about 60 m apart in places, so joining up the points may be misleading; however, the uniform temperature below the casing shoe depth is evidence of a downflow from formations at this level.

The deepest point in the temperature survey shows a slight rise, suggesting that the flow may have left the well and entered the formation above this level and the well surroundings at the bottom are hotter and probably less permeable, so the flow rate is very low there. Something complicated is happening in the region of the casing shoe which has prompted the more detailed temperature survey shown, but a geological log of the well and drilling records are required if this is to be interpreted. Permeability at the casing shoe (top of the slotted liner) can be the result of fracturing the formation by overpressurising while drilling the rest of the well. If it is simply due to the presence of a high-permeability formation, then

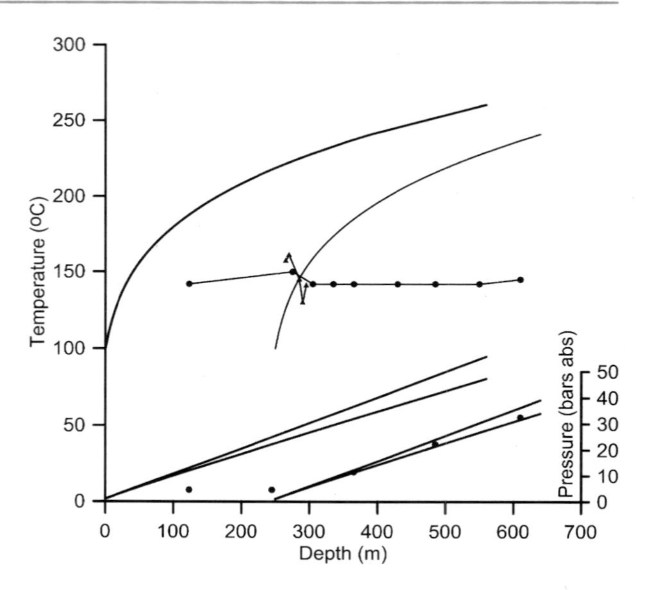

Fig. 6.6 Survey data for a well with a downflow from 300 to 600 m *(measurement data reproduced by permission of Century Drilling and Energy Services (NZ) Ltd)*

drilling problems are likely to have been experienced—loss of circulation and cementing difficulties in setting the production casing—all of which should be on record.

It seemed hardly worthwhile transferring the bpd curve since the downflow survey temperature data are clearly not going to cut it, but allowing the transfer to be governed by obtaining a match with the pressure survey lower in the well reveals that the downflow begins as saturated water, becoming subsaturated down the well as the pressure increases with depth.

For a non-discharging well, lengths of several hundred metres of uniform temperature in a resource area known to be hot are a sure sign that permeable formations are separated by impermeable material and that the pressure difference between them is not that of an equilibrium hydrostatic pressure distribution. The downflows may run for periods of years or may die away, depending on local conditions. In principle, they are damaging to the resource by reducing the temperature of formations at depth, and in Chap. 14 examples of the further potential for damage are illustrated.

6.3.3 A Well with the bpd Temperature Distribution

Figure 6.7 shows the surveys on Kawerau Bore 1 (New Zealand)—drilling it would have required care as the temperature distribution in the area follows the bpd from very close to the surface. The well was 457 m deep, cased to 305 m, and had been heating for 2 years when these surveys were made. The recorded wellhead pressure was 2 bars abs.

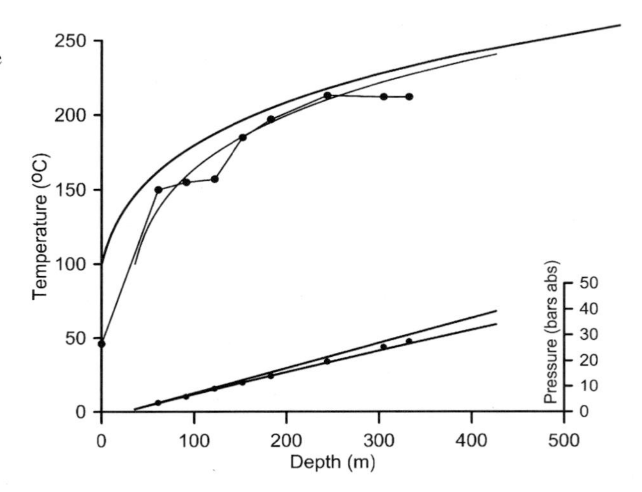

Fig. 6.7 Survey data showing the bpd temperature distribution and corresponding pressure distribution *(measurement data reproduced by permission of Century Drilling and Energy Services (NZ) Ltd)*

It was explained in Sect. 3.5 how the bpd temperature distribution might occur even though it is not a true equilibrium distribution. This well was probably on bleed, providing the necessary upward heat flux for the bpd distribution to occur, although the original graphs from which the data have been taken are silent on this issue. There is no doubt that the well temperature distribution is that of the surrounding area, which has hot springs at the surface, providing the upward flux over the whole area and also enabling the bpd distribution.

6.3.4 Water Loss Surveys

This and the following two examples are from the same set of surveys, carried out on a Kawerau (New Zealand) well over a period of 2 months following the completion tests in June 1977. The surveys shown in Fig. 6.8 were made without the benefit of a downhole flowmeter and were carried out in a single day; the graph shows temperature distributions at 2 flow rates, 7.6 and 25.0 kg/s. The well depth was 1,271 m and slotted liner started at 536 m below CHF, marked as the production casing shoe. The bpd curve is shown at wellhead simply to make the point that the flows were well below saturation temperature—cold water was being injected into a well which had already been cooled by drilling. In a well with no inflows, the injected flow shows a steady rise to some depth at which the water leaves the well and enters a formation. Below that depth the well temperature distribution does not change in response to changes in the injected flow rate. In Fig. 6.8 the injected cold water flows down the well with little temperature rise until a step change occurs in the open part (slotted liner). The height of the step evidently depends inversely on the flow rate, which is consistent with hot water entering the well from a formation or fracture zone somewhere between 750 and 850 m depth and mixing with the water being injected.

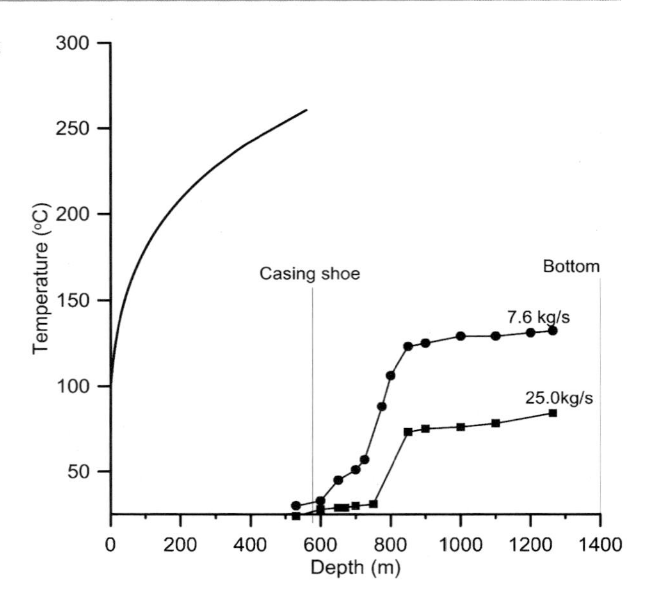

Fig. 6.8 Survey data during water injection—water loss surveys *(measurements reproduced by permission of Century Drilling and Energy Services (NZ) Ltd)*

The mixture then flows down the well to the bottom with only a small rise in temperature—this is what demonstrates that the water is actually flowing down to the bottom; if there was no flow, the temperature would be increasing and probably have a non-uniform distribution to match the formation. Two heat balance equations can be written, incorporating a mass balance, and there are only two unknowns, the mass flow rate and temperature of the inflow—that is, provided it is assumed that the inflow rate is constant for both injected flow rates. It may vary because of time—the pressure in the formation providing the flow may be declining as a result of the flow and the pressure in the well at the inflow point is likely to be higher for the higher injection rate. The pressure difference between the sandface and the undisturbed formation drives the flow, but the relationship will be non-linear. Nevertheless, with the information available, there is no alternative but to assume that the inflow rate is constant, allowing an equation for each flow rate to be written as

$$(\dot{m}.CpT)_{pump} + \dot{m}_{in}.Cp_{in}.T_{in} = \dot{m}_t.Cp_t.T_t \tag{6.2}$$

where $\dot{m}$ is mass flow rate (kg/s), Cp is specific heat (kJ/kgK), T is temperature (°C), and suffix *pump* indicates the pumped flow, *in* the inflow and *t* the total mixed flow.

This equation can be solved with the mass balance equation

$$\dot{m}_{pump} + \dot{m}_{in} = \dot{m}_t \tag{6.3}$$

Combining these equations and assuming Cp is constant give

$$(\dot{m}.T)_{pump} + \dot{m}_{in}.T_{in} = \left(\dot{m}_{pump} + \dot{m}_{in}\right).T_t \tag{6.4}$$

From the survey data the temperature of the flows just before and just after the inflow can be estimated as 55 and 125 °C for the flow of 7.6 kg/s and 30 and 73 °C for the flow of 25 kg/s and provides two equations:

$$7.6 \times 55 + \dot{m}_{in} \times T_{in} = (7.6 + \dot{m}_{in}) \times 125 \tag{6.5}$$

$$25.0 \times 30 + \dot{m}_{in} \times T_{in} = (25.0 + \dot{m}_{in}) \times 73 \tag{6.6}$$

which reveal an inflow $\dot{m}_{in}$ of 10.4 kg/s at a temperature T_{in} of 176 °C.

In general, a well being subjected to a water loss survey might have more than one inflow and more than one outflow. The mixing algebra is simple enough to allow a range of flow rates to provide enough information to determine the flows, in principle, as discussed by Bixley and Grant [1981], who noted that the formation pressure and productivity of each zone could be found. However the approach relies on the concepts of injectivity and productivity, with their arbitrary definitions, and they found that the method did not give acceptable results in all cases.

6.3.5 Heat-Up Surveys

Heat-up temperature surveys were taken on the Kawerau well of the last example, 1 day after the water loss surveys, then 4 days, 2 weeks and 4 weeks. Simultaneous pressure surveys were taken with the 4-day and 4-week surveys.

The wellhead pressure recorded was 4.1 bars abs after heating for 1 day, falling to 2.7 bar abs after 4 days and 1.7 bar abs after 4 weeks. Following the survey in week 4, the well was set on bleed for a further 4 weeks and P and T surveys were made. Only the temperature surveys are shown in Fig. 6.9. The lowest flow rate water loss survey from Fig. 6.8 is included for reference, and the figure shows how the temperature has recovered towards what can only be taken as the resource temperature. The upper curve with the data points is the survey taken after the well had been on bleed for 4 weeks; it comes close to the bpd curve in the 500–700 m depth region. The wellhead pressure for this survey is recorded as 21.6/27.2 bar abs, a large increase from the less than 2 bar abs recorded before the well was on bleed. The very small flow induced by opening the bleed valve has been sufficient to fill the well with near-saturated water, suggesting a powerful producer.

6.3.6 Pressure Surveys During Heat-Up and the Pivot Point

Figure 6.10 shows the temperature distribution during a water loss survey of another Kawerau well. The water loss surveys suggest two loss zones: at 450–500 m and 1,200 m.

The well is cased to 439 m and the bottom is at 1,281 m. The heat-up temperature surveys followed a similar pattern to those shown in Fig. 6.9, so Fig. 6.10 has been

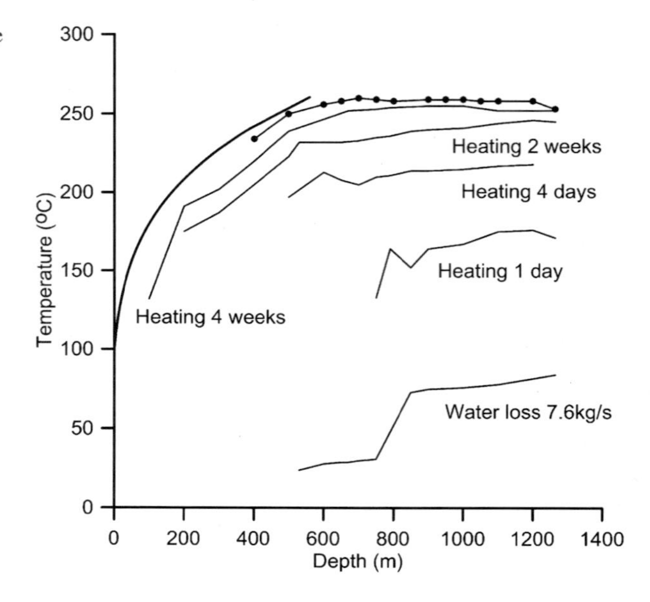

Fig. 6.9 Surveys during the heat-up period for a well *(measurements reproduced by permission of Century Drilling and Energy Services (NZ) Ltd)*

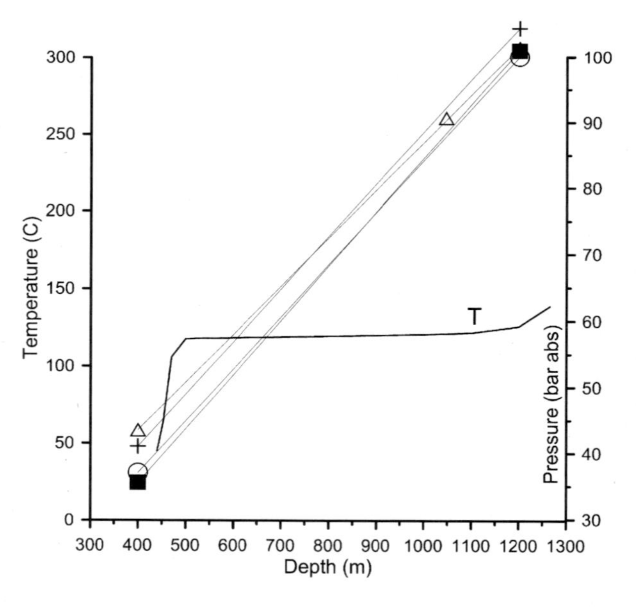

Fig. 6.10 Survey data in a well with a downflow, showing a pivot point *(measurements reproduced by permission of Century Drilling and Energy Services (NZ) Ltd)*

modified by expanding the pressure axis to see the difference between a set of pressure surveys which lie very close together. The only temperature survey included is marked "T".

The surveys were taken at regular intervals during the heat-up—the important detail is that the pressure surveys cross one another, at a depth between 700 and 900 m. The earliest pair of surveys, marked with a △ and a +, shows a clear crossing.

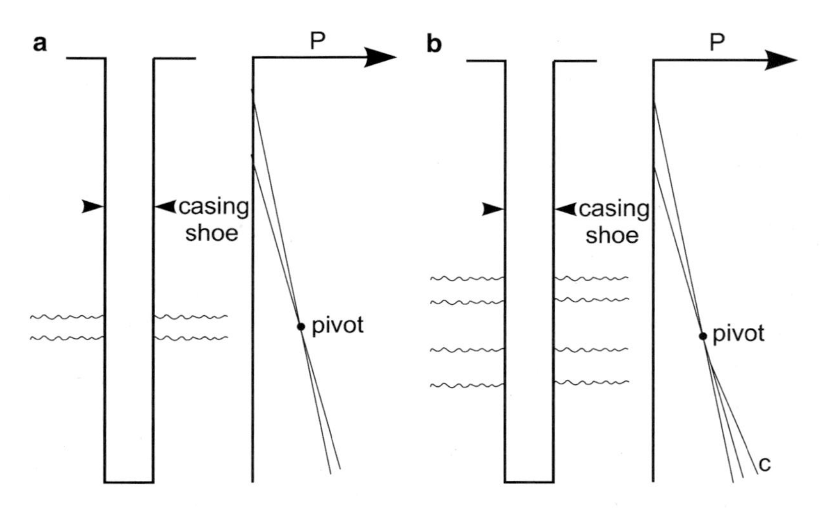

Fig. 6.11 Two wells illustrating the pivot point idea, (**a**) with a single producing formation and (**b**) with two, the lower one having the greater permeability

The later pressure distributions have shifted to the right, consistent with the wellhead pressure starting at atmospheric pressure at the beginning of heat-up, and rising to 3.4 bars at the end, 32 days later; the effective water level in the well has moved deeper. The later set shows a less distinct crossing because there is little change in the slope of the distributions, the result of the temperature distribution in the well not changing (although this is not demonstrated in the figure).

The crossover point is called the pivot point, a conceptually simple idea. Figure 6.11a shows a well filled with water and open to a single producing formation which is not at the bottom of the well. The hydrostatic pressure of the water column in the well balances the formation pressure, which is assumed to remain steady. The height of the water column depends on its temperature distribution, so if this varies, perhaps as the well heats up, the water level will vary but the pressure at the formation will remain constant. The well acts like a manometer. Any measured pressure distribution will pass through the pivot point marked. The pressure distribution below that depth in the well is also fixed at the pivot point if the well is impermeable below the depth of the formation so in principle a kink in the measured distributions at the pivot point is possible.

In Fig. 6.11b the well has intersected two producing formations, the upper one being less permeable than the lower one. It is possible, but extremely unlikely, that the undisturbed pressures of the two formations are such that the fluid filling the well between them provides exactly the right increase in hydrostatic pressure to leave them undisturbed, with neither flowing into the well—it is unlikely because the temperature distribution in a recently completed well changes as it recovers from the cooling effect of the drilling fluids. Before the well was drilled, the two formations may have been joined by a lengthy flow path, but both being in a geothermal resource means that they would not have been in static equilibrium because of the resource-scale natural circulation. The result is that an internal flow in the well is to be expected, one formation discharging and other receiving fluid. If the capacity to accept fluid of the lower formation is infinite, then the pressure at that point in the well is fixed at the undisturbed lower formation pressure and the well will act as a manometer for that level. All measurements will pivot about that point. A kink in the pressure distribution below the lower formation could occur—line C in the figure. In reality there will be no fixed pivot point; instead, it will be close to the formation with the greater permeability, as explained by Grant [1979], but will move as the temperature distribution in the well changes and the pressure in the formation with the lower permeability changes. It will change in response to the internal flow taking place in the well—pressure in the "lesser" formation will tend towards an equilibrium with the well and the greater formation, the internal flow will decrease and the system will move towards static equilibrium.

No convincing mathematical support has been offered to demonstrate this reasoning. The problem is relatively easy to specify for numerical modelling but the issue is mainly of academic interest.

6.4 Resource Pressure and Pressure Gradient Estimates from Individual Wells

When drilling a new resource, the hydrology and stratigraphy are unknown and every opportunity must be taken to build up an understanding. Dench [1980] noted that it is possible to have, above a resource, permeable formations with pressures that do lie on the regular increase with depth of the datum curves used above, and he explained how their individual pressures were measured during drilling. At every depth at which circulation was lost, drilling was halted and the pressure in the loss zone was measured by using the well as a manometer, measuring the mud density and the final static mud surface height above the loss zone. After the measurement the loss zone was sealed with mud and drilling continued to the next zone where the process was repeated. Dench showed the pressure distribution above 600 m depth in an Olkaria (Kenya) well; it was very non-linear with a maximum pressure at about 500 m depth that was much higher than the deeper steam-filled formations. The upper "aquifers" were described as being perched, although this usage does not imply being perched on a shelf formed by a discontinuous impermeable formation, but simply that the high temperature steam lies below an extensive water-filled layer with lower permeability. He then presented data from the Kawah Kamojang (Indonesia) resource which has this arrangement. The pressure distributions in the wells resembled those of Fig. 6.3, but there was no consistency about the apparent depth of the liquid surface.

Figure 6.12a represents the measurements that Dench reported, and at first sight each well appears to have been drilled into a liquid-filled resource, with widely varying water levels. The casing shoe of each of the wells was deeper than the apparent water level.

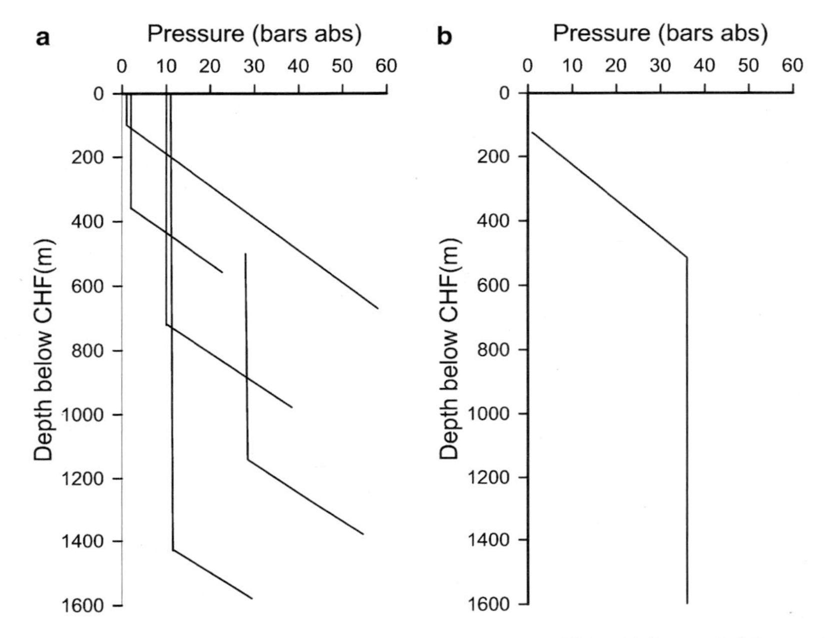

Fig. 6.12 Representing the measurements reported by Dench [1980] which revealed the presence of a steam-filled zone lying beneath a water-filled zone: (**a**) the individual well pressure distributions and (**b**) the actual resource pressure distribution

Using the method of formation pressure measurement described, he identified the formation with the highest pressure and plotted the results; Fig. 6.12b is a representation. This plot revealed that the resource is a steam-filled zone of great thickness, with a hydrostatic pressure corresponding to steam, over which a water-saturated zone lies—almost the exact inverse of what Fig. 6.12a suggested.

The vertical pressure distribution in the central area of liquid-dominated resources is sometimes plotted, but the value of doing this is open to debate. Grant [1987] explained that McNabb developed the technique of deducing the vertical pressure distribution in the New Zealand resources in the 1970s and used it with the surface discharge rate to estimate the vertical permeability. All of the parameters required for this estimate can only be roughly estimated—surface fluid discharge rate, area through which the upflow occurs, the pressure gradient itself and whether it varies across the resource. It was noted in Sect. 4.4 that in the case of the vertical flat plate arranged in a semi-infinite reservoir of fluid and heated to induce natural convection with a velocity and temperature profile as shown in Fig. 4.10, there is only one vertical pressure gradient of significance, that which existed everywhere before the plate was heated, and exists outside the heated region during convection. This is a hydrostatic distribution, the result of the setting of independent variables, but the vertical pressure distribution within the rising fluid adjacent to the plate, or in the plume of the resource, is not. It is the distribution which happens to accompany the upward natural convection, which is controlled by the heat source and resource composition and has no separate significance; there is a different vertical distribution for every distance from the plate surface. In contrast, the pressure gradient along the pipe of Fig. 4.7 in the fully developed section is an independent variable of great significance; it is an applied variable which drives the flow, whereas the pressure gradient in the resource is the result of the heat transfer. With reservoir simulation now able to provide detailed flow patterns and matches to well temperature and pressure surveys, there is no need to attempt to measure and interpret the vertical pressure distribution in the upflow resource-wide.

6.5 Making a Well Discharge

Some geothermal wells drilled with the best intentions will never discharge, others have to be prevented from discharging by keeping the valves closed, and in between there are wells which present varying degrees of difficulty in getting them to discharge. Very little has been published on the details of achieving a discharge, perhaps because it is not a common problem. However an increase in the use of geothermal energy is likely to result in exploration in more remote and mountainous sites of the same type that prompted some of the methods discussed here to be developed. Wells in the Palimpinon resource, Southern Negros, the Philippines, had been drilled from a high elevation, and the resource pressure was controlled by land at lower elevation. Even in already developed resources, there is an economic incentive to have production wells discharge as soon as the pre-discharge testing will permit. In addition, however, in the context of this book, the techniques provide an example of the analysis of events within a well. The examples given here are mainly based on the paper by Brodie et al. [1981].

Suppose a well has been cased to a single permeable formation at the bottom, 200 m below the surface, and after it has been left until it takes on the formation temperature distribution, the well does not discharge but stands open with a water level 20 m below the CHF. There must be a column of water 180 m long which is everywhere below saturation temperature. The formation pressure and the resource temperature distribution control water density and hence the water level. If a submersible water pump was inserted, the pump suction would take in water and reduce the water level, and in response the sandface pressure would fall and water would flow into the well from the formation. Alternatively, if the water density could be lowered somehow, the length of water column required to balance the formation pressure would be greater; if it could be made greater than the length from CHF to the formation, then the well would overflow and keep discharging. The greatest reduction in density occurs when the water is hot enough to flash when the pressure is reduced. Figure 6.13 shows some actual measurements on a well 2,100 m deep cased to 600 m, with 9 5/8″ production casing and 7 1/2″ slotted liner in an 8 1/2″ hole.

The datum curves would fit the measurements if moved about 250 m down the well. The change in slope of both temperature and pressure distributions at about 700 m marks the change from single- to two-phase conditions (the flash level); the lower part of the well is liquid filled and the upper part filled with a two-phase

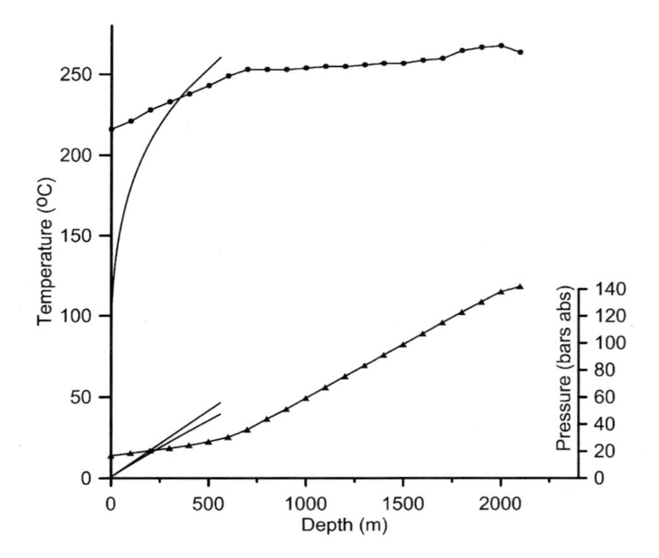

Fig. 6.13 Pressure and temperature surveys in a discharging well *(this material was created and used with permission by Energy Development Corporation, Philippines, which reserves all rights thereto)*

mixture. The mass flow rate and specific enthalpy of the discharge were measured at the surface as 13.8 kg/s and 1,170 kJ/kg, respectively. At the pressure of the first point of the survey, just below wellhead, 17 bar abs, the dryness fraction of the flow can be calculated as

$$X = (h - h_f)/h_{fg} = (1,170 - 872)/1,923 = 0.155$$

so the mean density is given by

$$\frac{1}{\rho} = \frac{(1 - X)}{859.58} + \frac{X}{8.57} \qquad (6.7)$$

The mean density has fallen to approximately 52 kg/m^3, so the pressure distribution above the flash level is now less steep. The temperature of the flow in the upper 200 m of the well declines as the pressure declines—together they represent local saturation conditions (Fig. 6.13).

It may not be possible to make a well discharge because the formation temperature and permeability may not be high enough. The rock properties can be modified, by either acid treatment or fracturing, using techniques which are not discussed in this book, but there are several methods of inducing a discharge that can be attempted before resorting to these measures. These are:

- Gas lifting
- Lowering the water level by air compression
- Steam heating of the casing
- Injecting the discharge from an already discharging well

They all represent steps to fill the upper part of the well with a low-density two-phase mixture and reduce the sandface pressure.

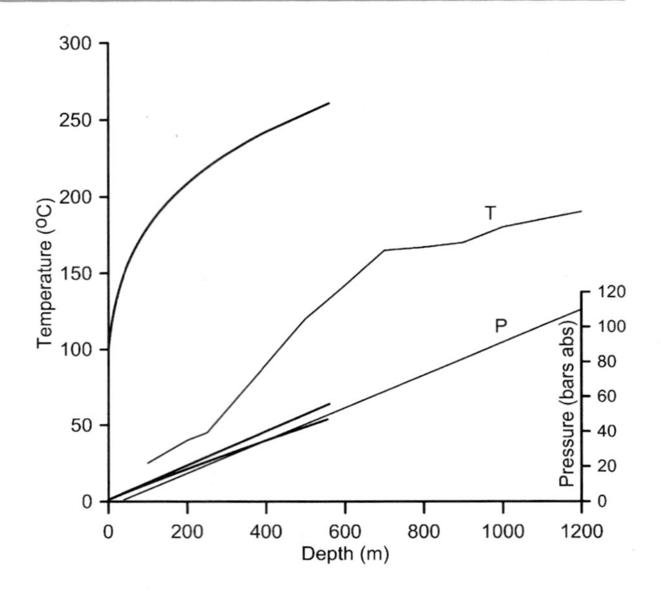

Fig. 6.14 Distributions in a well prior to discharge by airlifting *(this material was created and used with permission by Energy Development Corporation, Philippines, which reserves all rights thereto)*

6.5.1 Gas Lifting

A tube is inserted into the well to a depth below the water level, and compressed air or liquid nitrogen is supplied to it using an air compressor or liquid nitrogen tanker, respectively; the liquid nitrogen vapourises on its way down the supply pipe, which is stainless steel thin-walled tubing of about 1″ diameter, flexible enough to be coiled into a "coiled tube unit". The unit might be used for air; otherwise, a drilling rig can support a rigid, heavier pipe. The gas emerges from the bottom and rises in the well as a stream of bubbles. The mean density of the mixture is less than that of water so the water level in the well rises and if the tube has been inserted sufficiently deeply and the gas flow is adequate, the water level will overtop the well. Cold water is thus removed from the top of the well as long as the gas flow is maintained. The reduction in density of the mixture above the bottom of the tube reduces the pressure in the well at the level of the deeper formations and allows them to discharge into the well. Eventually, if the producing formations are above saturation temperature near the wellhead, the gaseous part of the two-phase flow will have increasing amounts of steam. If compressed air and rigid tubing is used, a drilling rig is required for safety.

Figure 6.14 shows an example of a well that was discharged by airlifting. The well, 1,200 m deep, was cased to 580 m, and substantial permeability had been found at 670–700 m, 870–910 m and 1,100–1,200 m, but there was a downflow from the upper zone. The measured temperature distribution, labelled T, is well below the bpd curve and the pressure distribution, labelled P, is consequently closer to the cold water datum than the saturated water datum (the datum curves have been left at the origin). The water level was about 40 m below CHF. Airlifting by

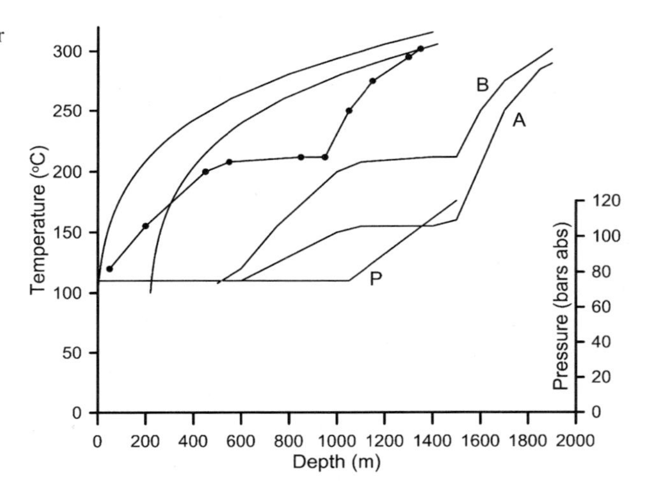

Fig. 6.15 An example of air compression *(this material was created and used with permission by Energy Development Corporation, Philippines, which reserves all rights thereto)*

inserting a rigid tube to below the water level and applying an 8 bar abs air compressor resulted in the well discharging in 20 min.

6.5.2 Lowering the Water Level by Air Compression

This method is used for wells with high downhole temperatures but with a low water level, because of the difficulty of installing long tubing and the size of compressor needed. The air is supplied at the wellhead. The well shown in Fig. 6.15 had a total depth of 1,965 m and was cased to 670 m. Drilling losses indicated permeability in the region of 1,125 m and the well had been drilled blind from 1,540 m to the bottom. Two successive heat-up surveys labelled A and B show evidence of a downflow from about 1,100 to 1,500 m, consistent with the drilling records. The standing water level in the well was very low at 220 m. Measured temperatures at the bottom of the well were high enough to suggest that the well would discharge if it could be filled with two-phase fluid. It was decided to use an air compressor capable of pushing the water level down to about 1,000 m, just above the source of the downflow, and the corresponding pressure distribution is shown in the figure, labelled P. It was not expected that this would halt the downflow, because the water level was determined by the greater permeability of the deeper production zones.

The aim of air compression is to push the cool water in the well back into the formation and hold it there until it has reached formation temperature. When the pressure is eventually released, the hot water enters the well, flashes and creates a high-volume two-phase mixture of sufficient volume to overtop the well, lowering the sandface pressure and establishing a continuous discharge. In this case the well failed to discharge. The reason can be seen by adjusting two of the temperature profiles in Fig. 6.15. The bpd curve is moved down to begin at the depth of the original water level, 220 m, and the 25-day shut temperature distribution, which

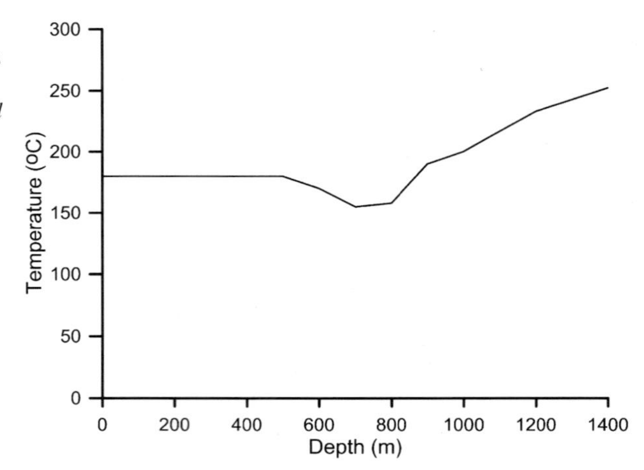

Fig. 6.16 Well temperature distribution after several days of injecting steam at 15 bars abs *(this material was created and used with permission by Energy Development Corporation, Philippines, which reserves all rights thereto)*

was the distribution when the air compression began, is moved up the well by the amount that the water level had been depressed, to show the temperature distribution in the water that re-entered the well when the pressure was released. The problem lies with the water between a depth of about 250 and 1,300 m, which remains below bpd temperature so does not flash. Flashing is indicated only in the relatively short length of well above 250 m depth.

Air compression can be a successful method of initiating a discharge, but the example chosen was clearly beyond the limit for this technique, despite the high bottom hole temperature. The continuation of the downflow throughout the procedure was probably also a significant factor in its failure.

6.5.3 Heating the Casing with Steam

It was demonstrated in Sect. 4.5 that the thermal response of steel casing is less than 10 s, so unheated casing can condense steam as the two-phase mixture rises towards the wellhead in any attempt to initiate a discharge. The results of preheating the casing by injecting steam were reported by Brodie et al. [1981]. The result of several days of injecting steam into a well is shown by the temperature distribution of Fig. 6.16. The standing water level in the well was 600 m below CHF and the steam raised the temperature of this empty section to about 180 °C. The steam pressure is reported to have been 15 bars abs (saturation temperature 198 °C) so it was essentially a combination of steam heating and the equivalent of air compression.

Brodie et al. [1981] carried out a heat balance on a rising column of two-phase mixture, considering the factors which would reduce its volume. Obtaining the largest volume mixture possible ensures the maximum amount of the mixture will flow out of the well by overtopping it, but the factors opposing that are the heat loss to the casing and the reduction in specific enthalpy as a result of work being done against gravity.

The steam heating of the casing was carried out using a portable boiler, which incurred the cost of transporting the boiler and its fuel to the site. This is the only means available for the exploration of a remote resource, but where there are nearby wells that already discharge, their output can be injected directly into the well in question. Siega et al. [2006] report that this method is used in Mahanagdong, Philippines, which is a high-elevation resource. The technique that has been developed is to connect a nearby well capable of producing a two-phase discharge when opened (the production well) to a silencer with a branch to the well to be discharged; the connection is to one of the 10″ gate valves using 10″ diameter piping, to provide minimum flow restriction. The production well is then discharged to the silencer, throttled to maximum wellhead pressure, and the discharge gradually redirected to the well to be discharged. Discharge occurs after typically 1 h. In contrast, discharging two-phase fluid through a 2″ pipeline and into a side valve was not successful even after injecting for 2 days. Temperature distributions in the well to be discharged resemble that of Fig. 6.16.

References

Bixley PF, Grant MA (1981) Evaluation of pressure-temperature profiles in wells with multiple feed points. In: Proceedings of the New Zealand geothermal workshop, University of Auckland, Auckland

Brodie AJ, Dobbie TP, Watson A (1981) Well discharge stimulation techniques in hot-water dominated geothermal fields. In: Proceedings of ASCOPE '81, ASEAN council on petroleum, second conference and exhibition, Manila, Philippines, Oct 1981

Clotworthy AW (1988) Complex feed zone observed at Wairakei. In: Proceedings of the 10th New Zealand geothermal workshop, University of Auckland, Auckland

Dench ND (1980) Interpretation of fluid pressure measurements in geothermal wells. In: Proceedings of the New Zealand geothermal workshop, University of Auckland, Auckland

Grant MA (1979) Interpretation of downhole measurements in geothermal wells. Report No. 88, December 1979, Applied Maths Division, Department of Scientific and Industrial Research, New Zealand

Grant MA (1987) Reservoir engineering of Wairakei geothermal field. In: Okanden E (ed) Geothermal reservoir engineering, vol 150, NATO ESI series, Series E: applied sciences. Kluwer, London

Grant MA, Bixley PF (2011) Geothermal reservoir engineering, 2nd edn. Academic, New York

Grant MA, Donaldson IG, Bixley PF (1982) Geothermal reservoir engineering. Academic, New York

Grant MA, Bixley PF, Donaldson IG (1983) Internal flows in geothermal wells: their identification and effect on the wellbore temperature and pressure profiles. Soc Pet Eng J 23(1):168–176

Johnston J, Adams LH (1916) On the measurement of temperature in bore-holes. Econ Geol 11(8):697–740

Kuster Company (2012). http://www.kusterco.com

Lighthill MJ (1953) Theoretical considerations on free convection in tubes. Q J Mech Appl Math 6(Pt4):398

Murgatroyd W, Watson A (1970) An experimental investigation of the natural convection of a heat generating fluid within a closed vertical cylinder. J Mech Eng Sci 12(5):354–363

Siega HC, Saw VS, Andrino RP Jr, Canete GF (2006) Well-to-well two-phase injection using a 10″ diameter line to initiate well discharge in Mahanagdong geothermal field, Leyte, Philippines. In: Proceedings of the 7th Asian geothermal symposium, July 2006
Wainwright DK (1970) Subsurface and output measurements on geothermal bores in New Zealand. UN Symposium of the development and utilization of geothermal resources, Pisa (Geothermics (1970) special issue 2)
Zarrouk SJ (1999) Numerical solution of near-critical natural convection in porous media with reference to geothermal reservoirs. MSc thesis, Department of Mechanical Engineering, University of Auckland, New Zealand

Phase-Change Phenomena and Two-Phase Flow

7

Phase change usually occurs as a result of heat transfer, but it can also take place in response to a change of pressure. Using heat to produce power relies on the water to steam phase change which occurs by boiling or flashing, and sometimes the phase change of fluids other than water. Boiling takes place in one of two possible modes, nucleate boiling, which involves surface tension, and homogenous nucleation. These are introduced first, followed by flashing, which is the change from liquid to gas resulting from pressure reduction, and then condensation. The nature and possible causes of thermal explosions and hydrothermal eruptions are briefly discussed, since phase-change phenomena are involved. An introduction to two-phase flow then follows, since it is empirically based and requires the introduction of many new variables to the governing equations established in Chap. 4. Finally, the physics of aqueous solutions of gas is explained.

7.1 Background

Geothermal engineering usually involves the extraction of hot water from a resource and flashing it in stages to produce dry steam. Flashing might appear today as a minor industrial activity, but in the early part of the twentieth century, storing saturated water in an "accumulator" and using it to produce extra steam when required was a standard practice in factories—see, for example, Lyle [1947] and books of similar vintage. Accumulators were insulated horizontal cylinders kept partly full of water at saturation conditions so that a large surface area existed from which steam could evaporate when the pressure was reduced in response to demand. Alternatively, steam could be supplied at higher pressure when there was a surplus, so that it condensed. The process of heat storage as a means of fuel economy was well established, and many of the steamfield practices of liquid-dominated geothermal resource use must have originated from this background.

A. Watson, *Geothermal Engineering: Fundamentals and Applications*,
DOI 10.1007/978-1-4614-8569-8_7,

Detailed studies of the boiling process intensified in the mid-twentieth century as the level of heat flux used in equipment increased. A natural heat flux of 800 mW/m^2 is regarded as high over a geothermal resource, but engineering equipment such as boilers, nuclear reactors and chemical processing plant are often designed to operate with heat fluxes of the order of 1 MW/m^2. Details of the flashing process have been less important to normal engineering operations, but very rapid flashing, boiling and condensation have all attracted attention in studies of explosions, both in natural and in engineering equipment.

7.2 Surface Tension

The saturation pressure and temperature form a unique set of conditions at which water and steam (or any liquid and its gas) are in equilibrium. Unfortunately, phase change creates a surface, the pressure on each side of which is only saturation pressure for a pure substance if the surface is flat. This is illustrated in Fig. 7.1—the liquid molecules at the surface are maintained in close contact by the short-range forces which act throughout the liquid including along the surface, giving it a "surface tension".

For water, the surface tension ζ (N/m) varies with saturation pressure and becomes zero at the thermodynamic critical point; the interface between liquid and gas disappears at the critical point, and watching this happen is a classic laboratory demonstration (though not easy to set up). The IAPWS [1994] equation is written in terms of saturation temperature; thus (Fig. 7.2),

$$\zeta = 235.8\Gamma^{1.256}(1 - 0.625\Gamma) \tag{7.1}$$

where $\Gamma = \left(1 - \frac{\Theta}{\Theta_C}\right)$ and the critical temperature $\Theta_C = 647.096\text{K}$.

The surface tension influences the shape of the interface between phases and may shift the local conditions away from the saturation pair. Consider a steam bubble of radius r and examine the force balance which must exist to keep its diameter constant, by imagining it cut in half—Fig. 7.3.

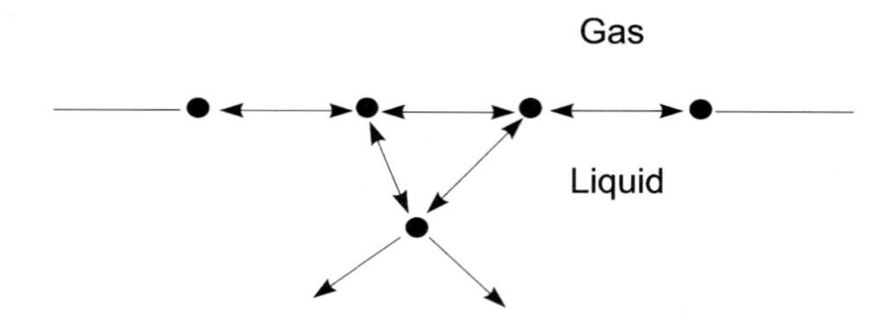

Fig. 7.1 Sketch of water molecules at a surface, to illustrate surface tension

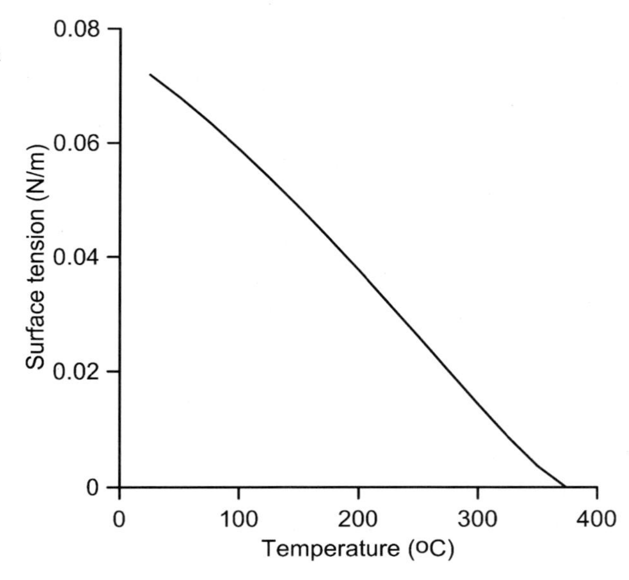

Fig. 7.2 Surface tension (N/m) for water as a function of saturation temperature calculated from IAPWS [1994]

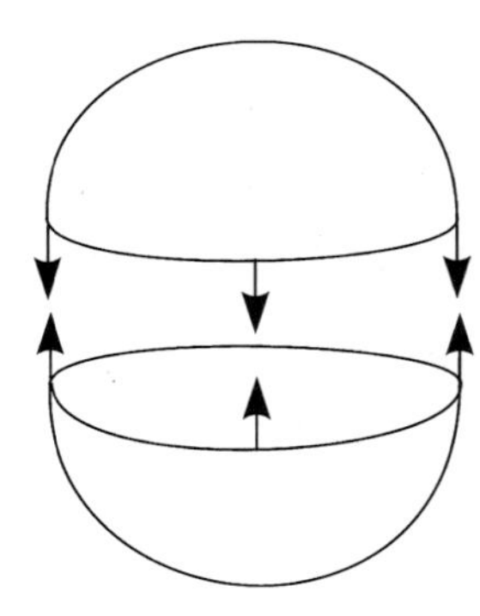

Fig. 7.3 The action of surface tension on a bubble

The surface tension around the perimeter keeps the two halves together but must be balanced by a pressure difference over the wall of the bubble to prevent it collapsing; thus,

$$\zeta .2\pi r = \Delta P.\pi r^2 \tag{7.2}$$

or

$$\Delta P = \frac{2\zeta}{r} \tag{7.3}$$

If the bubble and surrounding liquid are isothermal, then the liquid must be slightly superheated with respect to the steam in the bubble. A similar analysis can be carried out for a water droplet in steam.

7.3 Boiling

Quantifying the rate of heat transfer by boiling is seldom necessary in geothermal engineering, but the physical processes are worth understanding. Boiling studies are often reported in engineering literature as one of two types, pool boiling and flow boiling, depending on whether the liquid being heated is a fixed volume held as a "pool" in a container, or is flowing, as through a heated pipe. But the process is characterised by the way the phase change takes place, and three modes are commonly identified, nucleate boiling, film boiling and homogenous nucleation. The specific volume of steam is several orders of magnitude greater than that of water, so it is evident that steam generated at a surface has the potential to prevent replacement water from reaching it. In nucleate boiling the phase change occurs at nucleation sites on the heated surface, and bubbles are created which leave at regular intervals allowing new water to occupy the sites. If the heat flux at a surface is greater than a critical value, replacement water cannot reach the surface, which becomes blanketed by a layer of steam through which heat must pass in order to cause more evaporation at the steam–water interface. This is referred to as film boiling, in which the rate of heat transfer is very much less than in nucleate boiling. It is a mode that designers work to avoid—it is a failure condition of high heat flux equipment and is barely relevant to geothermal engineering, although it may arise in volcanic circumstances.

7.3.1 Nucleate Boiling

However smooth the surface of a container for boiling water appears, it will have microscopic pits and crevices. If the container can be evacuated first and then filled with water free of dissolved gas, the pits can be entirely filled with water, but normally air remains in the bottom when the water is added—see the sketch of Fig. 7.4a. The air pocket is compressed under hydrostatic head until it takes up a curvature which satisfies Eq. (7.3); it then becomes a pocket of air mixed with a small amount of steam, before the surface is heated. The air cannot condense, so even when the surface is cold, the pocket will not collapse. When the surface is heated, conduction ensures that the water adjacent to the surface and pit (more properly referred to as a nucleation site) is heated, and water molecules transfer across the interface from the liquid side. A bubble grows in the manner depicted by

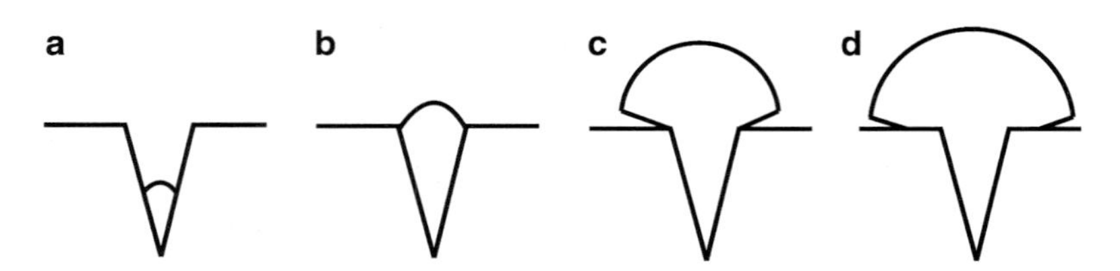

Fig. 7.4 Stages in the growth of a bubble at a nucleation site, from *left* to *right*

the remaining sketches of Fig. 7.4; it grows hemispherically, eventually leaving behind a thin layer of water on the heated surface as an annulus, with a dry patch around the nucleation site. Thermal conduction is sufficient to evaporate the thin layer (micro-layer). Relatively little evaporation takes place from around the dome of the bubble.

The surface and adjacent liquid are at a slightly higher temperature than the steam, providing heat transfer into the bubble. The growing bubble eventually becomes large enough to develop a buoyancy force sufficient to lift it off the heated surface, and it forms a neck which eventually breaks, leaving the bubble free to rise. The nucleation site then returns to its starting point, but the original air that filled it before heating started has been diluted and the new interface forms over mainly steam. Evaporation again takes place and the bubble growth and departure is repeated.

As the surface temperature is increased, the heat flux transferred increases, but eventually a stage is reached at which the rising bubbles obstruct the inflow of replacement water and the surface then becomes covered entirely by steam. In nuclear reactor fuel elements heated by fission and also in fossil-fuelled boiler tubes heated by thermal radiation from flames, the physical boundary condition of the surface is one of constant heat flux, not constant temperature. Given a heat flux entirely independent of temperature, it is possible for the surface to reach temperatures at which it will fail if the heat transfer rate is insufficient, and the term "burnout" is sometimes used to describe the critical heat flux condition.

There is a temperature level for any surface above which water sprayed onto it will not touch it but remains separated by a film of steam; the Leidenfrost drop is the name usually associated with this phenomenon.

Nucleate boiling also occurs in heated pipes through which liquid is flowing—called flow boiling; bubble growth can be very rapid, and the only difference from pool boiling is that the bubble is subjected to fluid drag as it gets bigger, so it is swept off the nucleation site instead of leaving it under its buoyancy force. The boiling crisis is likewise an issue with flow boiling. Some engineering equipment has nucleation sites deliberately created on the heated surfaces. The bumping in round-bottomed flasks in chemistry laboratories is caused by a lack of nucleation sites—it is traditionally cured by adding broken pieces of pottery, in other words, by adding nucleation sites. The pores of a permeable formation provide nucleation sites, as do particles suspended in a fluid. In the absence of nucleation sites, boiling does not occur with the few degrees of superheat that they permit, but instead the violent process known as homogenous nucleation takes place.

7.3.2 Homogenous Nucleation

Homogenous nucleation takes place within the body of the fluid when there are no nucleation sites. It is an explosive process, i.e. it is very rapid and the specific volume increase from water to steam produces high acceleration rates and hence pressure waves. The physical idea is that as the level of superheat increases, the amplitude of vibration of the liquid molecules increases—they are bound by short-range forces so movement is restricted but becomes greater with the increased kinetic energy associated with increasing temperature. A minimum linear dimension can be defined beyond which adjacent molecules will break free, and as superheat increases, spaces within the liquid will open up randomly and allow the formation of a bubble. In nucleate boiling the superheat provided at the surface is only a degree or two, but 50 °C or more may occur in the absence of nucleation sites. The rate of evaporation is correspondingly high. Homogenous nucleation is a rare phenomenon because nucleate boiling usually begins first at a container wall, but it can occur under very high rates of heating. Skripov [1974] set out the theory which was expanded upon by Blander [1979]; Hasan et al. [2011] have recently provided a literature survey. This topic is very relevant to the search for the causes of thermal explosions in industrial equipment and of hydrothermal eruptions and explosive volcanism.

7.4 Flashing

Flashing is the phase change from liquid to gas produced by reducing the pressure without adding heat, which distinguishes it from boiling. It is a common process both in natural geothermal flows and in wells and steamfield equipment. Hot fluids convect upwards in permeable ground and spring conduits and undergo falling pressure as a result, and they flow upwards in discharging wells.

If a mass of water at saturation conditions suffers a pressure reduction, the fraction of the water which will evaporate is determined by the absolute pressure level and the pressure change. How the evaporation takes place depends on the rate of fall of pressure and the surface area of the water from which steam can leave. Suppose the mass of water is contained in a cylinder by means of a piston and in the initial condition the entire volume is taken up by water. The pressure is then reduced by withdrawing the piston slowly. The only surfaces over which evaporation can take place are those in the bottom of the nucleation sites, as already discussed, which are water–air surfaces or water–air + steam surfaces. Bubbles would appear on the vessel walls. Hahne and Barthau [2000], who reviewed previous studies, carried out experiments of this type in which the starting point was a vessel part filled with liquid and the remainder filled with its gas (vapour), both at rest and in thermal equilibrium. Slow pressure reductions resulted in nucleation around the liquid line in the cylinder, while rapid reductions caused an evaporation wave to traverse down through the liquid, forming a mixture of droplets and gas behind the wave.

If a pressurised liquid is to be flashed to a lower pressure, then spraying it into the available space will help the process by maximising the liquid surface area, and this is the usual practice. Given the high specific volume of steam, it is possible that spraying water into a vessel would generate vapour faster than it could escape, so that the pressure in the vessel could increase and restrict the flashing process—in other words, the vessel must be large enough to accept the steam generated at the required pressure.

In discharging wells the liquid becomes superheated as it flows upwards because of the effective hydrostatic pressure gradient. The heat loss to the casing can be neglected if the discharge has been taking place for some time and is steady, and the flow will be approximately adiabatic. The distribution of the two phases will depend on the particular parameters of the flow, and this is the main issue in the study of flow in Sect. 7.7.

The flashing process is used in the desalination of seawater and chemical processes for concentrating solutes and recovering solvents. "Multistage flash" in geothermal terms means 2 or 3, but 25–30 stage flash evaporators are used to produce drinking water from seawater in the Middle East countries where there is ample fuel supply but little fresh water. There is a difference in focus between the two industries: in geothermal the aim is to produce steam from the available energy in the well discharge, and in desalination it is to produce the maximum amount of pure water from the seawater after heating it only once, at the inlet, and then conserving the heat by recovering the h_{fg} specific enthalpy change by means of a condenser in every stage. There is no economic advantage in increasing the number of flash stages in a geothermal plant because it would make the turbine too complicated, as will become clear in Chap. 11, although this is an opinion unsupported by any analysis. The literature on multistage flash evaporation is rather impenetrable, but papers by Lior [1986] and Khademi et al. [2009] provide an inroad in the vernacular of this book.

7.5 Condensation

Both steady-state condensation and very rapid transient condensation are relevant to geothermal engineering. The physical process is the reverse of phase change by boiling. It can be achieved by either decreasing the temperature of the gas or increasing its pressure, and it can take place homogenously or heterogeneously (via nucleation sites).

Homogenous condensation arises when the gas is supersaturated enough to cause molecules to come into close enough proximity to allow short-range forces to form a liquid droplet. Once a liquid surface is established, then condensation will occur via the capture of individual molecules, and the distribution of the condensed liquid controls the rate of further condensation via the surface area exposed. Nusselt (1916) developed the theory associated with the idea of a continuous film condensing on a vertical cooled surface. The film starts at the top and grows in thickness as it flows down the wall. If the surface is held at constant temperature by removing the heat given up in the condensation process, the rate of condensation decreased lower down the plate because of the increasing temperature difference required to drive the heat flow through the film to the wall—Fig. 7.5.

The patterns of condensation described so far occur when the gaseous form of a single pure substance condenses on a clean surface. The rate of heat transfer can be

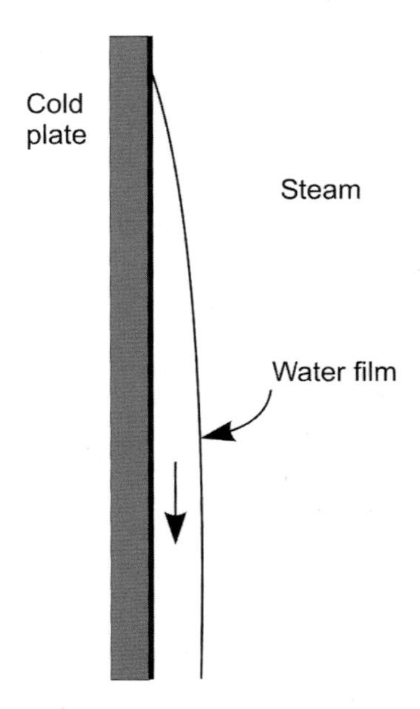

Fig. 7.5 Illustrating the process of film condensation

increased by the presence of other substances, just as it can be decreased by the presence of non-condensable gases. The surface can be treated with organic compounds which cause the condensation to form into drops (hence the name drop-wise condensation) which run off the surface quickly and leave it free for further condensation. If present, non-condensable gas (ncg) molecules mixed with the condensable gas molecules are carried to the cool surface but remain there after the liquid has formed. If the concentration of ncg molecules increases, they blanket the surface and prevent condensable molecules reaching it, thus reducing the condensation rate.

Transient condensation of very small bubbles has received a great deal of attention because it is the cause of the phenomenon known as cavitation, which occurs in high-speed flows. Many small bubbles form on the underwater surfaces of high-speed marine vessels and in the films of lubricant in high-speed rotating machines because even though the fluid is not at high temperature, local pressures may be transiently below saturation pressure under some flow circumstances—Bernoulli's equation applies and local pressure may fall considerably. The collapse of the bubbles causes metal fatigue and pitting of the surfaces. If the bubble is spherical, the condensation process proceeds fast enough for the liquid to accelerate towards the bubble centre symmetrically, resulting in a very high inertial force. Jones and Edwards [1960] reported peak pressures of more than 10,000 atm; Wang and Chen [2007] provide a review and the results of recent experiments.

A related phenomenon occurs in engineering plant when a steam cavity condenses. The maximum pressure achieved by the collapse is less than in a cavitation bubble because the geometry is usually not spherical, but the effect can in some circumstances be amplified by the local pressure increase travelling along a pipe until it rebounds and returns—the phenomenon known as water hammer. Chou and Griffith [1990] carried out experiments with the aim of developing protocols for the safe injection of cold water into pipework containing steam, in connection with water-cooled nuclear reactor safety studies. The avoidance of steam cavity collapse is a consideration in the design of geothermal pipelines, as will be discussed in Chap. 13.

7.6 Thermal Explosions and Hydrothermal Eruptions

Material used in engineered systems at ever higher temperatures and heat fluxes has given rise to explosive phenomena, the mechanisms of which remain to be established. Explosions also occur naturally in geothermal resources and volcanoes, some small and some very large. There are only a few fundamental physical processes which could contribute to these, yet they have so far defied explanation.

As regards hydrothermal eruptions, McKibben [2007] reviewed his previous work on events following the explosion itself but without identifying its origins, considering that the outflow of material may be the result of a failure of the containment of saturated water or other high-pressure fluids. Vessels filled with water at saturation temperature are known to produce considerable reactive forces when they burst. Ohba et al. [2007] examined the circumstances of a small isolated eruption on the flanks of a volcano, involving mud but not magma. Browne and Lawless [2001] presented evidence from the craters formed by eruptions from hundreds to thousands of years ago and addressed the terminology, which is poorly defined because the physics of the explosion have not been clearly identified. Phreato-magmatic and hydrothermal are two names often used, the first implying an explosion in which magma played a part and the second where no magma was present.

In engineering equipment the explosions are called steam explosions or vapour explosions, and they have been the subject of a great deal of laboratory experimentation. They occur, for example, when molten metal is dropped into water. Their history is reviewed by Hasan et al. [2011], and a recent numerical modelling study reported by Ursic et al. [2012] presents a picture very similar to the idea of a detonation wave. In the process called detonation, some combustible material is distributed throughout a volume of air which is traversed by a compression wave. Normally a wave of locally high-pressure fluid (air) will attenuate as it travels, but if it is fuelled by initiating combustion at or just behind the wave front, the energy release accentuates and accelerates the wave instead. The numerical model mentioned includes the molten metal being fragmented into very small drops which release their heat rapidly, in milliseconds, the result being an explosive expansion. Explosions of this type are of concern in foundries and also in the nuclear industry as the extreme outcome of fuel melting; the process also occurs in some types of volcanic eruption.

A connection between geothermal-related explosions and steam explosions is offered in a report by Stanmore and Desai [1993] of a total of 17 explosions which have occurred in fossil-fuelled boilers when dropping hot ash from the base of the boiler into a purpose-built pool of water beneath. Coal ash is primarily silica and alumina in particles of diameter less than 30 μm. A method of avoiding the explosions was recommended but the cause was not identified.

Watson [2002] identified and compared four physical processes which give rise to an explosion, namely, the failure of a vessel containing high-pressure fluid, a chemical explosion (TNT, gelignite, etc.), homogenous nucleation and the collapse of a vapour bubble. There are similarities in the craters caused by buried explosives and natural hydrothermal eruption craters and breccia pipes, but the natural events are so rare and so unpredictable that real progress in understanding will not be made without laboratory and small-scale field experimentation.

7.7 Two-Phase Flow

Two-phase flows occur in geothermal wells, pipelines, condensers and heat exchangers and also in the flows induced in a formation by the discharge from wells. They occur throughout the petroleum and chemical processing industries, often with more than one chemical species present. Only single-component (chemical species) two-phase flows are discussed here.

7.7.1 The Approach to Analysing a Two-Phase Flow

Some of the flow patterns that a steady mass flow rate of water goes through in changing into steam on its way through a vertical heated tube are illustrated in Fig. 7.6. An entirely liquid water flow enters at the bottom and gradually collects bubbles from the wall. Partway up the pipe, there is enough steam for it to collect as larger volumes separated by bodies of mainly water; still further up, the flow may become less organised before adopting an annular pattern, with most of the water flowing up the wall in an annular layer. The annular flow eventually dries out as the water evaporates. At any point in the flow, no matter what the frame of reference, the conditions vary randomly. There is no resemblance at all to the smooth variation of velocity with position in the single-phase flows described in Chap. 4.

The problems facing the designer of a two-phase flow pipe network are the same as for single-phase flow, usually calculating the pressure drop for a given pipe diameter and flow rate and optimising pumping power or energy loss. The fundamental approach of creating equations from the statements of conservation of mass, momentum and energy flow rates is still adopted, but a mainly new set of variables is required to express them. The mass flow rate through a pipe may be constant, but the fluid can change phase.

The flow rates of each phase are related—they add up to the constant total flow rate—and this introduces simple though clumsy algebraic relations into the equations. As in the treatment of turbulent single-phase flows, no solutions are possible without introducing experimentally determined relationships. In the case of two-phase flows, a set of experimental data must be provided for each of the flow patterns characterising the flow in Fig. 7.4 (there are more recognised patterns than

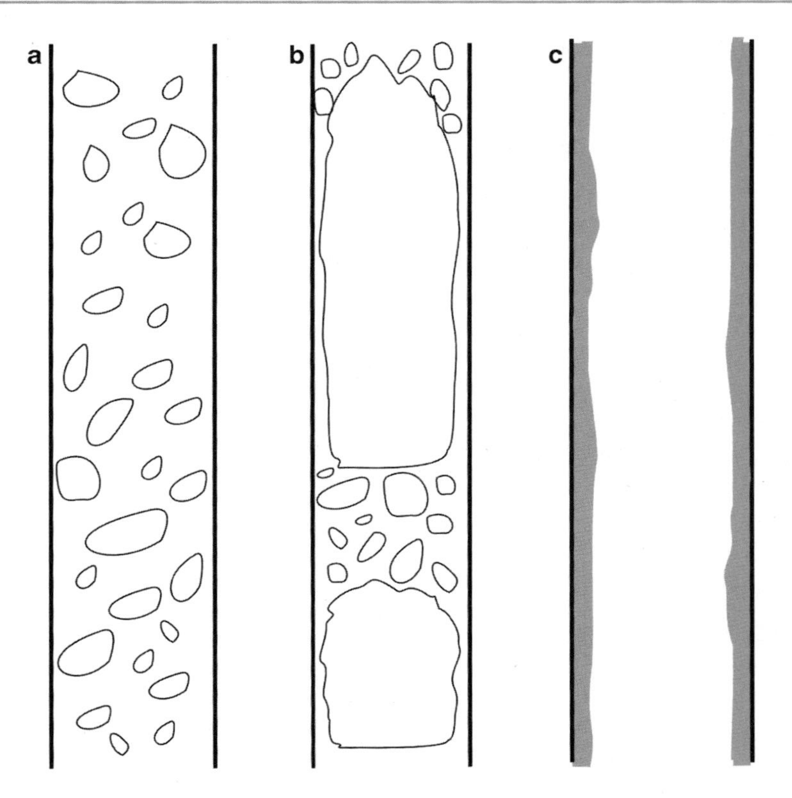

Fig. 7.6 Regimes of steady two-phase flow in a vertical, heated pipe; (**a**) bubbly, (**b**) slug and (**c**) annular

are shown here). The patterns have a measure of subjectiveness, which adds uncertainty to the outcome of solving the equations. The experimental data is provided as a number of correlations written in terms of the new variables already mentioned.

Quite separate from the flow patterns, there are three well-known methods of considering the flow in setting up the governing equations. Wallis [1969] described them as models, and they remain in use today—see Ghiaasiaan [2008].

They are:

- Homogenous flow, in which the two phases are well mixed and travel together at the same velocity. The flow can be represented by a single momentum equation because the phases are in thermodynamic equilibrium.
- Separated flow, in which it is accepted that phase change from liquid to gas induces a velocity change because of the change in specific volume, and hence a momentum equation for each phase is introduced.
- The drift flux model in which the phases travel as in the separated flow model, but the equations are expressed in terms of the relative velocity between the phases.

Only homogenous flow will be dealt with here as it is sufficient to illustrate the approach.

7.7.2 The New Variables

7.7.2.1 Void Fraction, α

Consider a pipe of cross-sectional area A and imagine the flow passing through a plane at some particular value of z, the direction of the pipe axis. At any instant, part of the cross-sectional area will be occupied by steam and part by water, and because of random shape and size of the "lumps" of water and steam, these areas will be constantly changing. The void fraction is the average fraction of A occupied by the steam phase:

$$\alpha = \mathrm{A_g}/\mathrm{A} \tag{7.4}$$

For dealing with the simple, steady-state homogeneous flow, the concept of a simple average is sufficient, but in general the phases can travel at a different velocity and the method of averaging requires careful consideration. By the definition of mean density, the void fraction is related to the phase densities as follows:

$$\overline{\rho} = \alpha\rho_f + (1 - \alpha)\rho_g \tag{7.5}$$

7.7.2.2 Mass Quality, x_m

If a 1 m length of the pipe being considered could be instantaneously removed, taking with it the flow that happened to be there at the time, and if the fluid content was weighed and the mass of steam measured, then the dryness fraction could be found. On the other hand, if the average flow rate of each phase could be measured at a cross section of the pipe, then the proportion of the flow that was steam could be calculated—this is the mass quality, which is a characteristic of a flow rather than a given mass.

$$x_m = \frac{\dot{m}_s}{\dot{m}} \tag{7.6}$$

The mean density can be written in terms of mass quality:

$$\frac{1}{\overline{\rho}} = \frac{x_m}{\rho_g} + \frac{(1 - x_m)}{\rho_f} \tag{7.7}$$

7.7.2.3 Mass Velocity G

The mass velocity is the mass flow rate per unit area, in other words a mass flux (kg/sm^2),

$$G = \frac{\dot{m}}{A} \tag{7.8}$$

It can be written for each phase.

7.7.2.4 Volumetric Flux or Superficial Velocity j

This is similar to the mass velocity, but is the volumetric flow rate per unit area instead of the mass flow rate. The volumetric flux is better known as the superficial velocity because it has the units of a velocity (m^3/s per m^2), and it may be defined for the total flow as well as each phase:

$$j = \frac{Q}{A} \tag{7.9}$$

7.7.3 The Governing Equations for Homogenous Flow

The flow is vertically upwards, steady state and one dimensional, varying only with z (positive upwards). It is also isothermal and finely mixed, whatever the proportions of steam and water. The phases travel at the same speed, and this has an influence on the averaging process and hence on the value of mean density. It is a single-component flow (H_2O) through a pipe of diameter d.

The continuity equation, (4.7), is most useful when written as a simple statement that the mass flow rate is constant:

$$\dot{m} = \bar{\rho} w \frac{\pi d^2}{4} \tag{7.10}$$

The momentum equations reduce to one only, which may be recognised as being similar to Eqs. (4.10) and (4.18) but with more basic terms since the flow is one dimensional:

$$\bar{\rho} w \left(\frac{\pi d^2}{4}\right) \frac{dw}{dz} = -\left(\frac{\pi d^2}{4}\right) \bar{\rho} g - \left(\frac{\pi d^2}{4}\right) \frac{dP}{dz} - (\pi d) \tau_w \tag{7.11}$$

On the right-hand side, the terms are gravitational, pressure gradient and wall shear stress acting on the pipe perimeter. This equation reduces to

$$\bar{\rho} w \frac{dw}{dz} = -\bar{\rho} g - \frac{dP}{dz} - \frac{4}{d} \tau_w \tag{7.12}$$

Since the flow is steady state, the energy equation is a version of the Steady Flow Energy Equation (3.6). However, in geothermal engineering, the flows are mainly adiabatic—at least, they are not heated and the heat loss is small, either through a well casing or through an insulated pipeline—so the energy equation provides no input to the set.

If Eq. (7.12) is rearranged as an expression for pressure gradient,

$$\frac{dP}{dz} = -\bar{\rho} g - \bar{\rho} w \frac{dw}{dz} - \frac{4}{d} \tau_w \tag{7.13}$$

it can be interpreted as a statement that the pressure gradient is the sum of components from three separate effects:

$$\frac{dP}{dz} = \left(\frac{dP}{dz}\right)_{grav} + \left(\frac{dP}{dz}\right)_{accel} + \left(\frac{dP}{dz}\right)_{fric} \tag{7.14}$$

To calculate the pressure gradient, each of the component parts must be provided: gravity, acceleration and friction. The gravitational contribution depends only on the mean density, which is a function of the flow regime and the details of mass flow rate, water and steam properties, etc., and must be found from experimental data. The acceleration component is the result of the volumetric flow rate changing as a result of heating or flashing and likewise depends on the mean density. The frictional component requires experimental data, very similar to the friction factor for single-phase turbulent flow.

7.7.4 Correlations for Obtaining Flow Parameters

Examining Eq. (7.12), the opportunities for introducing experimental data to help with the solution are only in the average density $\overline{\rho}$ and the wall shear stress τ_w.

With the flow to be analysed fixed in terms of a pipe diameter, mass flow rate, local quality, pressure and temperature, the mean density can be found from the definition, Eq. (7.10). The gravitational and acceleration pressure gradient components can then be calculated.

The wall shear stress can be considered to be proportional to the kinetic head of the flow, with a friction factor as the constant of proportionality, exactly as for single-phase fluids:

$$\tau_w = \frac{1}{2}\overline{\rho}w^2 f \tag{7.15}$$

This can be made more convenient by replacing the kinetic head by more practical parameters:

$$\frac{1}{2}\overline{\rho}w^2 = \frac{1}{2\overline{\rho}}\left(\frac{\dot{m}}{A}\right)^2 \tag{7.16}$$

But this simply defers the real issue of providing an appropriate friction factor. ESDU [2008] cites experimental data correlations by four different authors who adopted a common approach. The friction factor was assumed to be related to the Reynolds number in exactly the same way as for single-phase pipe flow, being represented by the experimental correlation for fully developed turbulent flow

$$f = 0.046\,\mathrm{Re}^{-0.2} \quad \text{for Re} > 20{,}000 \tag{7.17}$$

and the theoretical solution for fully developed laminar flow (Re < 2,000)

$$f = \frac{16}{\text{Re}} \tag{7.18}$$

But instead of using the fluid viscosity μ to calculate Reynolds number, each author provided a new definition of viscosity, mostly based on mass quality, and provided a formula for calculating it using the viscosities of water and steam at the appropriate pressure and temperature. Each formula was based on a separate set of experiments.

Adopting the homogenous flow model defines some of the flow parameters very tightly, leaving few options for experimenters to correlate the measured variables, but it is simple. The separated flow and drift flux models are less restrictive, providing a better chance of good predictions.

7.7.5 Two-Phase Flow in Permeable Formations

In Sect. 7.1 reference was made to factory accumulators, large vessels half filled with saturated water overlain by saturated steam. If the steam demand suddenly increased, a resulting sudden drop in pressure in the supply lines was avoided by the saturated water flashing; the process is very rapid. Each phase has its own compressibility, as defined earlier by Eq. (4.80):

$$c = \frac{1}{\rho}\left(\frac{\partial \rho}{\partial P}\right)_T \tag{4.80}$$

But when dealing with the response of a fixed volume of a mixture at saturation conditions to a pressure change, the volumes of each phase change because of not only their individual compressibilities but also the exchange of mass between the phases. The exchange of mass can be considered as an effective compressibility. Transient pressure testing of formation properties relies on measuring the pressure change induced by discharging a well or pumping water into it, and if the formation being tested contains two-phase fluid, then this effective compressibility is important. Work on this topic culminated in an important contribution by Grant and Sorey [1979] in the form of an expression for the effective compressibility:

$$c_{tp} = \frac{\left[(1-\phi)\rho_R C p_R + \phi S_f \rho_f C p_f\right]\left(\rho_f - \rho_g\right)}{\phi h_{fg}\left(\frac{dP}{dT}\right)_S \rho_f \rho_g} \tag{7.19}$$

In this expression the compressibilities of the individual phases have been neglected because they are small compared to the effective compressibility due to the phase change. The denominator contains the Clausius–Clapeyron equation given earlier as Eq. (3.16).

The saturation S appears in this equation, being defined as the volume fraction of a porous medium occupied by either of the phases; it is equivalent to the dryness fraction for mass fractions. The "rules" in respect of saturation are

$$S_f + S_g = 1 \tag{7.20}$$

and, for example, mean density is given by the following, with other properties similarly derived:

$$\overline{\rho} = \rho_f S_f + \rho_g S_g \tag{7.21}$$

7.8 Geothermal Liquids with Dissolved Gases

Geothermal liquids usually contain dissolved gases, and as they rise, the pressure falls and the gases come out of solution. The presence of gas in solution changes the pressure at which the water–steam phase change takes place for a given solution temperature, in other words the pressure at which bubbles form. The saturation line for pure water is now redundant, replaced by a new relationship involving the concentration of gas in solution. The presence of the gas changes the mechanics of the flow, both directly by forming bubbles and indirectly by altering the parameters at which phase change takes place; the bubbles contain both gas and steam. Fortunately most of the gas dissolved in geothermal water is CO_2 and it can be considered as the only solute so far as flow effects are concerned, for the present purposes.

An ideal gas is one in which the molecules move independently of each other, except for elastic collisions. The pressure (force per unit area) on the wall of the container is created by the change of momentum of the molecules hitting it. If two ideal gases are present as a mixture, red molecules and blue molecules, they still act independently, there is a pressure contribution from each colour, and the total pressure is the sum of the two. This was established by Dalton (1802) and is known as Dalton's law of partial pressures. Each pressure contribution is called the partial pressure of that species.

For some pairs of chemical species, the equilibrium vapour pressures formed throughout the entire range of mixture strengths, from a dilute solution of the red species in the blue as solvent, to the other way around, the total pressure above a liquid mixture is the sum of the partial pressures, and the partial pressure of each species is proportional to its concentration. This is an ideal gas-like behaviour and is referred to as Raoult's law, but it is not the general case. More often, the linear relationship of partial pressure to concentration occurs only for low-solute concentrations, when it is referred to as Henry's law

$$P_{pp} = K_H.C_M \tag{7.22}$$

in which
P_{pp} is the partial pressure of the gas in bars
C_M is the molar fraction of the gas in solution
K_H is Henry's constant in bars/molar fraction.

Not all chemical species behave as ideal gases, and a parameter known as fugacity is introduced as a measure of the departure from ideal. Fugacity is not needed here as Henry's law is an accurate enough representation for typical geothermal calculations.

Lu [2004] reviewed the experimental measurements of Henry's constant for dilute aqueous solutions of CO_2 and their correlation by Alkan et al. [1995]:

$$K_H = 406.41 + 47.088T + 7.6975x10^{-2}T^2 - 7.4695x10^{-4}T^3 \quad (7.23)$$

$$\text{for } 25 < T < 172 \ ^\circ C$$

Sutton [1976] described a method of calculating the equilibrium mass fraction of CO_2 in an aqueous solution over a range of pressures and temperatures, from which the new "boiling point" curve can be calculated for a given solute concentration. This will be referred to again in Chap. 8.

References

Alkan H, Babadagli T, Satman A (1995) The prediction of PVT/phase behaviour of geothermal fluid mixtures. In: Proceedings world geothermal congress, vol 3, Florence, Italy

Blander M (1979) Bubble nucleation in liquids. Adv Colloid Interface Sci 10:1–32

Browne PRL, Lawless JV (2001) Characteristics of hydrothermal eruptions, with examples from New Zealand and elsewhere. Earth Sci Rev 52:299–331

Chou Y, Griffith P (1990) Admitting cold water into steam filled pipes without water hammer due to steam bubble collapse. Nucl Eng Des 121(3):367–378

ESDU 04006 (2008) Pressure gradient in upward adiabatic flows of gas-liquid mixtures in vertical pipes. IHS, Feb 2008

Ghiaasiaan SM (2008) Flow, boiling, and condensation. Cambridge University Press, Cambridge

Grant MA, Sorey ML (1979) The compressibility and hydraulic diffusivity of a steam-water flow. Water Resour Res 13(3):684–686

Hahne E, Barthau G (2000) Evaporation waves in flashing processes. Int J Multiphas Flow 26:531–547

Hasan MN, Monde M, Mitsutake Y (2011) Model for boiling explosion during rapid liquid heating. Int J Heat Mass Transf 54(13–14):2837–2843

IAPWS (1994) IAPWS release on surface tension of ordinary water substance. Orlando, USA

Jones IR, Edwards DH (1960) An experimental study of the forces generated by the collapse of transient cavities in water. J Fluid Mech 7:596–609

Khademi MH, Rahimpour MR, Jahanmiri A (2009) Simulation and optimisation of a six effect evaporator in a desalination process. Chem Eng Process Process Intensif 48(1):339–347

Lior N (1986) Formulas for calculating the approach to equilibrium in open channel flash evaporators for saline water. Desalination 60(3):223–249

Lu X (2004) An investigation of flow in vertical pipes with particular reference to geysering. PhD thesis, University of Auckland, New Zealand

Lyle O (1947) The efficient use of steam. HM Stationary Office, London

McKibben R (2007) Force, flight and fallout: progress on modeling hydrothermal eruptions. In: Proceedings of the New Zealand 29th geothermal workshop, University of Auckland, Auckland

Ohba T, Taniguchi H, Miyamoto T, Hayashi S, Hasenaka T (2007) Mud plumbing system of an isolated phreatic eruption at Akita Yakeyama volcano, Northern Honshu, Japan. J Volcanol Geoth Res 161(1–2):35–46

Skripov VP (1974) Metastable liquids. Wiley, New York

Stanmore BR, Desai M (1993) Steam explosions in boiler ash hoppers. Proc Inst Mech Eng 207:133–142

Sutton FM (1976) Pressure-temperature curves for a mixture of water and carbon dioxide. N Z J Sci 19:297–301

Ursic M, Leskover M, Mavko B (2012) Simulation of KROTOS alumina and corium steam explosion experiments: applicability of the improved solidification influence modeling. Nucl Eng Des 246:163–174

Wallis GB (1969) One-dimensional flow. McGraw Hill, New York

Wang Y-C, Chen Y-W (2007) Application of piezo-electric PVDF film to the measurement of impulsive forces generated by cavitation bubble collapse near a solid boundary. Exp Therm Fluid Sci 32(2):403–414

Watson A (2002) Possible causes of hydrothermal eruptions. In: Proceedings of the 24th New Zealand geothermal workshop, University of Auckland, Auckland

The Discharging Well

8

Once discharge is achieved, the discharge flow rate and its temperature, pressure and specific enthalpy can be measured; several methods are available. The main parameters are presented as "discharge characteristics", which are required for power station control but also contain information about the producing formations. The chemical constituents of the discharge are measured at the same time as the discharge characteristics and provide further clues as to conditions in the producing formations. Some wells discharge at a steady flow rate, with perhaps a long-term decline as the formation pressure declines, but some have a periodic flow or even a regular intermittent discharge like a geyser. It is sometimes useful to have a means of predicting the details of the flow during discharge, and numerical discharge prediction methods have been developed. All of these matters are discussed in this Chapter.

8.1 The Discharge Characteristic

8.1.1 The Form of the Discharge Characteristic

The discharge from a well is equivalent to the discharge from a pump, and the well responds in a similar manner, the flow rate increasing as the wellhead control valve or pump outlet valve is opened. Pumps are selected for a task according to their characteristic, a graph of pressure difference over the pump plotted on the vertical axis versus the mass flow rate discharged plotted on the horizontal axis. This plot is useful because the resistance to flow in a piping system increases with the square of mass flow rate and can be calculated for a range of flows; a graph of resistance versus mass flow rate can be plotted on top of the characteristic. The intersection of the two curves marks the mass flow rate and pressure difference that will occur if the two are coupled together.

A. Watson, *Geothermal Engineering: Fundamentals and Applications*,
DOI 10.1007/978-1-4614-8569-8_8, © Springer Science+Business Media New York 2013

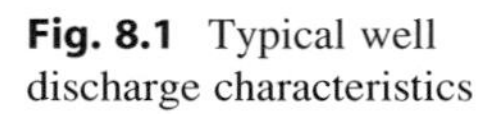

Fig. 8.1 Typical well discharge characteristics

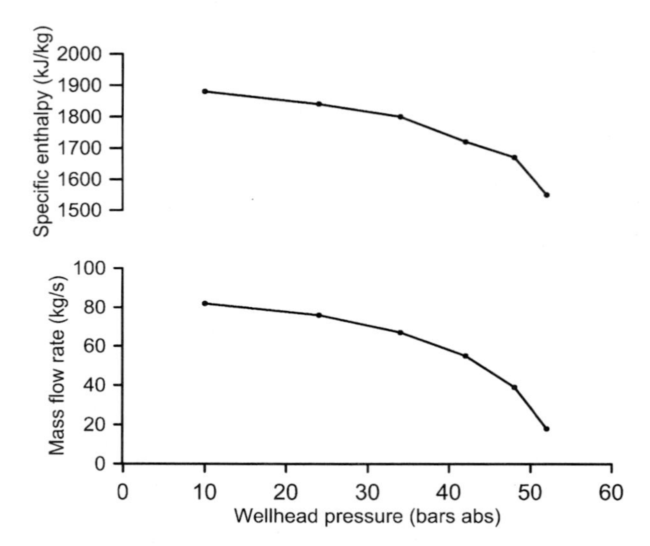

For geothermal wells a different convention has been adopted. Perhaps because the wellhead pressure is the controlled variable, it is plotted horizontally and the mass flow rate discharged is the vertical axis. Also, however, the discharge from a geothermal well needs two parameters to define it, mass flow rate and specific enthalpy, so two characteristic curves are necessary and are conveniently plotted as shown in Fig. 8.1 (which shows actual measurements and scatter). Like pumps, some mass flow rate characteristics have a maximum in wellhead pressure at low mass flow rates, curving back on themselves as zero mass flow rate is approached. A common practice is to fully open the well first until its discharge stabilises and then progressively throttle it to see if there is a clear maximum discharge pressure, which was not found in the well of Fig. 8.1.

Wells may be discharged for periods of weeks to determine production capability and resource characteristics such as geothermal gases and chemical species in solution. If the discharge rate is controlled by the wellhead control valve (never the master valve), the gate edges become very abraded and eventually will not seal. For long-term discharge it is common practice to place a less expensive restriction in the delivery pipe, usually a disc with a hole in it sufficient to pass the required flow rate, so the control valve can be fully opened, leaving the gates clear of the flow to avoid damage. A set of discs is made covering the range of flow rates to be tested, a trial and error exercise. Lovelock and Baltasar [1983], in discussing geochemical analysis of discharge samples, explain that PNOC-EDC (now EDC) discharge testing lasts for 4–6 weeks.

Because the pressure in the producing formations may fall gradually as the discharge continues, it is best not to change the wellhead pressure in uniform steps from high to low during the tests, but to randomise the sequence. By gradually opening the well so that the points on the characteristic form a regular sequence in time, any simultaneous reduction in formation pressure is hidden.

For a well discharging two-phase fluid at wellhead pressure P_{wh}, the specific enthalpy and mass flow rate of the discharge for a particular wellhead pressure can be read from Fig. 8.1. The dryness fraction can be calculated according to the equations given in Chap. 3 and in this way a third graph showing the mass flow rate of saturated steam versus wellhead pressure could be added to Fig. 8.1. Assuming a steam rate for the turbine, say 2.4 kg/s/MWe, a graph of power output versus wellhead pressure could also be added and also a graph showing the amount of separated water and condensate to be disposed of.

The measurement of the discharge characteristic is likely to be the first occasion on which the well has been discharged for a lengthy period, and it will be tested immediately afterwards to check for any damage. A calliper will be lowered into the well with the aim of checking that the production casing is still of uniform diameter, has not developed any holes and shows no evidence of solids deposition or acid attack from the aqueous solutions discharged. The calliper is a cylindrical instrument with several arms which make contact with the production casing and move in or out to follow any undulations in the casing wall, as already discussed in Chap. 5. The signal is recorded at the surface.

Before discussing how the discharge characteristics are measured, some more information about their use is appropriate.

8.1.2 Interpretation of Resource Behaviour from the Discharge Characteristics

The maximum wellhead pressure is a function of the resource and the fluid composition. The shape of the discharge characteristics together with the downhole measurements described in Chap. 6 can provide more information about the physics of the flow in the formations. In making this assessment the heat loss from the flow as it passes up the well can be ignored as a first estimate, as can the reduction in specific enthalpy by work done against gravity. Several circumstances can occur:

(a) The well produces from only one liquid-filled formation at a temperature significantly below saturation. The flow flashes on its way up the well, and this takes place over the whole range of wellhead pressures obtainable. The specific enthalpy of the discharge remains constant with wellhead pressure.

(b) The well produces from a formation at or only just below saturation temperature, so the fluid flashes in the formation as a result of the pressure reduction. The pressure reduction works its way radially outwards from the well, and as it does so the saturation temperature in the fluid falls. Whereas before production began the rock and fluid were at the same temperature, the fluid is now cooler, so there is the potential for the fluid to gain heat from the formation. The temperature difference driving this heat transfer is greatest at the sandface, where the pressure is the lowest. The specific enthalpy of the discharge will increase if heat transfer takes place, which accounts for the term "flowing enthalpy" to describe discharge specific enthalpy, a reminder that it may not be the specific enthalpy of the undisturbed fluid in the formation. Under these circumstances the well is said to produce "excess enthalpy", which is expected to be highest at low wellhead pressure.

(c) The fluid in the formation flashes as in (b), but the steam is able to move towards the well faster than the water—the steam (strictly, the combination of formation and steam) is said to have a higher relative permeability than the liquid water. The discharge has a higher specific enthalpy than the downhole pre-discharge measurements indicated for the producing formation and the well again has excess enthalpy.

Examples of wells exhibiting excess enthalpy were given by Menzies et al. [1982] and by Lovelock et al. [1982].

When the well has two or more production zones, the discharge characteristics are more difficult to interpret because the proportion of flow coming from each zone is a variable. The flow from each is governed by the pressure in the well at the zone. But these pressures are not independent; they form part of a non-linear distribution up the well which is linked to the specific enthalpy and mass flow rate of the discharge from the zones. This distribution changes in a complicated way with these four variables. Grant et al. [1979] give an example of a well penetrating two production zones, the upper one producing dry steam and the lower one being liquid filled. Suppose the lower zone has a high temperature and is capable of producing a high flow rate. The well stands shut with the upper section steam or gas filled because it is connected to the steam zone; below that depth the well is liquid filled. When the well is opened only enough to allow a small discharge rate, the pressure gradient in the well below the steam zone is unaffected—to enable the lower zone to discharge the liquid column must flash and reduce its density and hence the sandface pressure, and this calls for a major reduction in wellhead pressure. At low flow rates (wellhead pressure just a little below the shut value) the discharge will thus be from the steam zone only and the discharge specific enthalpy will be that of steam at the formation pressure. When the well is fully opened and the lower zone begins to produce, the discharge specific enthalpy will tend towards that of the lower zone liquid. The specific enthalpy discharge characteristic will change accordingly. This emphasises the importance of good well measurements and interpretation prior to discharge. In passing, note that in some wells of this type, the wellhead pressure cannot be lowered sufficiently to make the lower zone discharge by simply opening the valve without some encouragement of the type described in Sect. 6.5.

If some formations are steam filled, the question arises whether the flow from the formation or at the wellhead will be superheated. The thermal boundary conditions of the flow up the well need to be considered (wellbore heat loss) as well as the loss of specific enthalpy in the form of work done against gravity; in other words, if the exact degree of superheat is an issue, then greater precision in the analysis is needed.

8.2 Measuring the Discharge Characteristics

Several measuring methods are available to choose from, depending on the thermodynamic state of the fluid discharged and wider project issues. The alternatives are set out as a chart later in this section. Most methods combine several individual measurements made using equipment that is easy to describe and has been left until Sect. 8.3. The James lip pressure pipe is a more complicated "instrument" which is introduced first.

8.2.1 The James Lip Pressure Pipe

The lip pressure pipe was invented by James [1962, 1966], during the exploration and development of New Zealand resources in the late 1950s. James carried out experiments in which a two-phase flow was created by mixing two separate streams of water and steam at known rates and then discharging the mixture through a "lip pressure pipe", a piece of plain round pipe about 1 m long or less, flanged so that it can be fitted to the pipe delivering the flow. The discharge end is open to atmosphere and is cut off very precisely, normal to the axis, leaving sharp, right-angled edges. It has a pressure tapping fitted close to the end as shown in Fig. 8.2. James [1966] gave the results of experiments with the dimensions of the pressure tapping, which should be taken into account when designing a tube.

James recognised that so long as the emerging jet is supersonic, a simple correlation relates the lip pressure pipe reading to the mass flow rate and the specific enthalpy of the discharge. The correlation is

$$\frac{G h_D^{1.102}}{{P_{lip}}^{0.96}} = 22106 \tag{8.1}$$

$$\text{for } 400 < h_D < 2800 \text{ kJ/kg}$$

where

$G = \dot{m}/A(\text{kg/m}^2\text{s})$ is the mass velocity for a mass flow rate $\dot{m}(\text{kg/s})$ in a pipe of cross-sectional area $A(\text{m}^2)$

h_D = specific enthalpy of the discharge (kJ/kg)

P_{lip} = lip pressure (kPa abs)

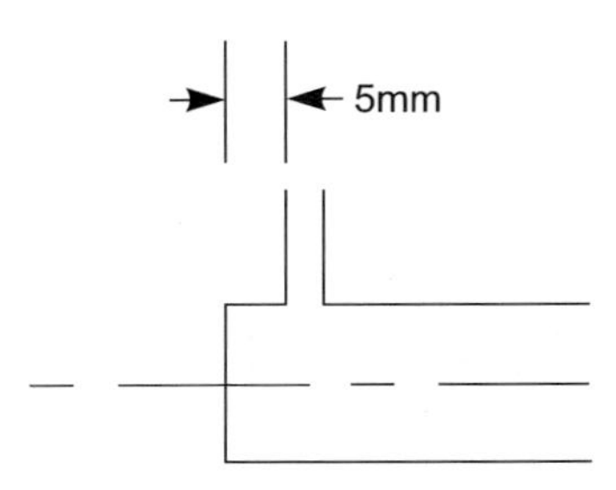

Fig. 8.2 Dimensions of the James lip pressure pipe

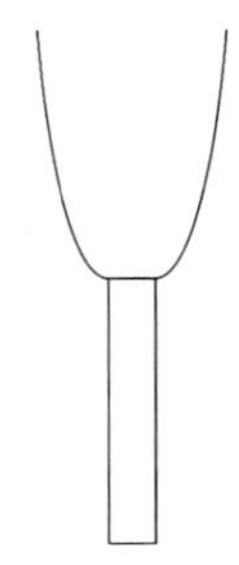

Fig. 8.3 Sketch of James lip pressure pipe flared supersonic flow

The constant on the right-hand side of the equation is experimentally determined and its value depends on the units used for the parameters in the equation. Constants for other units can be found in the literature. If the measurements are made using different units to those used in the formula, it is safer to change them to the formula units, carry out the calculation and convert the answer back to the required units. There is a further problem with this correlation; it is implicit in the variables to be determined, with one of them (specific enthalpy) raised to a power, so it cannot be solved simply. An example of its use is given below.

When testing a new well, the pipe diameter giving supersonic flow at maximum discharge is found by trial and error by watching the shape of the discharge, which has a characteristic flared shape when it is supersonic, as shown in Fig. 8.3. The velocity of sound in a two-phase mixture is lower than in either phase, see, for example, Kieffer [1977], so supersonic discharge can usually be achieved.

Measuring the complete discharge characteristic may take weeks and requires a good deal of equipment to be set up, so an initial estimate is often made with a lip pressure pipe attached directly to the wellhead. Used by itself, the lip pressure pipe provides insufficient information to determine mass flow rate, and an estimate of the specific enthalpy of the discharge from downhole measurements is necessary, which then allows mass flow rate to be determined.

There are practical problems with this method in terms of the disposal of the discharged fluid. In NZ, permits are usually obtainable to allow a short discharge vertically through a pipe attached directly to the wellhead and inclined at a slight angle to direct the flow away from the well but still having it fall within the drilling pad. The discharge is limited to an hour or so, on the grounds of noise and contamination. Wells are often in areas with natural surface discharge, so the extra contamination is tolerable.

8.2.2 The Available Methods of Measurement

The chart (Fig. 8.4) shows the main measurement methods available, characterising the state of the discharge as steam, two-phase or liquid, which dictates the level of complexity of the measurement method. One method has been left out of this diagram, namely, the calorimeter, which is only suitable for wells with a small discharge rate compared to that required for major power station projects.

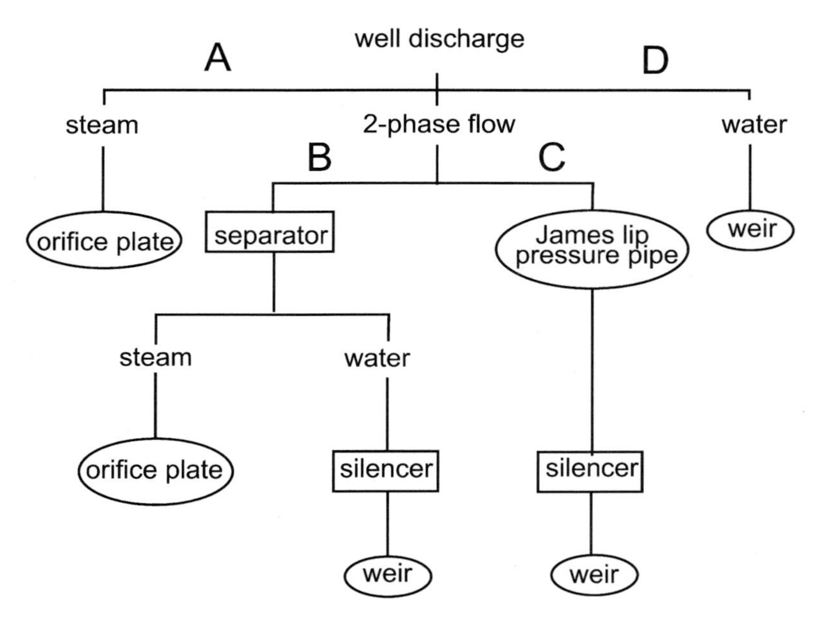

Fig. 8.4 Diagram showing method options

It consists of a thermally insulated tank of known volume, partially filled with a known volume of cold water at known temperature. The well is discharged into the tank for a measured length of time and the total volume and temperature of the mixed water are measured. Any steam discharged must be condensed in the tank, and the entry is submerged to encourage this. The mass flow rate of the well and the specific enthalpy of its discharge can be deduced from simple algebraic equations written for a mass and heat balance. The method is most often used for small diameter wells such as those supplying heat for domestic or small commercial use.

In the diagram are three measurement devices, namely, a single-phase orifice plate, a weir and the James lip pressure pipe, and these are used in combinations. The flow may have to be processed using a separator or an atmospheric pressure separator (otherwise known as a silencer), also shown in the diagram. Four routes through the diagram have been identified as A, B, C and D, of which route D requires only an explanation of the measurement weir, which is given in Sect. 8.3.

8.2.3 Route A: A Well Discharging Steam Only (Dry Steam)

The flow from a well producing steam only can be measured by directing it through a single-phase orifice plate, which is a regular obstruction in the form of a disc with a hole in it through which the flow passes, creating a small pressure drop—see Fig. 8.5. The disc must be manufactured and installed according to a standard, after which the mass flow rate can be calculated from the measured pressure drop and the

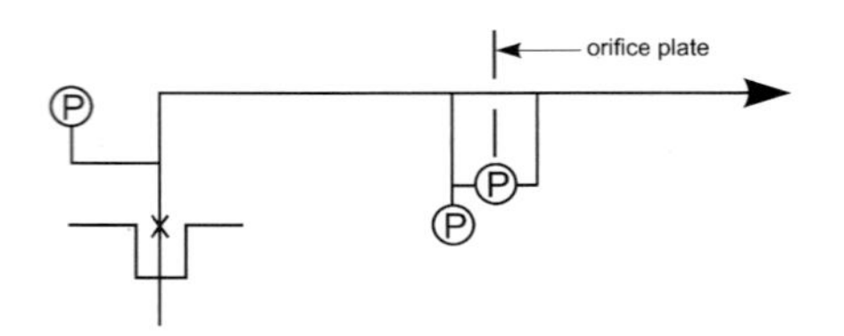

Fig. 8.5 Well discharging steam through an orifice plate—Route A of Fig. 8.4

pressure in the pipe just upstream of the orifice plate; it is described in detail in Sect. 8.3.1. The specific enthalpy of the well discharge can be found from the steam tables using the measured wellhead pressure. The flow through the pipe and orifice plate must be well below sonic velocity.

As a check on the wellhead pressure gauge and that the discharge is in fact saturated steam, a good estimate of the temperature of the flow can be found by attaching a thermocouple to the outside of the production casing or discharge pipe, making sure that it is thermally insulated over a circle around the thermocouple, perhaps 10 pipe wall thicknesses in radius.

8.2.4 Route B: A Well Discharging a Two-Phase Mixture

Route B is the most difficult (expensive) to set up in the field as a temporary arrangement during exploration, as it requires a separator and a silencer, but it should provide higher accuracy than Route C. The separator is essentially a closed vertical cylinder, higher than its diameter, in which water and steam are rotated at high enough speed that they separate under the centrifugal acceleration; a steel structure is needed to support it rigidly. A silencer is an atmospheric separator (essentially a separator vessel with no top) which may be as tall as the separator but is on a bigger base, so is more stable, and skid mounted moveable silencers are used during early exploration. On completed fields permanent separators and silencers are constructed. For the present purposes both separator and silencer can be assumed to separate the entering two-phase fluid into flows of saturated water and saturated steam, at the separator pressure and atmospheric pressure, respectively.

Figure 8.6 shows the arrangement and the relevant parameters. The specific enthalpies at the separator pressure and at atmospheric pressure are found from the steam tables at the measured pressures.

Recall that a two-phase flow has a dryness fraction X, the proportion of the total mass flow rate which is steam, which relates to specific enthalpy according to Eq. (3.18):

$$h = h_f + Xh_{fg} \tag{3.18}$$

There are two stages in Fig. 8.6 at which a two-phase flow separates into water and steam: in the separator, which receives the two-phase discharge from the well, and in the silencer, which receives separated water at separator pressure and flashes

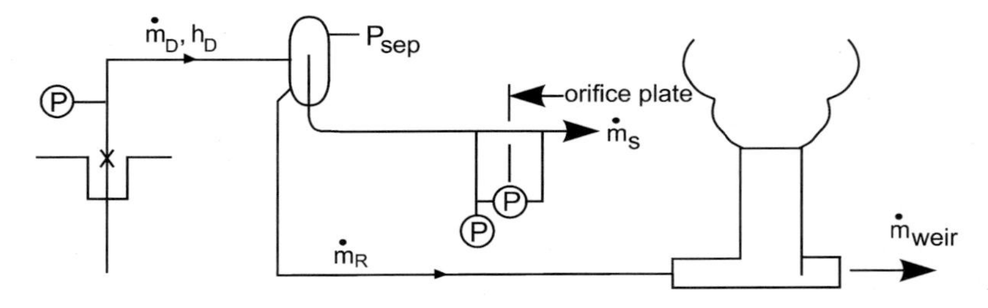

Fig. 8.6 Arrangement of equipment for Route B of Fig. 8.4

it to atmospheric pressure. Thus continuity of mass flow rate (a mass balance) can be written for the separator and reorganised, as follows:

$$\dot{m}_D = \dot{m}_R + \dot{m}_S \tag{8.2}$$

$$1 = \frac{\dot{m}_R}{\dot{m}_D} + \frac{\dot{m}_S}{\dot{m}_D} = \frac{\dot{m}_R}{\dot{m}_D} + X_{sep} \tag{8.3}$$

and similarly for the silencer,

$$\dot{m}_R = \dot{m}_{weir} + \dot{m}_{Satmos} \tag{8.4}$$

$$1 = \frac{\dot{m}_{weir}}{\dot{m}_R} + \frac{\dot{m}_{Satmos}}{\dot{m}_R} = \frac{\dot{m}_{weir}}{\dot{m}_R} + X_{sil} \tag{8.5}$$

The steam discharged from the silencer at atmospheric pressure is not measured, but the specific enthalpy of the mass flow rate $\dot{m}_R$ is known from the steam tables because the separator pressure P_{sep} is measured. Thus using Eq. (3.18) above,

$$h_R = (h_f)_{Psep} = (h_f + X_{sil}.h_{fg})_{atmos} \tag{8.6}$$

The atmospheric pressure is also measured, so X_{sil} for the silencer can be found. The mass flow rate of atmospheric pressure water is measured using the weir, which allows the mass flow rate of water entering the silencer, $\dot{m}_R$, to be found from Eq. (8.5). Equation (8.2) then allows the total discharge from the well to be found, since the mass flow rate of steam leaving the separator is measured, and also the separator dryness fraction. Continuity of energy (an energy balance) finally allows the specific enthalpy of the discharge to be calculated:

$$\dot{m}_D.h_D = (\dot{m}_R.h_f + \dot{m}_S.h_g)_{Psep} \tag{8.7}$$

The calculation proceeds upstream from the silencer, using the key fact that the flow entering the silencer is saturated water at a known pressure.

Example

- The measured parameters are the following:
  Separator pressure = 6.0 bar abs.
  Atmospheric pressure = 1 bar abs.
  Steam discharge rate from the separator, $\dot{m}_S = 6.2$ kg/s.
  Atmospheric pressure water flowing over the weir, $\dot{m}_{weir} = 35.3$ kg/s.
- Collect the properties required.

P (bar abs)	h_f (kJ/kg)	h_{fg} (kJ/kg)	h_g (kJ/kg)
6.0	670.5	2085.6	2756.1
1.0	417.4	2257.5	2675.0

- Apply the equations.

From Eq. (8.6)	$670.5 = 417.4 + X_{sil}.\ 2257.5$ (kJ/kg) giving $X_{sil} = 0.1121$ (it is advisable to carry four significant figures)
From Eq. (8.5)	$\dot{m}_R = 35.3/(1 - 0.1121) = 39.76$ kg/s
From Eq. (8.2)	$\dot{m}_D = 39.76 + 6.2 = 45.96$ kg/s
From Eq. (8.7)	$45.96\ .\ h_D = 39.76.\ 670.5 + 6.2\ .\ 2756.1$ giving $h_D = 951.85$ kJ/kg

8.2.5 Route C: An Alternative for a Well Discharging a Two-Phase Mixture

Route C addresses the same problem as Route B but avoids the use of a separator. Instead, a James lip pressure pipe is fitted where the fluid from the wellhead enters the silencer and the separated liquid passes over a measurement weir.

As before, the liquid flow rate over the weir is not the mass flow rate entering the silencer, as some of the discharged liquid is lost to atmosphere as steam. There is enough information to allow for this in the calculation.

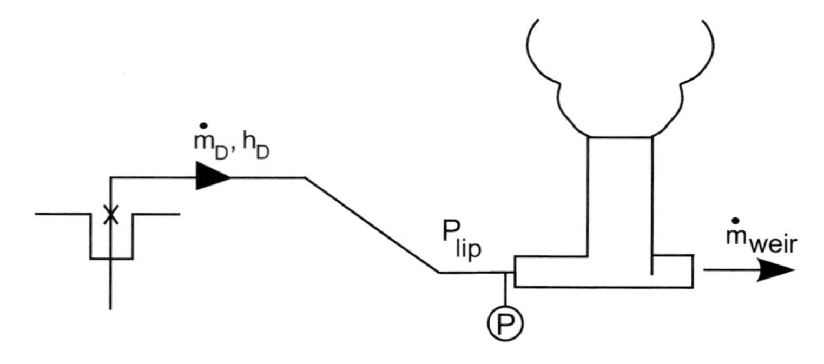

Fig. 8.7 Arrangement of equipment for Route C of Fig. 8.4

This time, the mass balance equation is

$$\dot{m}_D = \dot{m}_{Satmos} + \dot{m}_{weir} \tag{8.8}$$

and

$$\dot{m}_D = \frac{\dot{m}_{weir}}{(1 - X_{atmos})} \tag{8.9}$$

Equation (8.1) for the James lip pressure pipe is to be used:

$$\frac{G h_D^{1.102}}{P_{lip}{}^{0.96}} = 22106 \tag{8.1}$$

so the discharge mass flow rate must be written in terms of G:

$$G = \frac{4\dot{m}_D}{\pi d^2} = \frac{4\dot{m}_{weir}}{\pi d^2 (1 - X_{sil})} \tag{8.10}$$

The dryness fraction for the silencer, X_{sil}, is inconvenient and can be eliminated as follows:

$$(1 - X_{sil}) = \left(1 - \frac{(h_D - h_{fatmos})}{h_{fgatmos}}\right) = \frac{(h_{gatmos} - h_D)}{h_{fgatmos}} \tag{8.11}$$

With these substitutions, the correlation becomes

$$\frac{4\dot{m}_{weir}.h_{fgatmos}.h_D^{1.102}}{\pi d^2.P_{lip}^{0.96}(h_{fgatmos} - h_D)} = 22106 \tag{8.12}$$

This equation has only one unknown, h_D, but it appears twice. A simple approach for field work is to rearrange the equation so that the unknown appears on both sides:

$$\left(\frac{4\dot{m}_{weir}.h_{fgatmos}}{22106.\pi d^2.P_{lip}^{0.96}}\right).h_D^{1.102} = (h_{fgatmos} - h_D) \tag{8.13}$$

Then calculate the left-hand side and right-hand side and plot these on a graph against h_D repeating for a range of values until the two lines cross. The value of h_D at which they cross is the specific enthalpy of the discharge, from which the dryness fraction can be calculated, then G and finally $\dot{m}_D$. Iterative methods are available.

8.3 Further Details of the Measurement Equipment

8.3.1 The Single-Phase Orifice Plate

Several standards are available for orifice plates, ASME, ISO and BSI. The British Standard BS 1042 is used for the description here; it gives several options for construction. That referred to as the "D and D/2 version" shown in Fig. 8.7 and 8.8 is most often used with the flow from left to right.

The orifice plate is made with its edges accurately machined as shown. It is mounted between two flanges in a straight length of pipe of diameter D. The upstream pressure is measured at a wall tapping a distance D upstream of the plate and is an absolute pressure measurement used to determine the density of the fluid; call this location 1.

The other pressure of interest is where the cross-sectional area for the flow as it passes through the orifice is the smallest—call this location 2. Surprisingly, it turns out that the best position for location 2 is a distance D/2 downstream of the plate, so a wall tapping is placed there.

The formula relating pressure difference to mass flow rate is an empirical modification of Bernoulli's equation (4.28), which is for an ideal frictionless fluid (viscosity = 0). The equation can be reduced to

$$\frac{P_2}{\rho_2} + \frac{u_2^2}{2} = \frac{P_1}{\rho_1} + \frac{u_1^2}{2} \tag{8.14}$$

because gravitational effects are negligible over such a small arrangement, which is usually horizontal in any case. Rearranging,

$$u_1^2 - u_2^2 = 2\left(\frac{P_2}{\rho_2} - \frac{P_1}{\rho_1}\right) \tag{8.15}$$

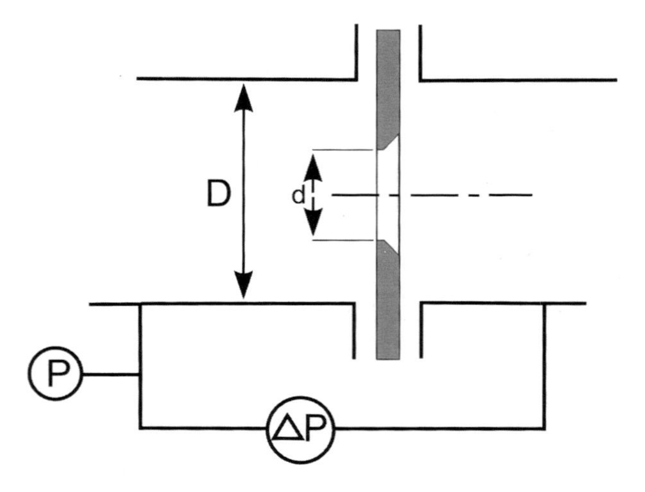

Fig. 8.8 Cross section of a single-phase orifice plate to BS 1042

Since $\dot{m} = \rho u A$ by definition, the left-hand side of the equation can be juggled to become

$$\frac{\dot{m}^2}{(\rho_2 A_2)^2}\left[\left(\frac{\rho_2 A_2}{\rho_1 A_1}\right)^2 - 1\right] = 2\left(\frac{P_2}{\rho_2} - \frac{P_1}{\rho_1}\right) \tag{8.16}$$

giving an expression for mass flow rate in terms of pressures and areas, which can be measured, and densities, which can be calculated if the fluid temperature is known. Despite the sites chosen for the pressure tappings, the areas A_I and A_2 are for the pipe and orifice plate, respectively. For a steam flow the pipe could be thermally insulated and the temperature measured near the orifice plate by a thermocouple attached to the insulated pipe wall or a thermometer pocket. Equation (8.15) can be rearranged as

$$\dot{m} = (\rho_2 A_2)\sqrt{\frac{2\left(\frac{P_1}{\rho_1} - \frac{P_2}{\rho_2}\right)}{\left(1 - \left(\frac{\rho_2 A_2}{\rho_1 A_1}\right)^2\right)}} \tag{8.17}$$

The equation used by BS 1042 for real fluids is

$$\dot{m} = C.\frac{\pi d^2}{4}\sqrt{\frac{2(P_1 - P_2)\rho_1}{\left(1 - \left(\frac{d}{D}\right)^4\right)}} \tag{8.18}$$

where C is called the discharge coefficient

The equations can be made more similar. If the pressure drop over the plate is small so that $\rho_1 \approx \rho_2$ and the value ρ_1 is used for both, and if the areas are expressed in terms of their diameters, then the equations differ only by the BS1042 form containing the discharge coefficient C. The discharge coefficient is necessary because of two real fluid effects. Firstly, the neck of the flow is not at the orifice plate but further downstream—the neck is referred to as the vena contracta and it is smaller in diameter than the hole in the plate. Secondly, there is a small energy loss between the upstream location and the neck because of eddies formed in the corners of the plate. These are illustrated in Fig. 8.9 which compares streamlines for an ideal (non-viscous) fluid and a real fluid.

The pressure is fairly uniform over the eddy formed after the plate, and the downstream tapping position has been chosen so that it is in this uniform pressure region. The value of C is of the order of 0.6 but is sensitive to a number of factors which BS 1042 takes account of, for example, it depends slightly on mass flow rate. The standard contains rules governing the required upstream length of straight pipe, the proximity to bends, etc., and these must be followed exactly if the measurement

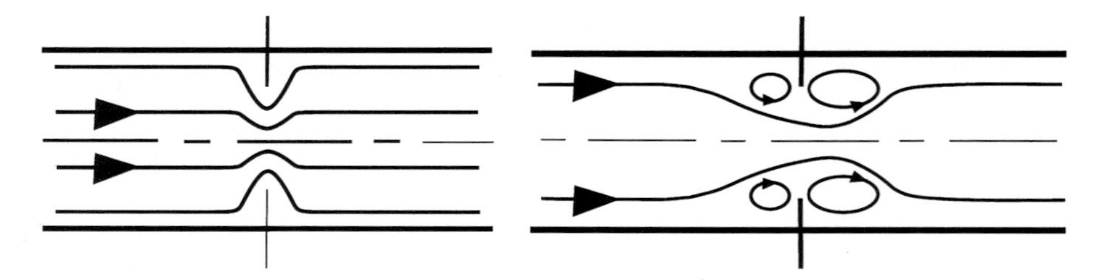

Fig. 8.9 Streamlines for ideal and real flow through an orifice plate

is to be to the prescribed precision. The plate itself must be accurately machined and free from burrs and defects along its sharp edge. (It should not be hung on a nail by its hole!) In addition, any pressure tappings made through the pipe must be of small diameter compared to the pipe wall thickness, say 1 mm diameter, and must have drilling burrs polished off on the inside leaving a square edged hole. Tappings welded on on-site are most unlikely to produce good results—the orifice plate is a laboratory technique and needs very careful attention to detail.

8.3.2 The Two-Phase Orifice Plate

The simplicity of placing an orifice plate in a flow attracted attention to its use for two-phase flows in various industrial sectors many years ago. It represents a very attractive option since the equipment arrangement shown in Figs. 8.6 and 8.7 would reduce to that of Fig. 8.5.

The problem of making predictions about the behaviour of any two-phase flow has already been explained in Sect. 7.7—experimental data is essential. A large number of parameters is required to define a two-phase flow, so a very large number of permutations must be examined in experiments if they are to provide data for a comprehensive correlation for any particular flow property of interest (in this case the pressure drop over an orifice). Adding to the problem is the uncertainty in knowing that the list of parameters is complete. Early empirical correlations were produced by Murdock [1962] and James [1965], the latter specifically for geothermal applications. Helbig and Zarrouk [2012] have reviewed them and others and have proposed a new correlation which they tested against field data. They note that the specific enthalpy of the flow is not an output of the orifice plate measurements and must be found separately.

8.3.3 The Thin-Plate Sharp-Edged Weir

There are even more standards governing weir flow measurement than orifice plates—ASME, ASTM, AWWA, BSI, ISO and so on; British Standard 3680 is used for this description. Like the orifice plate, which is an obstruction in a pipe flow,

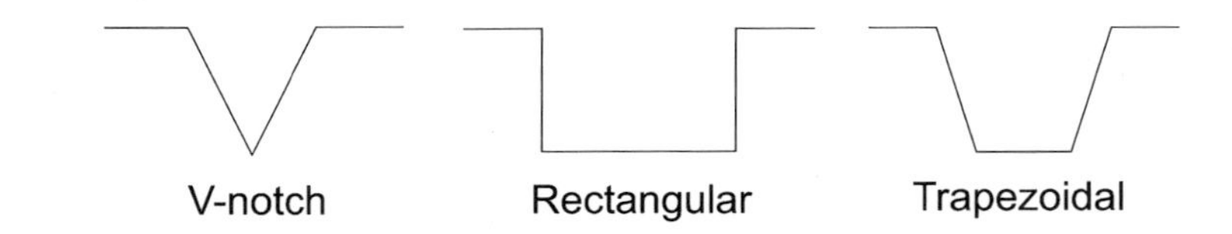

Fig. 8.10 Patterns of weir notches often used for geothermal measurements

the thin-plate weir is an obstruction in an open channel flow, and the fluid flow is disturbed in a controlled way to create a measurable effect that can be related to the mass flow rate. Weirs for measurement were developed extensively in the nineteenth century by hydraulics engineers, and many interesting details of their use can be found in older fluid mechanics textbooks.

British Standard 3680: part 4A sets out precise rules for the construction and installation of thin-plate weirs in the same manner as BS1042 does for orifice plates, although less manufacturing precision is required. For geothermal engineering, the hole, or "notch" in the weir, is usually shaped as either a V notch, a rectangular notch or a trapezoidal notch—Fig. 8.10.

A Cipolletti weir, sometimes used in geothermal well measurements, is a trapezoidal notch of particular shape, with a side slope of 1:4. The various notches have characteristics that influence the choice of which to use, but the differences are not complex. The plate must be vertical and the main measurement required is the water level above the bottom of the notch, although it is practically easier to deduce it by measuring below the top edge. The V-notch weir widens with height above the point, allowing increasing flow without a proportional increase in depth. This helps to keep the accuracy of the measurement uniform over the range of flow rates, whereas with the rectangular weir, low flow rates give a very small wide flow, the height of which is difficult to measure accurately. The trapezoidal weir is an alternative to the rectangular weir giving a smaller variation of depth measurement with flow rate.

Whichever notch is chosen, it will be accurate only if it is kept free of debris and chemical deposition (calcite and silica), is installed vertically and normal to the flow, and if the water flows over it correctly. The stream of water flowing over the weir is supposed to emerge as a jet—referred to as the "nappe"—which must make no contact with the downstream face of the weir plate. When the water forms a jet with clear air beneath it (Fig. 8.11) the nappe is said to be aerated and this is the requirement; a "drowned nappe" will not provide accurate measurements.

8.3.4 The Separator

Separators or, more precisely, cyclone separators are a normal part of steamfield equipment and are discussed in Chap. 13. Small units suitable for well testing can be constructed.

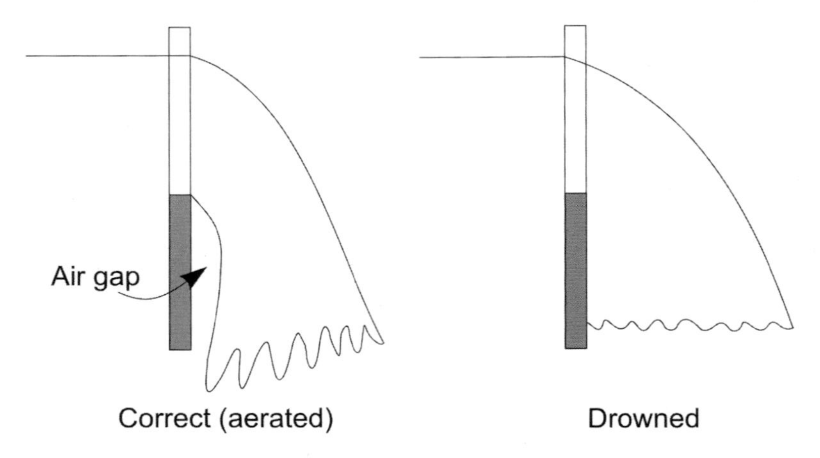

Fig. 8.11 Correct and incorrect flow patterns over notched weirs

8.3.5 The Silencer

A silencer is effectively a separator which has had its upper end dome removed, leaving it open to atmosphere. It works in precisely the same way as a separator, with a tangential entry that rotates the flow within the cylinder, but the steam and gas are allowed to discharge to atmosphere through the open top. Its purpose is to separate the water so that it can be measured and also to reduce the noise of the discharge. Many silencers are built as twin stack, with a single entry causing spinning in opposite directions and a single liquid exit—see, for example, Thain and Carey [2009] where it is referred to as an atmospheric separator. The water exits the bottom at the saturation temperature corresponding to the atmospheric pressure for the aqueous solution produced, but the difference between this and the properties of pure water are neglected and pure water properties are used. The mass flow rate of the water is measured with a simple weir; the mass flow rate of steam from the top of the silencer does not need to be measured, as already demonstrated in the examples given. The weir has already been explained, and usually the only problem, particularly in portable silencers, is arranging for the required length of undisturbed upstream flow. Some silencers are made as permanent concrete structures within easy piping distance of several wells, others from steel and some with wooden slat cylinders.

Most of the noise from a silencer is generated at the inlet nozzle, and the duct directing the flow into the silencer barrel (the tall cylinder) is often a thick-walled concrete construction to help reduce the noise level. The annular gap between the concrete pipe and the nozzle may be many cm wide, and air is dragged in, which also helps, but nevertheless these are noisy devices. The annular gap is usually covered by a loose-fitting steel plate, since the air entrained in the jet creates a suction that may be a hazard to operators.

8.4 Chemical Measurements During Discharge

Although it is not dealt with in any detail in this book, an understanding of the chemistry of the resource fluids is an important key to understanding the behaviour of the resource as a whole as an actively convecting, chemically reacting system. Sufficient understanding cannot be gained by examining the thermo-fluid dynamics of the resource in isolation. By adding the geochemistry it is possible to understand how the resource functions in the natural state and thus what to expect when fluid is removed by discharging wells and replaced (or not) with separated water at a lower temperature containing higher concentrations of the dissolved species in the original discharge, without the gas. The various large-scale effects that can result are discussed in Chaps. 13 and 14; their detection comes as a result of several types of geoscientific measurements, including geophysics (e.g. gravity changes), but perhaps most importantly, sampling of discharged fluids. Wells provide the only direct access to the resource, and by documenting changes in chemical species in the discharge, changes in the resource as a whole can be deduced.

From a purely mechanical engineering point of view, the gas present in the well discharge is important to the power station operation, as it is non-condensable and must eventually be pumped from the condenser at the cost of some power that could otherwise be sold as electricity. Some of the dissolved chemical species are also important, in particular silica, which often comes out of solution to form scale deposits in the pipelines and on the turbine blades. Various other chemical species may deposit in the well production casing, on first flashing, a phenomenon which is resource specific.

8.4.1 Sampling Arrangements During Discharge Measurements

Chemical samples are usually taken between the wellhead and the discharge measurement equipment and from just upstream of the weir attached to the separator. Ellis and Mahon [1977] state that water samples are best taken where the discharge is still at wellhead pressure, so that the volume of steam present is at its smallest, and a wellhead side valve is often used. Samples for gas content are best taken close to the silencer, downstream of any pressure restriction, where the water content is at its minimum. Geothermal gases do not all partition into the steam and their solubility in water cannot be ignored.

The samples are taken using a hand-carried cyclone separator, identified by those in the industry as a “Weber” separator (although Weber was responsible for the development of cyclone separators in general, which includes any separator used at a geothermal resource). The handheld device has a cooling system to condense the steam sampled.

8.4.2 Discharge from a Well Producing from Several Formations containing Chemically Different Fluids

Consider a well penetrating two producing formations, A and B, each with different chemical compositions, identified by different concentrations of a chemical species C and different specific enthalpy. Three balance equations can be written, for mass, energy and chemical species, in terms of mass flow rate $\dot{m}$, specific enthalpy h and species concentration C, as follows:

$$\dot{m}_A + \dot{m}_B = \dot{m}_D \tag{8.19}$$

$$\dot{m}_A.h_A + \dot{m}_B.h_B = \dot{m}_D.h_D \tag{8.20}$$

$$\dot{m}_A.C_A + \dot{m}_B.C_B = \dot{m}_D.C_D \tag{8.21}$$

where the suffixes refer to formations A and B and the total discharge D.

These form a set of linear algebraic equations, and the result of mixing the two sources in various proportions appears as a straight line on a graph of h versus C, for example. This is the basis of mixing models; h and C can be regarded as tracers provided they are passive. This approach was used by Pinder and Jones [1969] in examining the chemical composition of the total runoff of groundwater in terms of its individual sources and has been formalised as "endmember mixing analysis (EMMA)" according to Durand and Torres [1996]. The same approach was used by Fournier [1977], less formally in mathematical terms but with clear physical explanations, to demonstrate the use of geothermometers and mixing models to the examination of springs. These linear relationships provide a means of examining variations in total discharge composition with time and total discharge rate. Glover et al. [1981] used the approach to interpret the gas content in the total discharge of wells, citing examples from Krafla (Iceland) and Tongonan (Philippines) and drawing conclusions about the source of excess enthalpy. Lovelock and Baltasar [1983] illustrated the same general approach using a variety of examples from discharge tests in which the measurement method of Fig. 8.7 was used. The measurement uncertainty is a problem, as the uncertainty range as represented by an uncertainty bar on a point tends to lie along the mixing line that is the result, reinforcing the conclusions. However, a greater issue is that the parameters in the equations are not always passive, specific enthalpy varies with pressure and temperature, some species deposit when they reach saturation concentration and the concentration of species is increased as a result of flashing, the latter being the most significant.

8.4.3 Changes in Concentration of Dissolved Species as a Result of Flashing

The discharged fluid is likely to have a sufficiently high specific enthalpy to cause it to flash in the discharge measurement equipment, with the result that the concentration of dissolved solids increases; for the present purposes the concentration of silica (SiO_2) is the main interest. Silica dissolved in steam is of concern in fossil-fuelled power stations operating with high purity water at very high temperatures and pressures (550 °C and supercritical pressures), but in the range of pressures experienced by discharging wells it has negligible solubility in steam so is assumed to remain with the liquid phase. The dryness fraction, which defines the proportion of an original liquid which flashes and forms steam, can be used to calculate the concentration of the species in the remaining water. Thus Eq. (3.17),

$$h = (1 - X)h_f + Xh_g \tag{3.17}$$

states that the specific enthalpy of the original liquid is shared between the water and steam and shows that the mass flow rate of water is reduced to $\frac{1}{(1-X)}$ of the original. The chemical mass balance is

$$\dot{m}_D.C_D = \dot{m}_g.C_g + \dot{m}_f.C_f \tag{8.22}$$

where C is the species concentration, in the total discharge, and in steam and in water, respectively, reading from the left, and since the concentration in the steam is zero, the concentration in the remaining water is

$$C_w = \left[\frac{\dot{m}_D}{\dot{m}_f}\right].C_D = \frac{C_D}{(1 - X)} \tag{8.23}$$

The concentration measured in a liquid sample must be modified in this way according to how much of the original liquid remains after the flashing processes. The concentration increases as the liquid mass diminishes and may reach saturation, or supersaturation for a period of time until it deposits. Silica will be found to be a significant factor in steamfield design (Chap. 12).

8.4.4 Mass Flow Rate Measurement by Chemical Tracers

The same simple mixing algebra can be used to measure the mass flow rate of the two phases in a pipe carrying a two-phase mixture, and a measurement method using it was originally patented by Chevron Corporation [1988]. Lovelock [2006] explains that a tracer substance, isopropanol, is injected into the flow, and at some distance downstream sufficient for it to have become well mixed, it will have become distributed between the liquid and steam. At 180 °C the distribution is quoted as being about 5 % remaining in the liquid. The tracer is injected near the wellhead of a discharging well, and then at some distance downstream, samples of water and steam are taken. The steam is condensed and laboratory analysis can determine the concentration of the tracer in each phase. The mass flow rate of each phase is thus determined. Lovelock [2006] also describes the same technique but using sulphur hexafluoride (SF_6). He gives the results of tests using dry steam wells where the flow rate was already being measured with an orifice plate (see Fig. 8.5) and found an average difference between the two methods of only 1.2 %. A comprehensive set of test results is provided for two-phase flows comparing the two tracers. The distance from injection to sampling point was quoted as 10–15 m to obtain good results.

8.5 Surveying Wells During Steady-State Discharge and Predicting Pressure and Temperature Distributions

The downhole instruments described in Chap. 6 may be used while the well is discharging. The instruments partially block the flow in the liner, and particularly in the casing, which is transmitting the full flow, and an upward drag force is produced, capable of lifting the instrument up to the surface, damaging it and producing a tangle of steel wireline. Weights must be added to the instrument to counteract this. Measurements during discharge are helpful in locating production zones and the temperature of the fluid emerging from them. Measurements in the production casing are helpful in designing calculation procedures to predict details of the flow in the well

It is sometimes helpful to be able to predict the steady-state pressure and temperature distributions in the flowing well. For example, these would show the depth at which flashing first occurs in a liquid flow, which might be important in a discharge which deposits calcite. For a well with any type of axial tubular insert in the upper sections, it might be helpful to know how the discharge characteristics would be modified. The mixture produced at wellhead when more than one formation is producing depends on the flow from each, which is bound up with the sandface pressure of each, which both results from and partly controls the axial pressure distribution. A combined analysis of flow in the well and the formations could in principle be carried out. As a final example, the production casing diameter might be chosen based on pressure drop from formation to wellhead; King et al. [1995] presented a cost–benefit analysis of using different-sized production casings, using the commercial wellbore “simulator” WELSIM. For some reason, numerical predictions of this type have become known as wellbore simulation, despite “simulation” being almost universally reserved to describe the prediction of transients. Elmi and Axelsson [2009] show the application of a more recent simulator known as HOLA.

For single-phase flows the standard pipe flow solutions can be applied, with friction factor–Reynolds number correlations chosen for an appropriate wall roughness (see Sect. 4.3.4). For example, Leaver and Freeston [1987] produced a correlation from which the discharge characteristic of a steam-producing well could be determined.

Predicting heat loss from the flow is a problem because the thermal boundary condition on the production casing is ill-defined, and an average for the well must be used. Flow through the slotted liner can be reasonably represented by using an equivalent diameter, but an effective roughness is also needed and this is empirical and specific to each well, since rubble may obstruct the annular passage in places.

This problem was studied for geothermal wells at least as early as 1964 (Ryley [1964]—see Kestin [1980]) and earlier for petroleum wells. Research on two-phase flow for the nuclear industry started in roughly the same era. All calculations of pressure gradient in pipes carrying two-phase flow are essentially similar, using the approach described in Chap. 7. The explanation offered here uses a calculation procedure based on an ESDU compilation [1978] for water–air and water–steam mixtures; it was developed for application to geothermal wells by Brennand and Watson [1987] and later rewritten for teaching purposes. It has shortcomings as discussed by Karaalioglu and Watson [1999], which could be avoided by using a more recently available collection of two-phase correlations, ESDU [2008].

Equation (7.14) is the basis of the calculation, rearranged as

$$dz = \frac{dP}{\left(\left(\frac{dP}{dz}\right)_{grav} + \left(\frac{dP}{dz}\right)_{accel} + \left(\frac{dP}{dz}\right)_{fric}\right)} \tag{8.24}$$

The calculation proceeds by starting at a known level in the well, z, and calculating the pressure reduction over a short element, dz. The starting point can be either the top or bottom of the well, so the calculation steps down or up. However, instead of specifying dz, the pressure difference dP across the element is specified, leaving dz as the unknown. This approach allows the fluid properties at each end of the element to be calculated (pressure, temperature, and liquid and gas densities and viscosities), so the axial variation of properties, which produces variations in the flow via the correlations, is taken account of. Single-phase correlations are used if the flow is single phase; otherwise, two-phase correlations are used, which depend on parameters such as mass velocity G and volumetric flux j (defined in Sect. 7.7.2). The correlations provide a pressure gradient for each of the terms, gravity, acceleration and friction, one at each end of the element, allowing an average gradient to be determined. Thus using Eq. (8.24) the step length dz is the outcome of the calculation. The two-phase correlations used were based on homogenous flow, with the seven gravitational correlations and six frictional ones given in the ESDU compilation, and the acceleration component based on either homogenous or separated flow models.

The calculation procedure was tested against some data from Rotorua wells. These have small diameter production casing occupying almost the full drilled depth, so their measured mass discharge characteristics provide a valuable test because the flow through the casing is known at all depths and there are no step changes of diameter, slotted liner or increases in flow from several formations that require assumptions which would cloud the comparison. Furthermore, the producing formation was virtually at saturation point so there is a long length of two-phase flow. The details are given in Table 8.1 below.

The measured discharge characteristics represent the extreme right, low mass flow rate, high-wellhead pressure part of the characteristics shown in Fig. 8.1. They show a wellhead pressure maximum which, if it were a pump characteristic, would lead to flow instability—two mass flow rates corresponding to a single-wellhead

Table 8.1 Dimensions of the Rotorua wells for Figs. 8.12 and 8.13

Well no.	Depth (m)	Cased depth (m)	Formation temperature (°C)	Sp. enthalpy (kJ/kg)
715	122	100	165	610–720
901	149	129	183	730–790
703	206	193	199	860–900

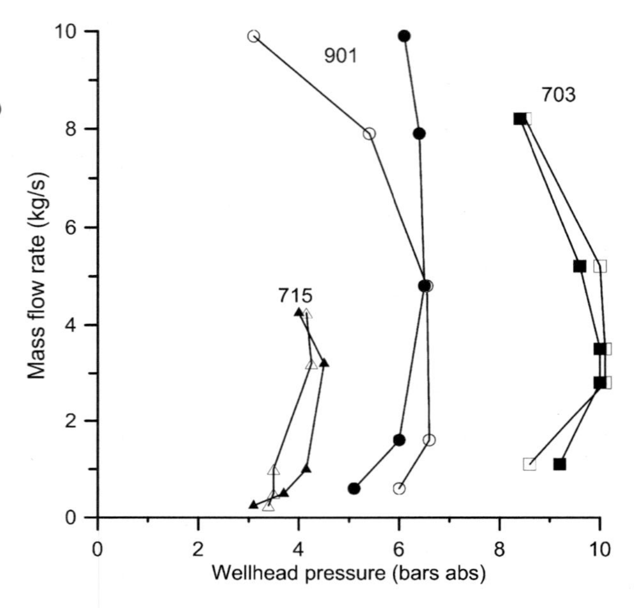

Fig. 8.12 Discharge characteristics of Rotorua wells of Table 8.1 showing measurements (*solid symbols*) and predictions (*hollow symbols*)

pressure—although the flow was apparently steady. The measurement data were given in the report of the NZ Ministry of Energy [1985] and are replotted in Fig. 8.12 together with the predictions of Brennand and Watson [1987].

For one of the points on the characteristic for well 703, the contributions to total pressure drop made by the three components are shown in Fig. 8.12. The frictional gradient makes the biggest contribution at the top of the well, because the mean density of the flow is least there, and the mean velocity highest—friction is generally proportional to kinetic head. The fluid is most dense near the bottom of the well, so the gravitational gradient is highest there. The acceleration component is small everywhere but increases up the well as the fluid is expanding.

For the comparisons of measurement and prediction shown, the roughness height was assumed to be 0.0003 m, which is much rougher than new commercial steel pipe, and the heat loss from the well was assumed to be zero because they had been in continuous use for a long time and the surrounding ground was warm. These assumptions were obviously suitable for well 703, based on Fig. 8.12, but not for well 901; to obtain a good match, the roughness height can be varied by trial and error, and this needs to be done for every well individually (it has not been

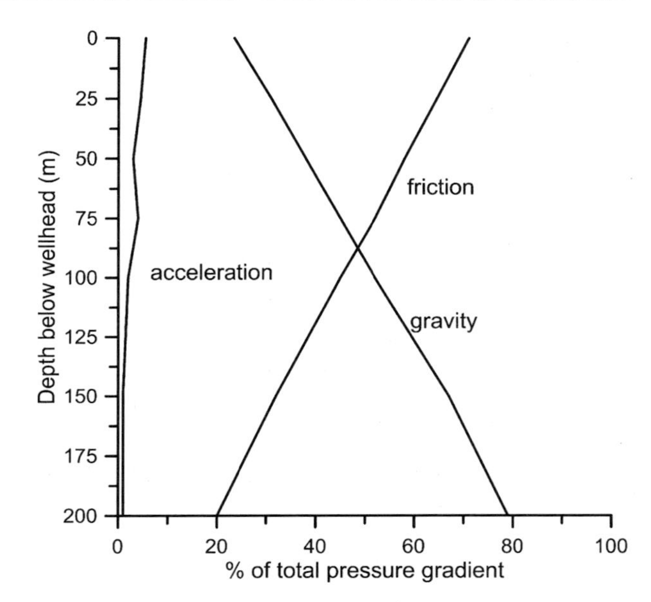

Fig. 8.13 Percentage contributions to total pressure drop made by friction, gravity and acceleration for Rotorua well 703

done here, to illustrate the point). Once a good match is obtained, then the effects of a modification to the well, such as inserting a sleeve to patch a hole in the casing, can be assessed by introducing the appropriate changes in diameter.

No provision was made for more than one production zone in the calculation procedure, because there are then too many degrees of freedom available which makes a match less reliable; the difficulties of specifying parameters for the slotted liner have already been explained. Thus the Brennand and Watson procedure is best suited to examining flow in the production casing only. Finally, the correlations used are not appropriate for near-sonic flows (choked flows).

Karaalioglu and Watson [1999] made comparisons of predictions and measurements for large diameter wells using the same procedure and noted the problem with using the ESDU [1978] correlations, which included heated flows. The acceleration component is greatly increased for the high heat flux situations of interest to fossil and nuclear boilers, for which the correlations were produced. More recently, a compilation of experimental data and correlations have been produced for adiabatic vertical flows, ESDU [2008], which would be applicable to geothermal wells.

8.6 Transient Discharge Measurements and Predictions

There are at least two different physical processes that lead to a periodically varying well discharge, which is what the title of this section refers to. The discharge is transient when the well is first opened and is being closed, but there has been little incentive to try to predict this for geothermal wells although it is an important research topic in the nuclear industry in relation to the rate at which water escapes through a break in the pressure vessel or pipes of a water cooled reactor.

One physical process involves two (or more) producing formations in a well. The interaction of drawdown of a zone with the phase distribution in the well offers the possibility of a major shift in wellbore pressure at the deeper zone. Consider a lower zone producing a fairly dry two-phase mixture but with low permeability which draws down, leaving the discharge from an upper good producer of liquid continuing. When the discharge from the lower zone reaches some particular low rate, the two-phase regime in the well collapses and becomes denser at the lower zone, halting or at least severely reducing the flow there and allowing the formation pressure to recover towards the undisturbed condition. The process repeats. This is entirely speculative of course, and Menzies [1979] offers a slightly different explanation for an impressive set of measurements of a Tongonan (Philippines) well with very regular periodic variations in discharge rate. Lovelock and Baltasar [1983] present convincing evidence of this type of behaviour in a Tongonan well, by chemically sampling the output at frequent intervals during the discharge, which had a period of 4 h, sufficiently long to allow sampling. The results were plotted on a graph of two components of the discharge, chloride and CO_2, and a mixing line of the type discussed in Sect. 8.4.2 above was obtained. The two formations were a high-chloride liquid-filled zone and a high-gas zone (presumably a shallower steam zone).

A different physical process occurs in some geysers. Geysers and geysering wells produce an extreme flow variation, an intermittent flow. Lu [2004] (see also Lu et al. [2005, 2006]) reviewed the history of investigation of geysers, which began in Iceland (the word is Icelandic), and also the occurrence of geyser-like behaviour in engineering equipment. Although natural geysers have been explained in terms very similar to those offered above for wells with multiple producing formations, geysering in engineered equipment is clearly not of this type. There exist a number of wells worldwide which exhibit the behaviour, and that at Te Aroha, New Zealand, was the subject of Lu's experiments. Like the Rotorua wells used for steady-state calculation comparisons, the Te Aroha well was cased to within a few metres of the bottom with a single 100 mm diameter steel tube; the well is 70 m deep so a long length of test section was available. It has a maximum temperature at the bottom of 83 °C and discharges at 70–75 °C, producing water with a high concentration of dissolved CO_2. Clearly, flashing of water to steam does not drive the phenomenon, and in this respect, it differs from natural geothermal geysers like those at Yellowstone and Rotorua. When shut it exhibits a wellhead pressure of 1.5–2.5 bars abs, depending on recent rainfall, and although it will discharge steadily when the discharge is restricted to a low flow rate, when fully opened, it discharges intermittently for approximately 120 s with a period of 700 s. The discharge is eruptive, with large amounts of gas produced with the water. Measurements were made by placing piezometric pressure transducers at various depths in the well. The void fraction was measured by placing two transducers a distance of 2–3 m apart and calculating the density between them.

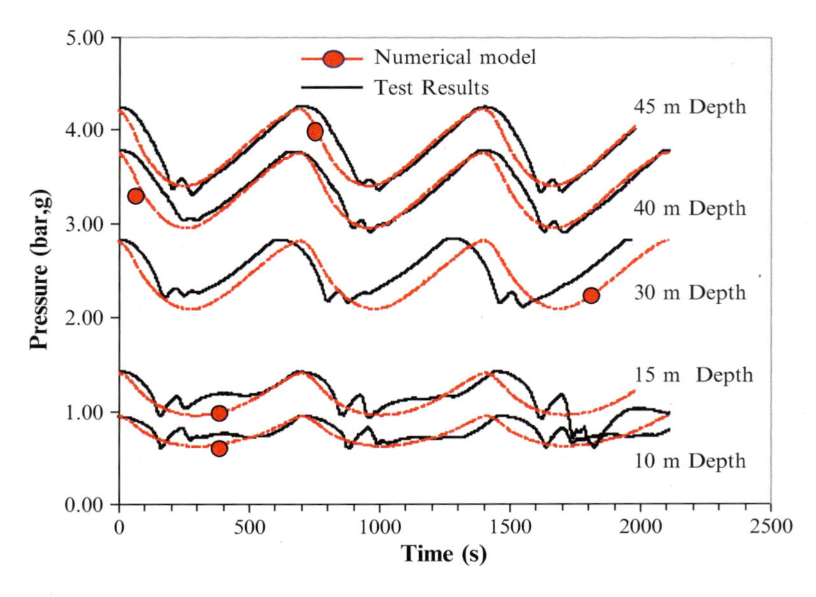

Fig. 8.14 Showing the cycling pressure variations (reproduced by permission of X.Lu [2004])

The equations governing the flow were continuity, expressed as one each for water and gas, with a source term representing the CO_2 coming out of solution, and also momentum in the form developed earlier as Eq. (7.14):

$$\frac{dP}{dz} = \left(\frac{dP}{dz}\right)_{grav} + \left(\frac{dP}{dz}\right)_{accel} + \left(\frac{dP}{dz}\right)_{fric} \tag{7.14}$$

The heat loss from the well was small so the energy equation was ignored on the grounds that energy exchanges did not significantly affect the flow. The solubility of CO_2 was described by Henry's law.

In the calculations (which represent a true simulation since they follow a time-dependent process), the gravitational term in the momentum equation was found using the calculated local mean density, and the frictional component assumed the homogenous model. Adopting separate conservation equations allowed the phases to have different velocities which in turn allowed the drift flux model to be used for the acceleration term in the momentum equation—despite homogenous flow being assumed for the frictional term. This was essentially based on an intuitive understanding of which component of pressure drop was having the greatest influence, gained from the experimental measurements. The process is accompanied by a periodic rise and fall of the bubble formation level (flash level) in the well, as Lu et al. [2006] demonstrated. The agreement between measurements and calculations is illustrated by Fig. 8.14.

The predictions compare very well with the measurements; those shown are for depths from 10 to 45 m. The agreement is good for period and amplitude of the main disturbance. Secondary measured regular disturbances, small at 45 m but quite large near the surface, are not represented by the calculations. It seems likely that a similar calculation approach incorporating the energy equation could be developed for application to natural geysers, but definition of the flow channel would present a problem.

References

Brennand AW, Watson, A (1987) Use of the ESDU compilation of two-phase flow correlations for the prediction of well discharge characteristics. In: Proceedings of 9th NZ Geothermal Workshop, University of Auckland

Chevron Corporation (1988) US Patent No 4788848

Durand P, Juan Torres JL (1996) Solute transfer in agricultural catchments; the interest and limits of mixing models. J Hydrol 181:1–22

Ellis AJ, Mahon WAJ (1977) Chemistry and geothermal systems. Academic, New York

Elmi D, Axelsson G (2009) Application of a transient wellbore simulator to wells HE-06 and HE-20 in the Hellisheidi geothermal system, SW-Iceland. In: Proceedings of 34th Workshop on Geothermal Reservoir Engineering, Stanford

ESDU (1978) Guide to calculation procedures for solving typical problems related to pressure drop in two-phase systems, ESDU 780018

ESDU (2008) Pressure gradient in upward adiabatic flows of gas–liquid mixtures in vertical pipes, ESDU 04006

Fournier RO (1977) Chemical geothermometers and mixing models for geothermal systems. Geothermics 5:41–50

Glover RB, Lovelock BG, Ruaya JR (1981) A novel way of using gas and enthalpy data. In: Proceedings of 3rd NZ Geothermal Workshop, University of Auckland

Grant, MA (1979) Interpretation of downhole measurements in geothermal wells, Report No. 88, Applied Maths Division, Department of Scientific and Industrial Research, NZ

Helbig S, Zarrouk SJ (2012) Measuring two-phase flow in geothermal pipelines using sharp-edged orifice plates. Geothermics 44:52–64

James R (1962) Steam-water critical flow through pipes. Proc Inst Mech Eng 176(26):741

James R (1965) Metering of steam-water two-phase flow by sharp edged orifices. Proc Inst Mech Eng 180:549–566

James R (1966) Measurement of steam-water mixtures discharging at the speed of sound to the atmosphere. NZ Eng 21(10):27

Karaalioglu H, Watson A (1999) A comparison of two wellbore simulators using field measurements. In: Proceedings of 21st NZ Geothermal Workshop, University of Auckland

Kestin J (ed) (1980) Sourcebook on the production of electricity from geothermal energy, US DoE, Contract No EY-76-S-4051.A002

Kieffer S (1977) Sound speed in liquid–gas mixtures: water-air and water-steam. J Geophys Res 82(10):2895

King TR, Freeston DH, Winmill RL (1995) A case study of wide diameter casing for geothermal systems. In: Proceedings of 17th NZ Geothermal Workshop, University of Auckland

Leaver JD, Freeston DH (1987) Simplified prediction of output curves for steam wells. In: Proceedings of 9th NZ Geothermal Workshop, University of Auckland

Lovelock B (2006) Flow testing in Indonesia using alcohol tracers. In: Proceedings of 31st Workshop on Geothermal Reservoir Engineering, Stanford University

Lovelock BG, Baltasar SJ (1983) Geochemical techniques applied to medium term discharge tests in Tongonan. In: Proceedings of 5th NZ Geothermal Workshop, University of Auckland

Lovelock BG, Cope DM, Baltasar AJ (1982) A hydrogeochemical model of the Tongonan geothermal field, Philippines. In: Proceedings of Pacific Geothermal Conference incorporating the 4th NZ Geothermal Workshop, University of Auckland

Lu X (2004) An investigation of transient two-phase flow in vertical pipes with particular reference to geysering PhD thesis, Department of Mechanical Engineering, University of Auckland, New Zealand

Lu X, Watson A, Gorin AV, Deans J (2005) Measurements in a low temperature CO2 driven geysering well, viewed in relation to natural geysers. Geothermics 34:389–410

Lu X, Watson A, Gorin AV, Deans J (2006) Experimental investigation and numerical modeling of transient two-phase flow in a geysering well. Geothermics 35:409–427

Menzies AJ (1979) Transient pressure testing. In: Proceedings of NZ Geothermal Workshop, University of Auckland

Menzies AJ, Gudmundsson JS, Horne RN (1982) Flashing flow in fractured geothermal reservoirs. In: Proceedings of 8th Workshop on Geothermal Reservoir Engineering, Stanford

Murdock JW (1962) Two-phase flow measurement with orifices. J Basic Eng 84:419

New Zealand Ministry of Energy (1985) The Rotorua geothermal field; technical report of the Rotorua Geothermal Task Force

Pinder GF, Jones JF (1969) Determination of the groundwater component of peak discharge from the chemistry of total runoff water. Water Resour Res 5:438–445

Ryley DJ (1964) Two-phase critical flow in geothermal steam wells. Int J Mech Sci 6(4):273

Thain IA, Carey B (2009) 50 years of geothermal power generation at Wairakei. Geothermics 38:48

The Transient Response of Formations to Flow in a Well: Transient Pressure Well Testing

9

Quantifying the ability of a geological formation to allow fluid to flow through it is an essential part of geothermal engineering, but fortunately it is equally important to groundwater and petroleum engineering and has been under development since the 1930s. Although it is a fluid mechanics-based topic, it could equally well be described as signal processing. This chapter introduces the fundamentals and then presents the relevant basic equations, their most often-used solution and the principle of superposition which allows the design of well tests. Ideal conditions for testing are defined and the historical development of methods of dealing with real effects such as skin and wellbore storage in single-phase flows is described. The chapter ends by describing the approach to analysing the flow in formations containing two-phase fluid and interference testing.

9.1 Introduction

Suppose interest is in a single, extensive producing formation, homogenous and of uniform thickness except that it is known to be intersected by a linear barrier to flow in the form of a fault. The formation beyond the fault has lost connection with the rest. Suppose two wells have been drilled, as shown in Fig. 9.1, and it is possible to send a signal from well W and receive it with some instrument in well M. The nature of the signal is not important at present.

Assume that the signal travels out radially from well W, reflects off the fault and is detected at well M. Only the distance r_p between the wells is known, and then there is insufficient information to draw the fault on the map. Two signals are received at M, the monitor well: first the direct signal from W and later the reflected signal, which is the same as would be received from a signal source in an imaginary well, labelled Image, situated such that a line joining it to W crosses the fault at right angles; this is because the signal travels in a straight line and rebounds from the fault at the same angle of inclination as it strikes it, as light from a mirror.

A. Watson, *Geothermal Engineering: Fundamentals and Applications*,

DOI 10.1007/978-1-4614-8569-8_9,

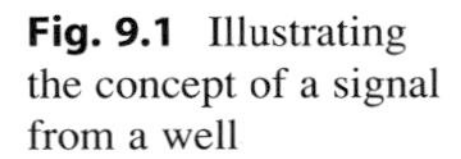

Fig. 9.1 Illustrating the concept of a signal from a well

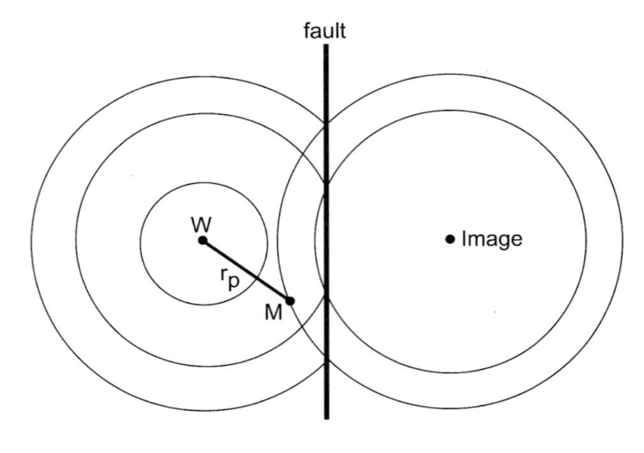

Fig. 9.2 Showing the deterioration of a pulse signal with time after generation and hence with distance from source

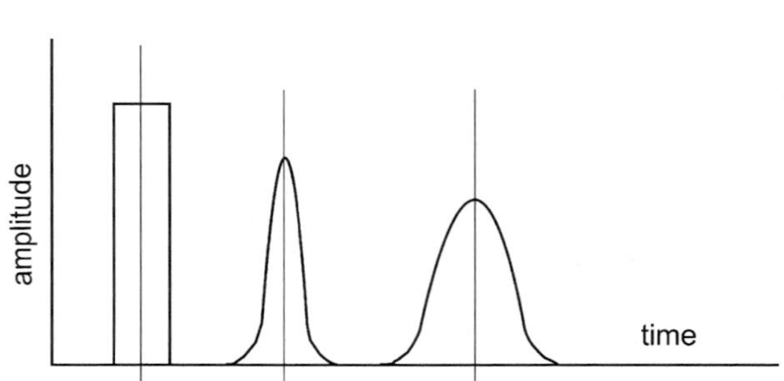

Neither the fault nor image can yet be drawn on a map. Vela [1977] produced an elegant proof that the image must lie on a circle centred on M and of radius length M to the image, and this shows that drilling another well, or preferably a few to reduce the experimental uncertainty, would allow the fault to be located.

The problem is the signal. It would be best to send a sharp pulse from W at a constant speed and have it received as a sharp pulse at M so that the time lag could be measured accurately. A pressure disturbance is the easiest way to generate a signal from a well, but fundamentally, the disturbance moves through the formation as a result of the expansion or compression of the fluid in every pore, and however much of a step change in time is generated at the point of origin, it very quickly loses its sharpness, as in the sketch of Fig. 9.2. The same would occur if the physical process was thermal conduction or any other "diffusive" process. The steady attenuation shown in the figure is much greater in radial geometry because the energy in the signal is distributed about a greater circumference as time passes.

Johnson et al. [1966] investigated the problem just described by creating a square edged pulse, by opening and then closing a well after times from 10 to 60 min and estimating the time lag from the received signal, distorted though it was. In addition, they made use of the distorted signal by matching its shape to a solution of the flow through the permeable formation, varying the formation properties by trial and error. Their application was petroleum engineering; however, the general technique arises in many other areas of physics, using sound, light or electromagnetic radiation, and falls into the category of signal processing. Although in the case of petroleum, groundwater and geothermal engineering the physical process is fluid flow, the details of the flow in the formation are dispensed with in a single step by using Darcy's law.

The majority of transient pressure well testing is carried out using a single well, which offers no chance of measuring the signal at the front of any disturbance. Consider a single well penetrating a single formation. The only measurement point available to investigate what is happening in the formation is in the well opposite the sandface, but it is left behind by the signal as it moves out into the formation. The sandface pressure measured as a function of time is the after-effect of the disturbance. The aim is to determine the formation properties, a much simpler use of the technique than the example used for the introduction. The signal is produced by opening or closing the well to allow discharge or injection as an individual step change in flow rate, a process aimed at producing as sharp a step as possible; the moving front of a step change can be visualised using Fig. 9.2 by considering the front half of the pulse only. The intended step change is inevitably rounded from the outset, because the sandface is separated from the wellhead valve by a large volume of compressible fluid contained in the well; the pressure change at wellhead travels down the well at the speed of sound, but the fluid in the well then expands or compresses slowly, at a rate dictated by the sandface pressure change. The process of interpreting the results consists of comparing the usually very slow and smooth variation of sandface pressure (inconveniently slow and smooth as will be seen) with a set of graphs or measuring the slope of the variation.

The petroleum engineering literature on transient pressure well testing is truly enormous. Such is the value of petroleum products that there has been continuous development in the topic, in instrumentation and in computerised data analysis. Improvements in instrumentation for petroleum wells are not always transferable to geothermal wells, for both economic and technological reasons. Petroleum fluids have calorific values of the order of 40 MJ/kg, and they are transportable and release heat at very high temperatures, allowing high efficiency conversion to work. In contrast the specific internal energy of steam is less than 1/10th of that released by the same mass of petroleum, it is available at only relatively low temperatures so less of it can be converted to work, and it must be used where it is produced. Setting aside economics, in petroleum wells it is possible to measure the flow rate in the well at or close to the producing formation and to use electronic instrumentation. Measurements in geothermal wells are less informative because the formations interact, being all open via the slotted liner, and instrumentation is less sophisticated.

The bulk of this chapter deals with single-phase fluids, i.e. the analysis of field tests is carried out using solutions to the governing equations obtained by assuming single-phase fluid. Where the fluid is a steam–water two-phase mixture, its physical characteristics require special attention, as explained further in Sect. 9.6. Finally, interference tests involving more than one well are discussed.

9.2 The Governing Equation in Axially Symmetric Coordinates and Its Solution and Application

9.2.1 The Preferred Solution

The solution described here is the "preferred solution" in the sense that it is the basis of transient pressure testing. To analyse the flow between a well and a formation of uniform thickness, the equivalent of Eq. (4.83) for a fluid with small, constant compressibility is required in axially symmetric geometry (Fig. 9.3):

$$\frac{\partial P}{\partial t} = \left(\frac{k}{\phi \mu c}\right) \frac{1}{r} \frac{\partial}{\partial r}\left(r \frac{\partial P}{\partial r}\right) \tag{9.1}$$

The formation is assumed rigid and isotropic—only the fluid responds to a change of pressure. The figure shows a segment only, for convenience. The well radius is r_w. The pressure over the formation is assumed uniform, and the sandface pressure P_w can be measured by hanging an instrument in the well at the average depth of the formation, which is usually thin enough for gravity not to have a significant effect although thick enough for variations in local volumetric flux to occur, which are assumed to average out. The formation is confined, meaning that fluid can only leave or enter via the sandface, and the well intersects only one formation. This equation appears in thermal conduction and electric circuit theory, so has received a great deal of attention since the early twentieth century, and there are several solutions for the basic problem described but with different boundary conditions. Several are for a defined well radius and outer formation radius, with various ways of specifying flow rate or pressure at each. A simpler solution is for a

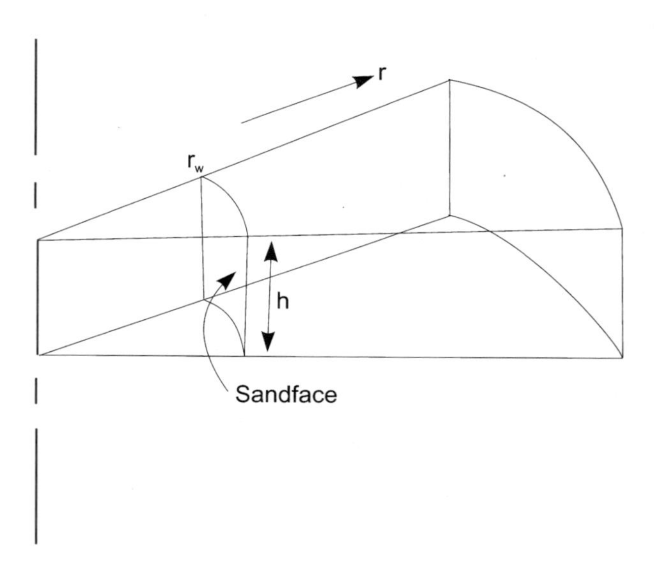

Fig. 9.3 The control volume for Eq. (9.1)

well represented by a line source or sink of constant volumetric flow rate, equivalent to a well of zero radius in a formation of infinite extent. This problem was addressed by Theis [1935] in connection with groundwater wells, and later by Polubarinova-Kochina (1962)—see Ramey [1966]. The solution is

$$P_i - P(r,t) = -\frac{Q\mu}{4\pi kh}\left(Ei\left(\frac{-\phi\mu cr^2}{4kt}\right)\right) \tag{9.2}$$

Although Ei is referred to as the exponential integral, it is a special case of the latter, which is defined more generally (Temme [1996]) as

$$E_n(z) = \int_1^\infty \frac{e^{-zt}}{t^n} dt \tag{9.3}$$

for the real part of $z = x + iy$ and $n = 1, 2$, etc. Interest here is with the real part only, $z = x$, and with $E_1(x)$ where $E_1(x) = -Ei(-x)$. Abramowitz and Stegun [1965] provide a series expansion form of E_1:

$$E_1(x) = -\gamma - \ln(x) - \sum_{n=1}^{\infty} \frac{(-1)^n x^n}{nn!} \tag{9.4}$$

in which γ is Euler's constant, 0.5772. Making these substitutions into Eq. (9.2) and neglecting the summation, which turns out to be small, gives the following equation for the pressure difference, valid for $\frac{4kt}{\phi\mu cr^2} > 100$:

$$P_i - P(r,t) = \frac{Q\mu}{4\pi kh}\left(\ln\left(\frac{kt}{\phi\mu cr^2}\right) + 0.80907\right) \tag{9.5}$$

or

$$P(r,t) = P_i - \frac{Q\mu}{4\pi kh}\left(-\ln\left(\frac{\phi\mu cr^2}{4kt}\right) - 0.5772\right) \tag{9.6}$$

Abramowitz and Stegun also provide polynomial approximations to $E_1(x)$ which are useful for numerical work.

Equation (9.5) is the building block for a variety of tests to measure the product of permeability and thickness, kh, known as the transmissivity. It defines the "cone of depression" illustrated in Fig. 4.12, which is the shape adopted by the step change disturbance. If the formation thickness is known from drilling, then k can be found. Equations (9.5) and (9.6) are loosely referred to in this text (as elsewhere) as the Theis solution. They provide the response of the formation pressure in space and time, $P(r,t)$, when a single well penetrating it is suddenly opened and fluid is either discharged or injected.

9.2.2 Conversion of the Equation and Solutions to Dimensionless Variables

The general case for introducing dimensionless variables was introduced in Sect. 4. 3.3; the process begins by identifying characteristic dimensions of the system. The only suitable characteristic length in the case of a well penetrating a formation is the well radius, since no variables change in the vertical direction, so r in the equation must be replaced by the dimensionless radius r/r_w. There is no obvious time to be used for comparison so the following variables are adopted:

$$r_D = r/r_w \quad \text{and} \quad t_D = t/\tau \tag{9.7}$$

where τ is a characteristic time yet to be determined. Making these substitutions Eq. (9.1) becomes

$$\frac{1}{\tau}\frac{\partial P}{\partial t_D} = \left(\frac{k}{\phi\mu c}\right)\frac{1}{r_w^2}\left(\frac{1}{r_D}\frac{\partial}{\partial r_D}\left(r_D\frac{\partial P}{\partial r_D}\right)\right) \tag{9.8}$$

Although only two of the variables are dimensionless, the equation is still correct because of the substitution process. Inspection shows that

$$\tau = \frac{\phi\mu c r_w^2}{k} \tag{9.9}$$

gives the equation its simplest appearance:

$$\frac{\partial P}{\partial t_D} = \frac{1}{r_D}\frac{\partial}{\partial r_D}\left(r_D\frac{\partial P}{\partial r_D}\right) \tag{9.10}$$

Only P remains with dimensions, and the clue to what to compare P with comes from the boundary condition of constant total volumetric flow rate over the sandface:

$$Q = v.2\pi r_w h \tag{9.11}$$

where v (m/s) is the velocity in the radial direction.

Darcy's law has already been "used" in the derivation of Eq. (9.1), but it now provides an alternative expression for v in the boundary condition:

$$Q = -\frac{k}{\mu}\left(\frac{\partial P}{\partial r}\right)_{r=w} 2\pi r_w h \tag{9.12}$$

Thus

$$\left(\frac{\partial P}{\partial r}\right)_{r=w} = -\frac{Q\mu}{2\pi r_w kh} \tag{9.13}$$

Introducing dimensionless radius this becomes

$$\left(\frac{\partial P}{\partial r_D}\right)_{r_D=1} = -\frac{Q\mu}{2\pi kh} \tag{9.14}$$

The group on the right-hand side has the dimensions of pressure and is constant for the conditions of the solution, which require Q to remain constant. So the dimensionless pressure could be defined as P divided by this group. However, the formation pressure at time $t = 0$ is P_i everywhere, so it has become usual for dimensionless pressure to be defined as

$$P_D = \left(\frac{P_i - P}{\left(\frac{Q\mu}{2\pi kh}\right)}\right) \tag{9.15}$$

With this substitution, the dimensionless equivalent of Eq. (9.10) is

$$\frac{\partial P_D}{\partial t_D} = \frac{1}{r_D}\frac{\partial}{\partial r_D}\left(r_D\frac{\partial P_D}{\partial r_D}\right) \tag{9.16}$$

to be solved with the boundary condition:

$$\left(\frac{\partial P_D}{\partial r_D}\right)_{r_D=1} = -1 \tag{9.17}$$

The solution is the exponential integral as before, for which the logarithmic approximation is appropriate:

$$P_D(r_D, t_D) = \frac{1}{2}\left(\ln\left(\frac{t_D}{r_D^2}\right) + 0.80907\right) \tag{9.18}$$

and in particular, at the sandface, where $r = r_w$ and $r_D = 1$

$$P_{wD} = \frac{1}{2}(\ln t_D + 0.80907) \tag{9.19}$$

where P_{wD} is the sandface dimensionless pressure, which is the pressure in the well at the formation. Equations (9.5) and (9.19) have the same form but in the latter the dependent variables are multiplied by constants, and this leaves the shape of graphs unaltered, although the axis variables are different; this is important in the use of "type curves", which are introduced later.

In converting the most commonly used solution into dimensionless terms by considering the real problem of a well in an infinite formation and working with the approximate solution, an issue has been overlooked. The Theis problem assumes a well of zero radius and an infinite outer boundary, so there is no characteristic length by which to make radius dimensionless. Instead a parameter combining the general coordinates r and t is adopted, which actually turns out to be dimensionless and which conveniently reduces the partial differential equation to an ordinary one. This is sometimes referred to as the Boltzmann transform but it is reasoned on purely dimensional grounds by Grigull and Sandner [1984] in connection with transient thermal conduction. This substitution was used by Polubarinova-Kochina in the solution presented by Ramey [1966].

9.2.3 Pressure Distributions in the Formation During a Simple Drawdown Test

Equations (9.5) and (9.19) have the form of a straight line, $y = mx + c$, so plotting the sandface pressure linearly and time logarithmically will result in a straight line with a slope related to the formation properties—a "semi-log plot". As already mentioned, the fluid mechanics of events in the formation have been left behind, but the flow regime that produces the linear variation is known as the infinite acting period of response because the disturbance has not reached any boundary that can affect the sandface pressure variation. It is easy to examine the details of the flow, such as the radial distribution of volumetric flux, if required.

A test carried out on an undisturbed formation by opening a well at constant flow rate and measuring the sandface pressure is called a drawdown test, and under the ideal circumstances assumed for the solutions above, a semi-log plot of pressure versus time should produce a straight line. Measuring its slope leads to kh. The solutions can be examined in more detail by calculating the pressure distribution in the formation, $P(r,t)$, if the formation and fluid properties are known.

Assume a well intersects a single formation with the following properties:

$$\text{thickness } h = 150 \text{ m}, \text{ porosity } \Phi = 0.10 \text{ and permeability } k = 5.0\text{E}{-}14 \text{ m}^2$$

and is filled with water with the following property values:

$$\text{density } \rho = 958 \text{ kg/m}^3, \quad \text{viscosity } \mu = 279.0\text{E}{-}6 \text{ kg/ms} \quad \text{and} \quad c = 4.74\text{E}{-}10 \text{ Pa}$$

The mass flow rate of the well measured at the surface, $\dot{m}$, is 80 kg/s, and since $Q = \frac{\dot{m}}{\rho}$ the volumetric flow rate at the initial pressure is 0.0835 m^3/s. To gain a rough estimate of the fluid velocity in the formation, assume that the volumetric flow rate above applies as the total over the formation cross-sectional area for the flow, which at a radius of 50 m is $2\pi rh = 47.13\text{E}3 \text{ m}^2$. This gives a velocity of 1.8E−6 m/s; this is the mean velocity, considering the full formation cross section is available for flow, when in fact only the pore cross section is available. The flow accelerates towards the well, but even at a radius of 1 m, the estimated mean velocity is only 0.1 mm/s, illustrating that the rate of progress of a signal through the formation is very slow.

9.3 Superposition and Its Use in Designing Tests

Simple drawdown tests are not usually carried out on petroleum or geothermal wells because the formations are rarely left undisturbed long enough, due to project priorities other than testing. Tests have been designed to fit in with other activities using the principle of superposition, otherwise referred to as Duhamel's theorem (1833) which he derived in connection with heat conduction—see Carslaw and Jaeger [1959]. The principle states that solutions of linear ordinary or partial differential equations can be superimposed. If the full solutions can be superimposed, then so can their approximations.

9.3.1 The Pressure Buildup Test

In a pressure buildup test, the well is discharged for a time, during which the sandface pressure declines. The well is then closed and the recovery or buildup of pressure is measured.

Consider superimposing two solutions labelled A and B in Fig. 9.4.

Solution A is the sandface pressure variation following the application of a step volumetric flow rate Q_p at the start of production, $t = 0$. This information is shown by the left-hand graph, and the right-hand graph shows the pressure falling from P_i at $t = 0$ to lower values as time proceeds, according to the Theis solution.

Solution B is for the same well and formation, but with injection at rate $-Q_p$ beginning as a step change at $t = t_p$ and continuing with time. The right-hand plot shows the pressure P_w increasing to higher values after t_p.

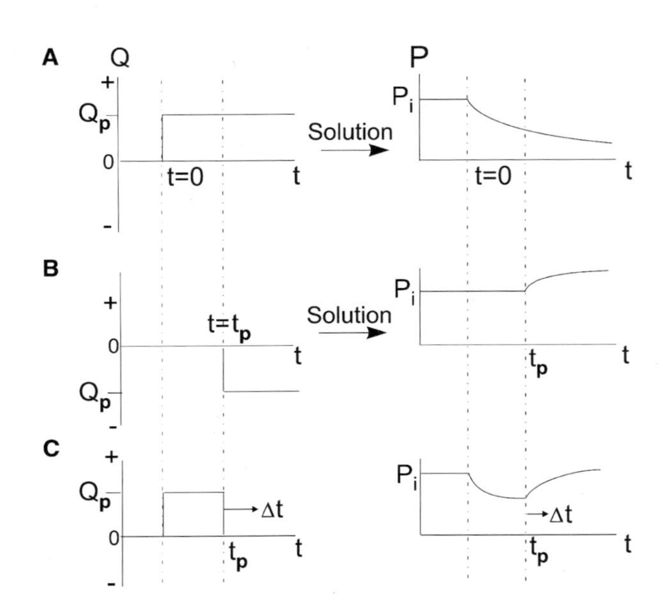

Fig. 9.4 Design of a pressure buildup test (PBU) using the principle of superposition

The combined effect of these two patterns of flow superimposed on the same well produces solution C; the flow begins at $t = 0$, stays uniform at Q_p until $t = t_p$ then stops instantaneously. The wellhead valve is opened at $t = 0$ and closed at $t = tp$. The superimposed sandface pressure variation is shown on the right in C; a decline after $t = 0$ proceeds until $t = t_p$ after which it recovers towards the initial pressure P_i. For this reason this test is known as a pressure buildup test (PBU), and the time period used in the analysis of results is that beyond $t = t_p$. The result being sought is the permeability thickness of the formation (kh), which is the slope of a semi-log plot.

The following algebra shows how this arises. With P_w as the sandface pressure, the solution for A is

$$(P_i - P_w)_A = \left(\frac{Q_p\mu}{4\pi kh}\right)\left(\ln\left(\frac{kt}{\phi\mu cr_w^2}\right) + 0.80907\right) \tag{9.20}$$

and for B is

$$(P_i - P_w)_B = \left(\frac{-Q_p\mu}{4\pi kh}\right)\left(\ln\left(\frac{k(t - t_p)}{\phi\mu cr_w^2}\right) + 0.80907\right) \tag{9.21}$$

The addition of the two pressure variations to give the pattern shown in Fig. 9.4 as C for times beyond $t = t_p$ is assisted by introducing $t = t_p + \Delta t$, in other words, denoting the time after $t = t_p$ as Δt; thus,

$$\begin{aligned}(P_i - P_w)_A + (P_i - P_w)_B &= (P_i - P_w)_{actual} \\ &= \left(\frac{Q_p\mu}{4\pi kh}\right)\left(\ln\left(\frac{k(t_p + \Delta t)}{\phi\mu cr_w^2}\right) + 0.80907\right) \\ &\quad - \left(\frac{Q_p\mu}{4\pi kh}\right)\left(\ln\left(\frac{k\Delta t}{\phi\mu cr_w^2}\right) + 0.80907\right)\end{aligned} \tag{9.22}$$

which reduces to

$$(P_i - P_w)_{actual} = \Delta P = \left(\frac{Q_p\mu}{4\pi kh}\right)\left(\ln\left(\frac{t_p + \Delta t}{\Delta t}\right)\right) \tag{9.23}$$

This equation has the form of a straight line $y = m.ln(x)$, so plotting the results on a semi-log graph allows the slope to be measured and kh deduced provided the volumetric flow rate at the formation can be determined from measurements at the surface. This form of graph is called a Horner plot (after Horner D.R.), and it is worth noting that the track of the plotted curve is from right to left as $(t_p + \Delta t)$ increases—this is the result of the ln term in Eq. (9.23). It often occurs that a PBU test is carried out on a well which has been discharging for a very long time and t_p is unknown. In this case the measurements can be plotted as a simple semi-log plot of

$(P_i - P_w)_{actual}$ versus $\ln(\Delta t)$, with Δt being the time measured from the time at which the discharge is halted; this is because the solution A curve after a long discharge is almost a horizontal straight line on the graph for times greater than t_p, as can be judged from Fig. 9.4, the superimposed solutions at C.

9.3.2 The Pressure Falloff Test

In the pressure fall-off test (PFO), fluid is injected into the well for a period, and the sandface pressure increases as a result. The well is then closed (injection terminated) and the sandface pressure is measured as it declines back towards the original formation pressure. The mathematics of superposition is used as above, resulting in a linear semi-log plot in terms of a modified time variable.

9.3.3 The Detection of an Outer Formation Boundary

As a final example of the use of superposition with the ideal solutions, consider an ideal formation with an outer impermeable boundary at $r = r_e$ penetrated by a well which is subjected to a simple drawdown test at a volumetric flow rate of Q_p m^3/s. This is the situation of Fig. 9.1 but without the monitor well. Before the disturbance to initial pressure has reached the boundary, the sandface pressure follows the approximate solution of Eq. (9.5) giving a semi-log plot with slope of $Q_p\mu/4\pi kh = m_{NB}$, say, the subscript indicating "no boundary". The pressure variation is that of solution A of Fig. 9.4. Once the disturbance has reached the boundary, the effective combined pressure disturbance affecting the sandface pressure is that of the real well plus that of the image. The radius of operation of the image well is $2r_e$—the distance from it to the real well—and the variation of sandface pressure is the sum of the pressure variations for both wells:

$$\begin{aligned} P_i - P_w &= \left(\frac{-Q_p\mu}{4\pi kh}\right)\left(\ln\left(\frac{kt}{\phi\mu c r_w^2}\right) + 0.80907 + \ln\left(\frac{kt}{\phi\mu c\left(2r_e^2\right)}\right) + 0.80907\right) \\ &= m_{NB} 2\ln t + \text{constant} \end{aligned} \tag{9.24}$$

The semi-log plot thus changes slope from m_{NB} to $2m_{NB}$ when the disturbance reaches the boundary or, more precisely, a sudden doubling of the slope after the test has been running long enough for the reflected disturbance to reach the measurement point. This demonstrates that despite the limitation of only a single well, which gives measurements long behind the disturbance front, an outer formation boundary is detectable.

9.4 Formation Testing in the Presence of Real Effects

The solutions presented so far in this chapter are for a carefully prescribed set of conditions which are not usually met with in practice. Several departures from the ideal can be accommodated in the theory, leaving the general method for well testing outlined above still useable, and they are:

- Skin effect
- Wellbore storage
- Lack of stratigraphic uniformity of the formation
- The presence of faults and related fracturing
- Radial temperature and phase variations

The effect of the first two can be illustrated by the results of a pressure falloff test on a well in the Tongonan field, as reported by Andrino [1998]. Figure 9.5 shows the resulting semi-log plot—the last decade of time difference shows a straight line, but there are earlier data to which a straight line could also be fitted over parts. Which data should be used and why is there such a slow non-linear approach to the linear parts? The later straight line turns out to be the variation of the infinite acting part of the ideal solution, the part with an interpretable slope, and the slow approach is due to the skin and wellbore storage effects.

9.4.1 The Skin Effect

The drilling process must be expected to cause changes to the fine structure of the formation at the sandface. The permeability might be reduced due to pores being blocked with drilling mud or being compacted as a result of high pressure applied

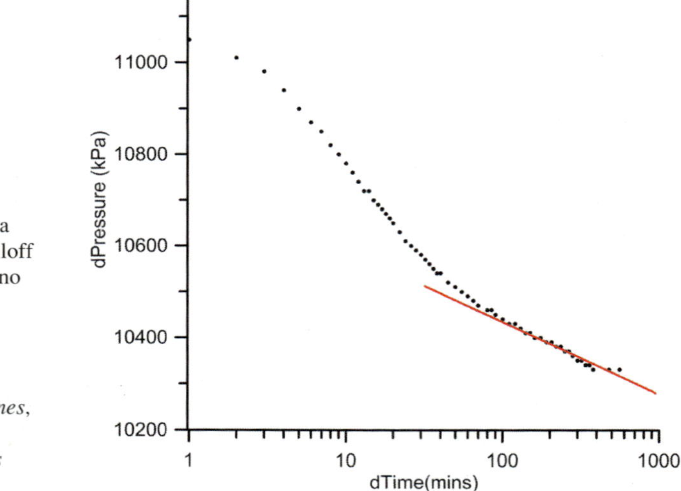

Fig. 9.5 Example of a Tongonan pressure falloff test reported by Andrino [1998] (*This material was created and used with permission by Energy Development Corporation, Philippines, and by Mr. Romeo P. Andrino, who reserves all rights thereto*)

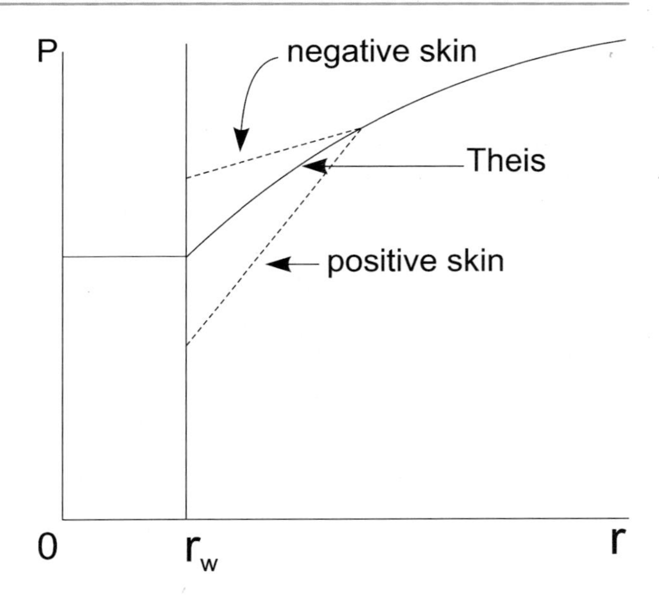

Fig. 9.6 The definitions of skin effect as variations of the Theis solution

by the mud column. Alternatively, the permeability of some formations might be increased as a result of material being washed out to leave bigger, better connected pores, and wells are sometimes treated with concentrated acids to rectify permeability reduction by inducing this process.

The result of any modification to the natural structure is to modify the pressure distribution adjacent to the sandface. Figure 9.6 shows the pressure variation of the Theis solution and the modifications for positive and negative skin effects, defined so that reductions in permeability (giving increased pressure drop) result from a positive skin and increases (decreased pressure drop) from a negative skin.

In a simple drawdown test, the sandface pressure changes from the initial pressure P_i by an amount $\Delta P = P_i - P_w$, which is modified by an amount ΔP_s if a skin is present. Thus with a positive skin

$$P_i - P_{act} = (P_i - P_{sf})_{Theis} + \Delta P_s \tag{9.25}$$

which can be written

$$(P_i - P_{act}) = \frac{q\mu}{4\pi kh}\left(-\ln\left(\frac{kt}{\phi\mu cr^2}\right) + 0.80907 + \Delta P_s\left(\frac{4\pi kh}{q\mu}\right)\right) \tag{9.26}$$

where P_{act} is the actual pressure at the sandface. Translating this into dimensionless terms gives

$$P_{wD} = \frac{1}{2}(\ln t_D + 0.80907) + s \tag{9.27}$$

where s is ΔP_s made dimensionless according to Eq. (9.15). Note that the radial thickness of the skin does not enter the discussion at all, only the modification to the Theis solution pressure change. However, the difference from the pressure variation of the Theis solution can be considered to be due to the addition of a layer of formation material, in other words, a modification of the wellbore radius from r_w to an effective radius r_{eff} as explained by Horne [1995], which is

$$r_{eff} = r_w e^{-S} \tag{9.28}$$

Assume for the sake of discussion that the skin has uniform permeability throughout its thickness (however unlikely this is). If the opening of the wellhead valve actually resulted in a step change in Q at the sandface, for a very short time the sandface pressure variation with time would be that of a formation with the permeability of the skin. Not until the pressure disturbance has moved out into the formation beyond the skin would the change in P_w begin to be influenced by the original formation properties. When the disturbance has moved sufficiently far out into the formation that ΔP_s is becoming small relative to P_i–P_w, then the slope of the semi-log plot will be dominated by the proper formation properties. This is the reason for the statement that the skin affects early times in a drawdown test. In the same way, if the flow rate is changed after a period of discharge, e.g. in a Horner test, the change in flow rate at the sandface will first produce a change in pressure within the skin, and because the fluid is compressible, Q at the sandface will not be identical to Q at the outer radius of the skin. This change in fluid storage in the skin is negligible; however, the same effect occurs in the well, which has a large volume and is certainly not negligible; its effect is known as wellbore storage.

9.4.2 Wellbore Storage and the Introduction of Type Curves

When a geothermal wellhead valve is opened, the fluid occupying the well flows first and expands as the pressure falls; flow through the sandface builds up gradually. The well "unloads", providing a cushioning effect which rounds off the corner of the attempted step change in Q. With the passage of time, the wellhead and sandface mass flow rates tend to equalise, and the sandface volumetric flow rate remains almost constant if the wellhead volumetric flow rate is constant—"almost" is essential here because the sandface pressure will keep declining so long as the well is flowing, so the average pressure in the well must also decline, allowing more unloading. However, this small slow change is accommodated by the small constant compressibility assumption (Sect. 4.6.3), and it is a reasonable assumption that the wellbore storage effect will be negligible after the sandface volumetric flow rate has caught up with the wellhead flow rate. How long this takes depends on the depth of the well and its diameter. Geothermal wells are left open to all formations below the casing shoe depth, so the volume available as storage is large. Petroleum wells are often completed with small bore tubing inside the main well casing, so the volume involved is less. The skin and wellbore storage effects take place soon after a step change in Q, and their combined effect produces the non-linear data before the straight line fit referred to in Fig. 9.6. Ramey [1970] presented a review of this problem at about the same time as the publication by Agarwal et al. [1970] of what became known as the Ramey type curves, and they provided examples of their use with real measurements (it is noteworthy that the name H.J. Ramey Jr. is hidden within "et al." in many references in this chapter, as it is in this case). The type curves, a terminology that was already in use in groundwater studies, are parametric plots on logarithmic axes of P_D versus t_D for various values of s and a dimensionless wellbore storage coefficient C.

Horne [1995] analysed wellbore storage in a petroleum well completed in a particular manner that is different to geothermal wells. For a geothermal well, consider a single production zone at depth L, and define the following terms:
Q = constant volumetric flow rate at wellhead, beginning at t = 0.
Q_w = volumetric flow rate at the sandface, which lags behind Q.
ρ_{wh} = fluid density at the wellhead.
ρ_w = fluid density at the sandface.
The volume of fluid in the well is $V_w = \pi r_w^2 L$, so a mass balance can be written:

$$Q\rho_{wh} = Q_w\rho_w + V_w\left(\frac{d\overline{\rho}}{dt}\right) \tag{9.29}$$

At the beginning of the process, the sandface volumetric flow has not started so

$$Q\rho_{wh} = V_w\left(\frac{d\overline{\rho}}{dt}\right) \tag{9.30}$$

An expression for the rate of change of mean density is needed:

$$c = \frac{1}{\rho}\frac{\partial\rho}{\partial P} \tag{9.31}$$

and

$$\frac{d\overline{\rho}}{dt} = \frac{d\overline{\rho}}{dP}\cdot\frac{d\overline{P}}{dt} \tag{9.32}$$

so

$$\frac{d\overline{\rho}}{dt} = \overline{\rho}c\frac{d\overline{P}}{dt} \tag{9.33}$$

and using this the mass balance, Eq. (9.30) can be written:

$$Q\rho_{wh} = \overline{\rho}cV_w\frac{d\overline{P}}{dt} \tag{9.34}$$

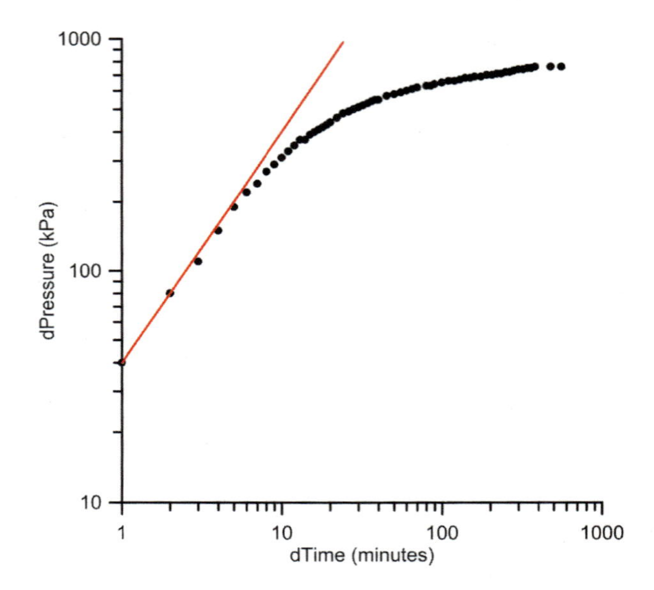

Fig. 9.7 Log–log plot of the data in Fig. 9.5 (*This material was created and used with permission by Energy Development Corporation, Philippines, and by Mr. Romeo P. Andrino, who reserves all rights thereto*)

or

$$\frac{d\overline{P}}{dt} = \frac{\rho_{wh} Q}{\overline{\rho} c V_w} \approx \text{constant} \tag{9.35}$$

This shows only that the initial rate of decline of mean wellbore pressure is linear, which could be the case in Fig. 9.6, but a long period elapses between the departure from this and the beginning of the infinite acting flow straight line. A solution for this period was presented by Agarwal et al. [ibid], the application of which resulted in a rule of thumb that the straight line of the infinite acting period begins 1 ½ decades after the data leave the 45° slope. Applying this to the data of Fig. 9.6 plotted as Fig. 9.7, this time using log–log axes on which 1 ½ decades can be visualised, it can be seen that the infinite acting straight line can be fitted after about $dt = 200$ min. The unit slope at the beginning of the plot is shown, lasting for a very short time of a few minutes only.

Although this rule of thumb apparently works, the solution of the wellbore storage problem carried out by Agarwal et al. warrants examination. They wanted to be able to interpret short-term well test data dominated by wellbore storage, to avoid interrupting production by longer tests. To include the transient expansion of the fluid in the well from the time of first valve opening is a very difficult mathematical problem. They adopted the solution to a thermal conduction problem solved earlier by Jaeger [1956], the problem of a disc with a hole in it, adiabatic except for the perimeter of the hole, at which heat was removed by a convecting fluid. Jaeger idealised the problem by assuming a fluid with infinite thermal conductivity. Agarwal et al. were able to superimpose Jaeger's solution on the

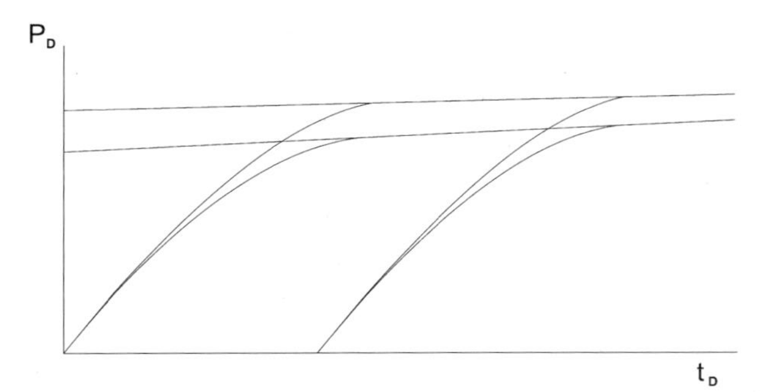

Fig. 9.8 A sketch showing the general form of the Ramey type curves, a family of curves for a range of skin and wellbore storage factors

Theis solution because both equations were linear as a result of adopting the small, constant compressibility assumption. However, the equations governing the transient flow in a large diameter geothermal well are non-linear, and the suitability of Jaeger's solution cannot be taken for granted, even if the fluid is a liquid—many geothermal discharges flash in the well, making the problem analytically impossible. Miller [1980] discussed the physics of this issue in detail, specifically with reference to geothermal wells, and presented a numerical solution of the transient flow in the well itself. In a simple drawdown test at times after the well opening which might be considered long enough for the sandface pressure variation to have become dominated by the infinite acting behaviour of the formation, the flow in the wellbore may still be in a transient stage, for example, a thermal transient due to heat loss.

Returning to Eq. (9.27), a set of solutions could be plotted as graphs of P_{wD} versus t_D for a wide range of values of s. If a range of possible wellbore storage effects was introduced also, the set would be what is referred to as type curves. The Ramey type curves look like Fig. 9.8, a family of curves plotted for a range of skin factors and wellbore storage coefficients.

An expression for the wellbore storage coefficient can be defined in terms of the volume of the well and compressibility of the fluid in it, and it can be converted into a suitable dimensionless form, but the value for a geothermal well is more often the outcome of curve matching. The Ramey curves are plotted with the following range of variables:

$$10^2 < t_D < 10^7, \quad 0 < s < 20 \quad \text{and} \quad 0 < C_D < 10^5$$

Ramey [1970] explains how the type curves are used practically—they can be used digitally, but he describes the manual method of the era, which will be described here also as being more easily understood. The aim is to plot the experimental measurements on transparent graph paper with the same axes as the printed version of the type curves. The well measurements for a pressure buildup test, for example, are a set of ΔP_w versus Δt, the changes since the flow was terminated. The data could be converted to dimensionless variables by the simple multiplication of each value by a constant, but the two constants are unknown—finding them is the objective because they are made up of the formation properties being sought. Because only a simple multiplication is required, the shape of a plot of ΔP_w versus Δt is exactly the same as that of a plot of P_{wD} versus t_D if plotted on the same grid, usually log–log. An easy way to proceed is to have a paper copy of the type curves, place over it a piece of tracing paper and use the type-curve grid to plot the measured data on the tracing paper. Keep the tracing paper firmly in place. The number associated with each decade line on the type curves and the curves themselves are ignored, and the grid is used just as a new (blank) piece of log–log graph paper—this is what would be used except that it is usually not transparent. When the data are finally plotted and a smooth curve drawn, the tracing paper is freed and moved about until the drawn curve matches one part of one of the type curves. The general nature of the Ramey type curves—Fig. 9.8—is such that there are a range of positions of the tracing paper that apparently give a match. This is a serious problem which was addressed as will be explained below, but regardless, once a match is obtained, the tracing paper is fixed in that position. The coordinates of a single point on the data curve are read from the tracing paper axes and the type curve axes and are equated; the benefit of using a point on the curve is that its coordinates are known. The definitions of dimensionless pressure and time are then used from Eqs. (9.7), (9.9) and (9.15):

$$t_D = t/\tau = t/\left(\frac{\phi\mu c r_w^2}{k}\right) \tag{9.36}$$

$$P_D = \Delta P/\left(\frac{Q\mu}{2\pi kh}\right) \tag{9.37}$$

and putting in the property values and both pairs of coordinates from the curve match allows kh to be determined. This is the main parameter of interest, but the particular curve of the family over which the experimental data fit has values of s and C_D, and these are the values for the well, if they are required. Ramey's definition of C_D is

$$C_D = \frac{C}{2\pi h\phi c r_w^2} \tag{9.38}$$

A new set of type curves was introduced by Gringarten et al. [1979] on the grounds that the Ramey curves were so similar that a superimposed set of data plotted on tracing paper could not be placed with confidence on the "correct" curve. These authors used the same solution for the response of the wellbore as had Agarwal et al., and it was in the choice of variables that the method differed and was claimed to be superior. The curves were plotted as dimensionless pressure versus (t_D/C_D), forming a family in terms of the single parameter $C_D e^{2S}$, but the method was superseded by incorporating the derivative of the solutions making up each type curve in what has become known as the derivative plot.

The derivative plot was introduced by Bourdet et al. [1983], who showed that interpretation was greatly assisted by plotting $t_D dP_D/dt_D$ versus t_D on the same graph as dP_D versus dt_D. The infinite acting part of the variation now appears as a horizontal straight line. Horne [1995] provides many examples. Being able to make use of the gradient of pressure variation with time is dependent upon accurate measurements, and appropriate advances have been made in petroleum engineering but not significantly in geothermal engineering. Calculations of local slope of a graph are made using differencing methods, and even with the more sophisticated instrumentation, Horne emphasises the importance of data smoothing. He recommends that given a data set x_i where i is 1, 2, 3, the local gradient is not determined at i by using the differences from immediately adjacent points, $i - 1$ and $i + 1$, but points some "distance" away, namely $i + k$ and $i - j$, where k and j are determined by trial and error but within the limits:

$$0.2 \leq \ln\left(t_{i+j}\right) - \ln(t_i) \geq 0.5 \quad \text{and} \quad 0.2 \leq \ln(t_i) - \ln(t_{i-k}) \geq 0.5$$

Since

$$t\left(\frac{\partial P}{\partial t}\right) = \left(\frac{\partial P}{\partial \ln t}\right) \tag{9.39}$$

he recommends smoothing as follows:

$$\begin{aligned}\left(\frac{\partial P}{\partial \ln t}\right) = &\left(\frac{\ln(t_i/t_{i-k})dP_{i+j}}{\ln\left(t_{i+j}/ti\right)\ln\left(t_{i+j}/t_{i-k}\right)}\right) + \left(\frac{\ln\left(t_{i+j}t_{i-k}/t_i^2\right)dP_i}{\ln\left(t_{i+j}/t_i\right)\ln(t_i/t_{i-k})}\right) \\ &+ \left(\frac{\ln\left(t_{i+j}/t_i\right)dP_{i-k}}{\ln(t_i/t_{i-k})\ln\left(t_{i+j}/t_{i-k}\right)}\right)\end{aligned} \tag{9.40}$$

Employing this method, Andrino [1998] analysed well measurements made by his colleagues in PNOC-EDC (now EDC), Philippines, using Kuster mechanical instruments; this organisation has a great deal of experience in well testing and interpretation. His work demonstrates that the derivative method is applicable to transient testing of geothermal wells using traditional instrumentation. Figure 9.9 shows the derivative plot using Andrino's data for the same well as Figs. 9.5 and 9.7.

The most significant features of the derivative plot are as follows:

- Wellbore storage begins as a unit slope, as in the log–log plot, but develops into a hump as time progresses.
- The height of the hump increases with increasing skin factor.
- A sandface pressure variation which would result in a sloping straight line on a semi-log plot produces a horizontal straight line. Thus, a formation satisfying the Theis solution produces a horizontal straight line at a value of $(dP_D/dt_D)(t_D/C_D) = 0.5$.

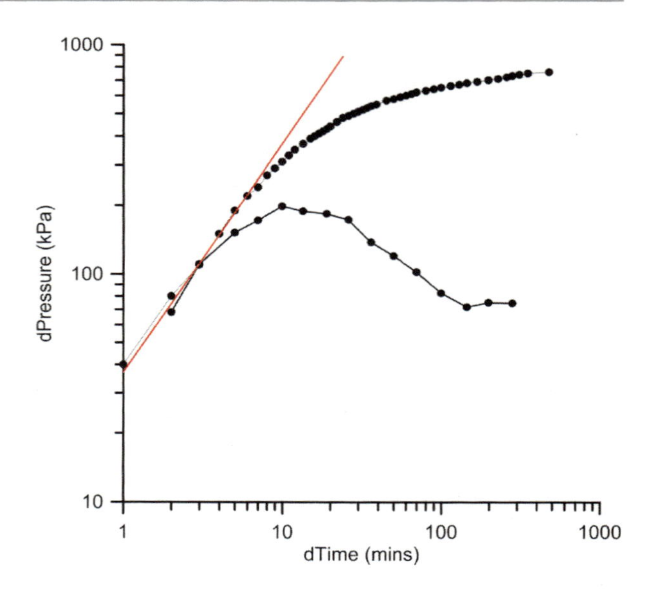

Fig. 9.9 Derivative plot for the well of Figs. 9.6 and 9.8 (*This material was created and used with permission by Energy Development Corporation, Philippines, and by Mr. Romeo P. Andrino, who reserves all rights thereto*)

9.4.3 Departure from Infinite Uniform Thickness Formation Geometry

Returning to the identification of real effects which prevent the direct application of the Theis solution, most high-temperature geothermal resources occur in regions subject to tectonic movements, and their strata are usually nonuniform in thickness and of limited radial extent in some directions. Irregular boundaries (i.e. not symmetric about the well axis) may be present in the form of sealed (offset) faults, illustrated by the sketch of Fig. 9.10, or a change in formation, perhaps the front of a volcanic lava flow of impermeable material. For a formation which satisfies the Theis assumptions except for the presence of such an imperfection, the sandface pressure transient will follow the infinite acting form until the disturbance reaches the boundary, and various authors have addressed this. Britto and Grader [1988] provided a survey.

Geothermal wells are open to all formations within the slotted liner part of the well, and there are usually several (a dual completion well which segregated production zones was constructed at Ngawha, New Zealand, but this was an exception). A transient pressure test on a geothermal well is usually a test of several different formations arranged in parallel. Because the kh values for each one are likely to be different, the flow and pressure at each sandface will change at different rates. The observation of a pivot point (Sect. 6.3.6) is an outcome of this. Placing the measuring instrument at the pivot point, assuming it can be accurately found, partly deals with this problem, but does not provide a way of identifying the kh value for any particular formation, as required for numerical reservoir simulation (Sect. 13.3).

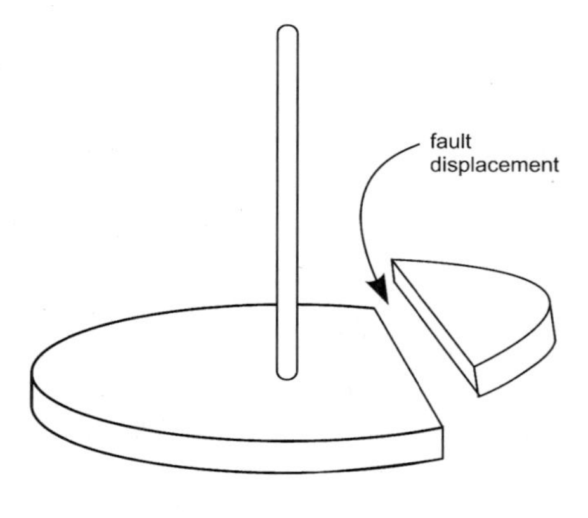

Fig. 9.10 A sketch of a sealed fault

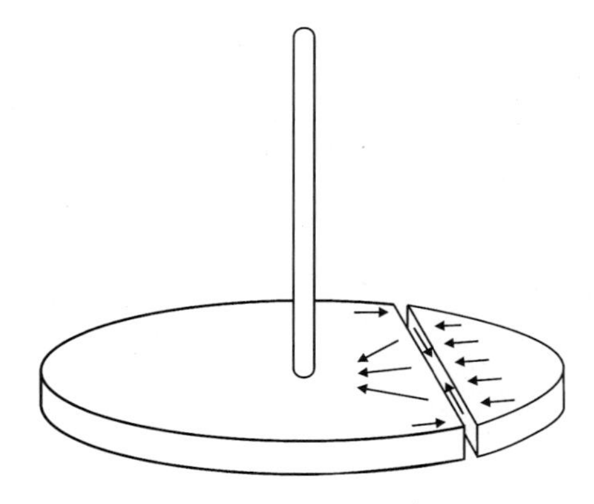

Fig. 9.11 A sketch of a fault and its effect on fluid flow in the formation of interest

9.4.4 The Influence of Fractures in the Formation

If the sealed fault of Fig. 9.10 had been an open fault with no displacement, it would have provided a short circuit for flow to the well. Instead of flowing tortuously through the pores of the formation, a lower drag flow passage is available—Eq. (4.37) in two-dimensional rectangular coordinates would describe the flow. The outcome is suggested by the sketch of Fig. 9.11, the arrows indicating flow directions.

The distinction between faults and fractures may appear unimportant to the engineer. In this book "fracture" is used to describe a planar break in a formation; fractures vary in scale over orders of magnitude, from single-crack splitting of huge blocks to hand-sized pieces, although at that scale they must be closely fitting pieces like a regular pile of bricks without mortar, and not rubble or a heap of pebbles. The term "fault" is used to describe a break on a large scale as a result of earth movements of geological origin (uplift, rifting, etc.). Fault types have been extensively analysed, and that which occurs depends on the state of stress of the material.

As a result of the granular, brittle nature of most rocks, natural faults are accompanied by fracturing. Although faults are recognisable at the surface as step changes in level, for example, and may appear to be a distinct planar break from a distance, so far as fluid flow is concerned, they are a broad band, perhaps tens of metres wide, of closely fitting pieces of fractured rock. The fit is close, but the frictional drag offered to a flow is very much less than through permeable material, and unless they have been sealed by chemical deposition, faults offer a short circuit to flows within a resource. Unintentional fracturing may result during drilling, sometimes at the casing shoe in geothermal wells. Deliberate fracturing is commonly practised as a way of improving the flow of fluids in permeable formations, and it is the source of all permeability in enhanced geothermal resources; it has attracted the attention of environmental concern groups at the time of writing. It was studied in detail many years ago—see Howard and Fast [1970] who explained that the orientation of the fracture produced in a well by pressurising to the point of material failure depends on the in situ stress field. The failure of rock does not comply with the theories applicable to steel. If the locality is not subject to any horizontal stress, the fracture will tend to be vertical, and the flow in a vertical fracture has received attention analytically, being a more easily defined problem than that shown in Fig. 9.11. It is clear from the origins of Darcy's law that flow in the fracture (a 2D channel instead of a tube) can be imagined to be flow in a permeable formation. Some of the treatments of fractured flow do not clearly describe the physical ideas; however, Horne [1995] approaches the problem by describing the flow boundary conditions from the formation into the fracture and along the fracture into the well for two cases and the variation of well pressure with time that results. McLure and Horne [2011] demonstrate the approach to analysing the measurements from the deep geothermal well into fracture rock at Soultz-sur-Forêts, NE France.

Da Prat et al. [1982] ascribe one method of representing fractured media to Barenblatt and Zheltov who in 1960 represented a heavily fractured block of permeable material as the superposition of two porous media with different sized pores. Returning to the idea of fractured material as an uncemented pile of closely fitting bricks, if the pile has a waterproof exterior and is flooded then allowed to drain through a single hole drilled through that exterior, water will drain from the fractures quickly. Once they are emptied, or the pressure has declined, water will begin to seep from the entire surface area of the porous (permeable) bricks. Both flows are governed by the same form of equation, regardless of geometry, Eq. (4.72)—that was Darcy's original idea—and so long as the small compressibility approximation is valid, both flows might be guessed to follow a semi-log pressure decline. The theory confirms this, and Andrino [1998] included an example of a well test demonstrating it, shown here as Fig. 9.12; a fractured formation with this type of behaviour is referred to as having double porosity. (The semi-log straight lines have been approximately fitted by eye.)

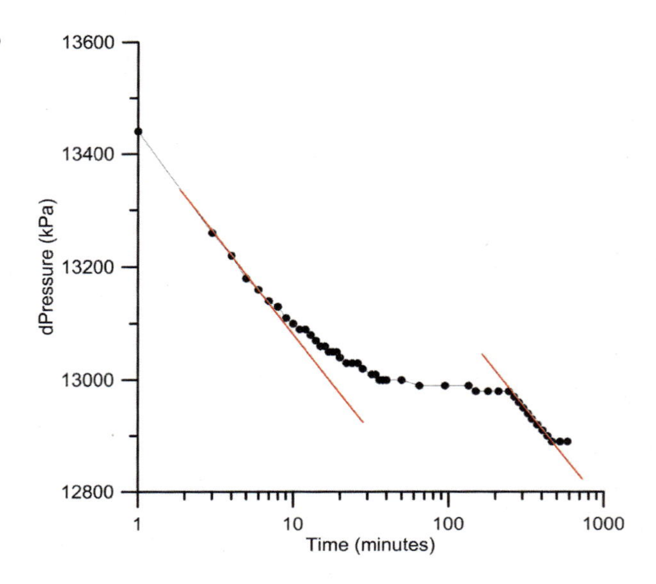

Fig. 9.12 Results of a PFO test demonstrating double porosity behaviour, Andrino [1998] (*This material was created and used with permission by Energy Development Corporation, Philippines, and by Mr. Romeo P. Andrino, who reserves all rights thereto*)

9.5 Transient Pressure Testing of Formations Filled with Steam

If the fluid is a gas following the relationship P/ρ = RT, which includes steam unless the conditions everywhere are very close to the saturation line, then a particular form of the governing equation can be derived. Returning to Eq. (4.78):

$$\phi\mu\frac{\partial\rho}{\partial t}=\frac{\partial}{\partial x}\left(k\rho\frac{\partial P}{\partial x}\right)+\frac{\partial}{\partial y}\left(k\rho\frac{\partial P}{\partial y}\right)+\frac{\partial}{\partial z}\left(k\rho\frac{\partial P}{\partial z}\right) \tag{4.78}$$

and rewriting it for the Theis geometry, it becomes

$$\phi\mu\frac{\partial\rho}{\partial t}=\frac{1}{r}\frac{\partial}{\partial r}\left(k\rho r\frac{\partial P}{\partial r}\right) \tag{9.41}$$

Incorporating the definition of c from Eq. (4.80),

$$c=\frac{1}{\rho}\left(\frac{\partial\rho}{\partial P}\right)_T=\frac{1}{P} \tag{9.42}$$

which with the equation of state becomes

$$\begin{aligned}\frac{\phi\mu}{RT}\frac{\partial P}{\partial t} &= \frac{1}{r}\frac{\partial}{\partial r}\left(k\rho\frac{\partial P}{\partial r}\right)\\ &= \frac{1}{r}\frac{\partial}{\partial r}\left(\frac{kP}{RT}\frac{\partial P}{\partial r}\right)\end{aligned} \tag{9.43}$$

Cancelling RT and reorganising the expression, this becomes

$$\phi\mu\frac{\partial P}{\partial t} = \frac{1}{2r}\frac{\partial}{\partial r}\left(kr\frac{\partial P^2}{\partial r}\right) \tag{9.44}$$

If k is constant, the familiar group $\frac{k}{\phi\mu c}$ can be formed by introducing c, leaving

$$\frac{\partial\left(P^2\right)}{\partial t} = \left(\frac{k}{\phi\mu c}\right)\frac{1}{r}\frac{\partial}{\partial r}\left(r\frac{\partial P^2}{\partial r}\right) \tag{9.45}$$

Thus in plotting transient tests on steam wells (wells drilled into steam zones), semi-log plots should be made using P^2 and not P. Economides et al. [1980] set out the fundamentals.

9.6 Two-Phase Flow in the Formation

The transient pressure tests described so far have been for single-phase fluids with small, constant compressibility. Discharging a geothermal well in a liquid-dominated reservoir inevitably causes flashing in the formation. When the fluid in the formation flashes as a result of pressure decline to form a two-phase mixture, the effective compressibility is not small—it was deduced by Grant and Sorey [1979] in the form already given in Chap. 7. In addition the phases do not move through the formation under a pressure gradient in the same proportions that exist when there is no pressure gradient and the mixture is stationary, and this is the cause of changes in specific enthalpy of the discharge. To pick up from Sect. 7.7, a method of analysing transient tests is required.

The nature of the porosity in any particular material influences the way the two-phase mixture flows towards the well. The two phases flow at different volumetric fluxes, and this is detectable in the well discharge characteristics—the specific enthalpy of the discharge rises when steam flows to the well faster than water. The formation effectively has a different value of permeability for each phase, and this is accommodated in calculations by defining a relative permeability for each one. Darcy's law was expressed as Eq. (4.72):

$$q_v = -\frac{k}{\mu}\frac{dP}{dx} \tag{4.72}$$

where q_v is the volumetric flux which becomes mass flux q_m when multiplied by the fluid density and allows Eq. (4.71) to become

$$q_m = -\frac{k}{\nu}\frac{dP}{dx} \tag{9.46}$$

where ν is the kinematic viscosity μ/ρ.

The total mass flux is made up of liquid and steam:

$$q_m = q_{mg} + q_{mf} \tag{9.47}$$

and the Darcy's law relationship is written for each phase:

$$q_{mf} = -\frac{k.k_{rf}}{\nu_f}\frac{dP}{dx} \tag{9.48}$$

$$q_{mg} = -\frac{k.k_{rg}}{\nu_g}\frac{dP}{dx} \tag{9.49}$$

Many formulations have been proposed for the relative permeabilities, mainly from the petroleum industry, and several are listed by Pruess [1987] including those of Corey which are often quoted for geothermal calculations. Darcy's law is itself empirical and a further layer of empiricism has now been added, which is unsatisfactory but inevitable. However, relative permeabilities are an integral part of reservoir simulation, to be introduced in Chap. 13, which can be made to produce good results, so the empiricism works.

To draw the above ideas together, consider a procedure for interpreting a simple drawdown test, in which the discharge enthalpy (the flowing enthalpy) has been measured together with the sandface pressure as a function of time. The flowing enthalpy can be expressed in terms of the relative permeabilities using the expressions developed above—it is defined as

$$h_{flow}.q_m = q_e \tag{9.50}$$

The energy flux is

$$\begin{aligned} q_e &= h_f q_{mf} + h_g q_{mg} \\ &= -k\frac{dP}{dx}\left(h_f\frac{k_{rf}}{\nu_f} + h_g\frac{k_{rg}}{\nu_g}\right) \end{aligned} \tag{9.51}$$

and expressions for the mass flux have been developed above [Eqs. (9.47), (9.48), (9.49)].

Thus,

$$
\begin{aligned}
h_{flow} &= \frac{q_e}{q_m} \\
&= \left(h_f \frac{k_{rf}}{\nu_f} + h_g \frac{k_{rg}}{\nu_g}\right) \Big/ \left(\frac{k_{rf}}{\nu_f} + \frac{k_{rg}}{\nu_g}\right)
\end{aligned}
\tag{9.52}
$$

This equation can be further rearranged as

$$
\frac{k_{rg}}{k_{rf}} = \frac{\nu_g\left(h_{flow} - h_f\right)}{\nu_f\left(h_g - h_{flow}\right)} \tag{9.53}
$$

A relationship often quoted for fractured geothermal formations but of unknown origin (to the author) is

$$
k_{rf} + k_{rg} = 1 \tag{9.54}
$$

and with steam table values and the measured flowing enthalpy, Eqs. (9.53) and (9.54) allow the relative permeabilities to be calculated. There is one further step required before any of the semi-log or curve matching methods already described can be applied. Writing Eq. (9.47) in terms of its parts, Eqs. (9.44), (9.48) and (9.49), gives

$$
\frac{k}{\overline{\nu}} \frac{dP}{dx} = \frac{k.k_{rf}}{\nu_f} \frac{dP}{dx} + \frac{k.k_{rg}}{\nu_g} \frac{dP}{dx} \tag{9.55}
$$

and thus

$$
\frac{1}{\overline{\nu}} = \frac{k_{rf}}{\nu_f} + \frac{k_{rg}}{\nu_g} \tag{9.56}
$$

An early application of this approach was given by Whittome [1979].

9.7 Tests Using More than One Well-Interference Testing

This chapter was introduced by describing an interference test, which is the name given to measurements carried out with two or more wells simultaneously. The tests described so far have had the aim of determining the formation properties close to a single well, sometimes restricted to short times to avoid production being interrupted. Measuring *kh* over wider areas and identifying barriers to flow is important for planning the drilling of a resource, and while it is possible to detect the presence of a boundary using a single well, it is not possible to locate it. For these purposes it is usual to provide a disturbance using a single well, but measure

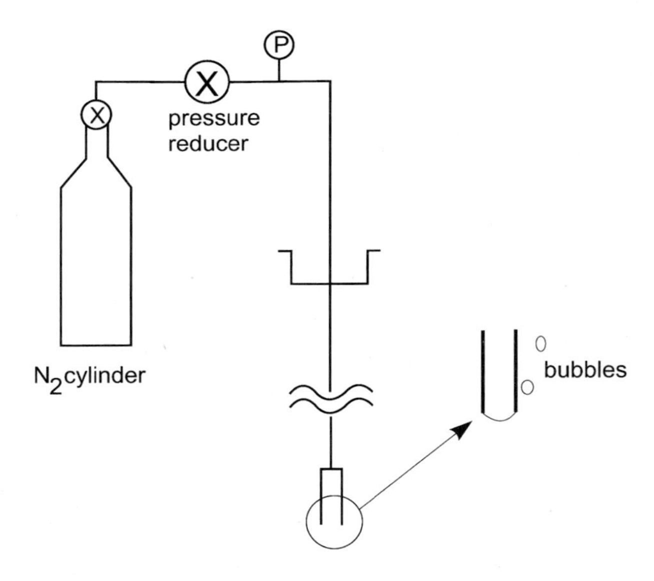

Fig. 9.13 The arrangement of capillary tubing for long-term small-amplitude pressure variations. The cylinder is marked "Nitrogen" but other gases may be used

its effect at several other wells which have remained closed so that the disturbance to formation pressure can be more accurately measured. Because there is little fluid movement in the formation near a monitor well, or in the bore of the monitor well, wellbore storage and skin effects are minimised and usually neglected. On the other hand, the disturbance to formation pressure may be very small, small enough that lunar tidal changes to formation pressure sometimes have to be allowed for, which requires longer recording. In addition, the time for a disturbance to reach a monitor well will be long and perhaps expensive if field operations are held up. The need to use existing wells rather than use purpose-drilled monitor wells may add to these difficulties if they are far apart. There may be problems with limited clock times if mechanical instruments are being used, but capillary tubing provides a remedy for this.

An accurate method of downhole pressure measurement was developed using small bore stainless steel of typically 6 mm outside diameter, although Inconel tubing down to 3 mm diameter has also been used. The arrangement is shown in Fig. 9.13.

The technique is to fit the lower end of the tubing into a short length of heavy walled tubing, which acts as a weight to help keep the tubing at the required depth opposite the sandface, and presents a larger interface between the liquid in the well and the gas which flows slowly down the capillary tube and out into the well.

The gas, usually nitrogen, is supplied from a gas cylinder through a pressure controller, downstream of which a pressure gauge is fitted. The principle of operation is that the hydrostatic head of the gas in the tubing is negligible compared to the pressure at the sandface; provided the gas flow rate is constant and very small, with negligible frictional pressure drop due to flow, then changes in the pressure measured at the surface represent changes at the tube end. Starting with the tube full of water, at least in its lower part, if the gas pressure is increased incrementally the water will be driven back into the well. When the pressure no longer increases as the valve is opened, the tube is full and the gas is bubbling out, and the gas flow rate can be minimised. If the gas–water interface is within the tube rather than at the end, then the gas pressure measured at the surface is not the absolute pressure in the well at the foot of the tube less the hydrostatic head of the gas column. However, pressure variations in the well will be accurately represented by gas pressure variations, so long as the gas–water interface stays at the same level, preferably close to the tube end. In practice it is difficult to know exactly where the downhole end of the tube is, because the tubing is stiff and is curved as a result of being rolled on a drum for storage, but the same reasoning applies, and pressure variations can be accurately detected.

The history of the development of interference testing was set out by Leaver et al. [1988] briefly, but stating the crucial publications along the way, and they analysed sets of two-well interference tests made at Ohaaki, New Zealand, determining average reservoir properties and locating the presence of a permeability barrier.

If the formation and fluid compressibility satisfy the Theis solution assumptions, then Eq. (9.5), which is

$$P_i - P(r,t) = \frac{Q\mu}{4\pi kh}\left(\ln\left(\frac{kt}{\phi\mu c r^2}\right) + 0.80907\right) \tag{9.5}$$

holds for every well with injection or discharge, and the pressure disturbances from each one can be superimposed. Thus, for two wells, W_1 and W_2, distant r_1 and r_2 from a monitor well, which begin flowing simultaneously at time $t = 0$ at volumetric rates Q_1 and Q_2, respectively, the change in pressure at the monitor well is

$$\Delta P = \frac{Q_1\mu}{4\pi kh}\left(\ln\frac{4kt}{\phi\mu c r_1^2} + 0.80907\right) + \frac{Q_2\mu}{4\pi kh}\left(\ln\frac{4kt}{\phi\mu c r_2^2} + 0.80907\right) \tag{9.57}$$

In writing this expression, it has been assumed that the formation properties are the same throughout the formation. A semi-log plot of ΔP versus $ln(t)$ would in principle provide a straight line with a slope from which kh could be found, in the same manner as for single-well tests.

A method of locating faults using a set of semi-log type curves was first designed by Stallman [1952], but Sageev et al. [1985] modified Stallman's curves to improve ease of use and to allow shorter test times.

The introduction to this chapter described pulse tests, as introduced by Johnson et al. [1966], which have been significantly developed for petroleum engineering use but appear to have been rarely used for geothermal resources. Pulse testing at the Sumikawa resource, Japan, was described by Nakao et al. [2003]. The pulse length varied from 3 to 13 h and was integrated into the normal operation of the system of production and injection wells. It appears possible to estimate fracture permeability to help in defining a dual porosity model for the resource. Pulse tests have also attracted attention in association with enhanced geothermal systems, involving high-pressure pulses in low-permeability systems (see Itoi et al. [1994]).

9.8 Concluding Remarks

Transient pressure testing is a time-consuming and relatively expensive activity, expensive in terms of the facilities required and human resources. The optimum level appropriate for any geothermal project is difficult to estimate intuitively and may be very different from that for petroleum engineering, although even this generalisation would warrant some study before it could be asserted—it will depend on the level of development of the reservoir model for a particular resource. In the author's experience, the activity successfully promotes serious study of a resource, but a cost–benefit analysis might be worthwhile. For instance, the information from transient testing assists reservoir modelling, which is also a time-consuming activity, but a single measured value of permeability for a well open to several formations leaves the kh value for each formation to be estimated and is likely to be discarded fairly readily by modellers if they have more certain data with which to test their models. As a further example, the extent to which a very early estimate of well production capacity, as gained from completion tests, is influential in directing a project has always been uncertain to the author, but the tests are usually carried out.

References

Abramowitz M, Stegun IA (eds) (1965) Handbook of mathematical functions. Dover, New York

Agarwal RG, Al-Husseini R, Ramey HJ Jr (1970) An investigation of wellbore storage and skin effect in unsteady liquid flow: I. Analytical treatment. Soc Pet Eng J: 279–290; Trans SPE, vol 249

Andrino R (1998) A review of pressure transient analysis using the pressure derivative method, with application to ten wells of the Leyte Geothermal Power Project, the Philippines. University of Auckland Geothermal Institute, New Zealand, Report No. 98.04

Bourdet D, Whittle TM, Douglas AA, Pirard YM (1983) A new set of type curves simplifies well test analysis. World Oil 196:95–106

Britto PR, Grader AS (1988) The effects of size, shape and orientation of an impermeable region on transient pressure testing. SPE Form Eval 33:595–606

Carslaw HS, Jaeger JC (1959) Conduction of heat in solids. Oxford Science, New York

Da Prat G, Ramey HJ Jr, Cinco-Ley H (1982) A method to determine the permeability –thickness product for a naturally fractured reservoir. JPT, June 1982

Economides MJ, Ogbe DO, Miller FR, Ramey HJ Jr (1980) Geothermal steam well testing: state of the art. SPE9272

Grant MA, Sorey ML (1979) The compressibility and hydraulic diffusivity of a steam-water flow. Water Resour Res 13(3):684–686

Grigull U, Sandner H (1984) "Heat conduction", International series in heat and mass transfer. Springer, Berlin

Gringarten AC, Bourdet DP, Landel PA, Kniazeff VJ (1979) A comparison between different skin and wellbore storage type curves for early transient analysis. Soc Pet Eng SPE 8205:1

Horne RN (1995) Modern well test analysis. Petroway Inc., Palo Alto, CA

Howard GC, Fast CR (1970) Hydraulic fracturing. Monograph vol 2 of the Henry L. Doherty series. Society of Petroleum Engineers, New York

Itoi T, Hirose Y, Hiyashi K (1994) Measurement of in-situ hydraulic properties from the pulse test for the case of unknown in-situ pore pressure and its application to HDR model fields. Geoth Res Counc Trans 18:445

Jaeger JC (1956) Conduction of heat in an infinite region bounded internally by a circular cylinder of perfect conductor. Aust J Phys 9(2):167–179

Johnson CR, Greenkorn RA, Woods EG (1966) Pulse testing: a new method for describing flow properties between wells. J Pet Tech 18:1599–1604

Kuster Instruments. http://www.kusterco.com

Leaver JD, Grader A, Ramey HJ Jr (1988) Multiple well interference testing in the Ohaaki geothermal field. SPE Form Eval 3(2):429–437

McLure MW, Horne RN (2011) Pressure transient analysis of fracture zone permeability at Soultz-sur-Forêts. GRC Trans 35:1487–1498

Miller CW (1980) Wellbore storage effects in Geothermal wells. Soc Petroleum Engineers Jnl 555–566

Nakao S, Ishido T, Hatakayama K (2003) Pulse testing analysis for fractured geothermal reservoir. Geoth Res Counc Trans 27:807–809

Pruess K (1987) TOUGH user's guide. Lawrence Berkeley Laboratory, University of California, LBL-20700

Ramey HJ Jr (1970) Short-term well test data interpretation in the presence of skin effect and wellbore storage. J Pet Tech 22:97–104

Ramey HJ Jr (1966) Application of the line source solution to flow in porous media – a review. SPE 1361, SPE-AIChE symposium, Dallas, Feb 1966

Sageev A, Horne RN, Ramey HJ (1985) Detection of linear boundaries by d rawdown tests: a semi-log type curve matching approach. Water Resour Res 21(3):305–310

Stallman RW (1952) Non-equilibrium type curves modified for two-well systems. Geological Survey Groundwater Note 3. US Department of the Interior, Washington, DC

Temme NM (1996) Special functions: an introduction to the classical functions of mathematical physics. Wiley, New York

Theis CV (1935) The relationship between the lowering of the piezometric surface and rate and duration of discharge of a well using groundwater storage. Trans Am Geophys Union 2:519–524

Vela S (1977) Effect of a linear boundary on interference and pulse tests – the elliptical interference area. J Pet Tech 29(8):947–950

Whittome AJ (1979) Well testing in a liquid dominated two-phase reservoir. Geoth Resour Counc Trans 3:781–784

Economic Issues Relating to Geothermal Energy Use 10

This chapter begins with a general discussion of why an economic analysis is required and whether it should be called economic or financial. The costs and cash flows for a single isolated geothermal project are then introduced, firstly without discounting the cash flow, to illustrate methodology, and then incorporating it, as is essential for the economic comparison of projects. The various parameters used for project economic comparison are explained, and sensitivity analysis discussed. The role of a geothermal power station in a network is considered next, followed by the need for a more widely based energy analysis. The chapter closes with a discussion of steam sales contracts.

10.1 Introduction

The scales of geothermal energy extraction vary widely. At one extreme lies the single-family dwelling which makes use of a nearby spring without the need to drill at all, and at the other, the government organisation or multinational company with hundreds of MWe of generating capacity installed at a cost of hundreds of millions of dollars. At the lower end, the funds are simply drawn from the family bank account if they are available, but at the upper end, economic analysis is required, and the decision-making process is complicated.

The upper end of the scale may be the efforts of the government to increase the national supply of electricity by making use of its geothermal resources, thus avoiding further dependence on fuel imports. The fundamental economic problem facing all countries is how to allocate limited resources (labour, capital, natural resources and foreign exchange) to a variety of different uses in such a way as to maximise the net benefit to society. Turvey [1968] deals with the problem of valuing the electricity that might be produced to see if the investment of resources and funds is justified. For developing countries, in which some of the best geothermal resources are situated, funding is provided by the development banks, e.g. the World Bank and the Asian Development Bank. Also at this upper end of the

A. Watson, *Geothermal Engineering: Fundamentals and Applications*,

DOI 10.1007/978-1-4614-8569-8_10,

scale lie government decisions to invest in geothermal resource use in response to global warming concerns. Such decisions will in part be politically motivated, but must be supported by economic analysis to some extent. It is not always necessary for scientists and engineers to be very close to the economic analysis at this level. In the early stages of considering the use of geothermal energy at a particular location, it may be that the exploration and feasibility studies necessary to support the economic analysis form a well-defined package of information which can be passed on without further involvement, but eventually one particular project will be decided upon for further investigation, and wider involvement is necessary.

A project may be defined in general terms as the investment of funds to construct a facility that will turn out goods, the sale of which generates a stream of profits sufficient to recover the initial investment and more. There might be several projects in contention, representing alternative ways in which the goods might be produced, and it must be decided which to invest in. Well-established methodologies exist to provide numbers by which the projects may be differentiated, but they are accounting calculations in which only money is recognised. Other benefits and disadvantages must be considered in some other way.

In the engineering industries, chemical engineers seem to have made use of the methodology ahead of other branches, perhaps because chemical plants are often stand-alone facilities reliant on specific markets for their output. Power stations for bulk electricity generation are different. In the commercial world in general, manufacturers can control the supply of their goods to maximise their profit; however, electricity supply is regarded by governments as essential for the social and economic well-being of their nation, and they seek to limit manipulation by private enterprise. One way of achieving this is by having the government own the electricity generators, which until the 1980s was the case in the UK (the Central Electricity Generating Board) and in New Zealand (the New Zealand Electricity Department), examples both large and small. Alternatively, governments can provide regulations to limit the profits from the sale of electricity to a level that suits their policies, and in this way, the revenue stream to the generating companies is controlled, directly or indirectly. The significance of this here is that the standard methodology produces a comparison of projects in terms of their level of profitability, so it must be modified if profitability is regulated. It turns out that the modification is in viewpoint rather than numerical procedure.

The numerical procedures themselves are very simple—the time value of money, which most of us understand intuitively via the concept of interest, is the only arithmetic process beyond simple addition and subtraction. However, difficulty arises because the focus of attention is cash flow, which is modified by taxation and the financial rules and customs of the particular country or company concerned. It is impossible to perform an economic analysis of a real project without incorporating these rules, for example, depreciation, but both the rules and the terminology are foreign to most engineers. Fortunately, they are not fundamental to understanding the methodology, so there is no necessity to include them here, and the project examples have been stripped down to basics to illustrate the effects of time on the interpretation of cash flows. Economic analysis without the full accounting rules still has value at project management level.

This is an appropriate stage at which to discuss terminology—are the issues of this chapter economic or financial? The assessment of engineering projects for profitability is traditionally called economic analysis—thus Allen [1991] uses the title "Economic evaluation of projects" in his book which is applied to the chemical engineering industry. The book by Marsh [1980] is entitled "Economics of Electric Utility Power Generation" and includes "financial simulation" as one of the methods of analysis available. Turvey and Anderson [1977] define the difference; thus,

> *"A 'financial' appraisal of a project is made either to determine its capacity to service debt and contribute to subsequent investment by the borrower, or to determine the return to the investor. An 'economic' appraisal or 'cost-benefit analysis' is aimed at determining whether or not the project is in the national interest".*

A definitive answer is unimportant here, and the term "economic analysis" will be used, although the subject matter would be included in what is defined above as financial appraisal. No attempt will be made to explain purely accounting matters such as ownership (shareholding), taxation and depreciation. Neither are the various means by which funds can be raised for geothermal projects discussed. The complexities of getting together many parties to fund a project are illustrated by Ogryzlo and Randle [2005] in their history of events leading up to the financing of a Nicaraguan project Scientific and engineering matters play a part in this, for example, in the form of an independent opinion (due diligence examination) on the likelihood of the project producing the steam required and generating electricity at the predicted cost, but such studies are well defined and, as mentioned earlier, form discrete packages of work.

A power generation project in isolation is examined first, assuming that its main components have been selected by the organisation after considering alternatives. The aim is to illustrate the general pattern of cash flow and how it can be affected by decisions and events that occur throughout the project life. Methods of quantifying project performance which are needed to decide between options are then addressed. The remainder of the chapter deals with related economic issues relevant to scientists and engineers.

10.2 A Single Isolated Project

A geothermal power project is a long-term undertaking involving the outlay of a large sum of money which generates no income over a period of several years during construction, followed by income from the sale of electricity as smaller sums over a much longer period. Figure 10.1 shows a spreadsheet with the first column representing the year of the project and the other columns representing, for every year, the cost incurred, revenue earned, the sum of these and the debt remaining, assuming all earnings are used to repay debt at the end of each year. In reality, the project plan,

	A	B	C	D	E	F
1	Year	Cash Flow items	Costs	Revenues	Annual net	Cumulative
2					cash flow	cash flow
3	0					
4	1	exploration and planning	3.50	0	-3.50	-3.50
5	2	drill 3 exploration wells	15.00	0	-15.00	-18.50
6	3	drill 3 wells, preliminary design, permitting	17.00	0	-17.00	-35.50
7	4	drill 2 wells, pay 30% pipeline and plant costs	28.60	0	-28.60	-64.10
8	5	pay balance of construction costs	43.40	0	-43.40	-107.50
9	6	first year only half capacity output	5.00	6.26	1.26	-106.24
10	7		2.20	12.52	10.32	-95.93
11	8		2.20	12.52	10.32	-85.61
12	9		2.20	12.52	10.32	-75.29
13	10		2.20	12.52	10.32	-64.98
14	11		2.20	12.52	10.32	-54.66
15	12		2.20	12.52	10.32	-44.34
16	13		2.20	12.52	10.32	-34.03
17	14		2.20	12.52	10.32	-23.71
18	15		2.20	12.52	10.32	-13.40
19	16		2.20	12.52	10.32	-3.08
20	17		2.20	12.52	10.32	7.24
21	18		2.20	12.52	10.32	17.55
22	19		2.20	12.52	10.32	27.87
23	20		2.20	12.52	10.32	38.19
24						
25						
26		**Data**				
27		Installed capacity (MWe)	20			
28		Full load net output (MWe)	18.8			
29		Annual load factor	0.95			
30		Operations and maintenance costs million$/yr	2.2			
31		Total annual output (kWh)	1.56E+08			
32		Electricity selling price c/kWh	8.0			
33						
34						
35						
36						
37						
38						
39						
40						
41						
42						
43						
44						
45						
46						

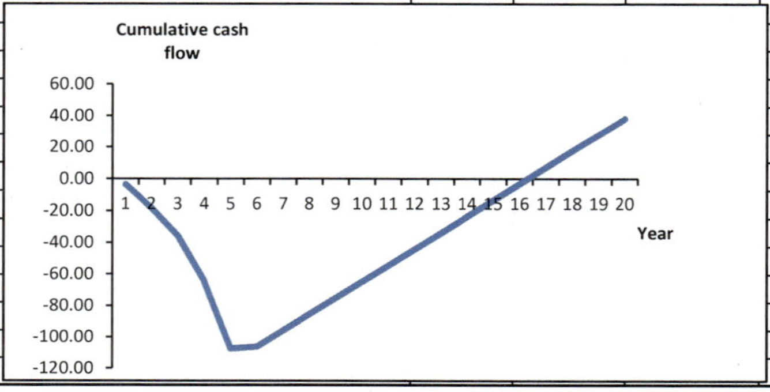

Cumulative cash flow

60.00
40.00
20.00
0.00
-20.00
-40.00
-60.00
-80.00
-100.00
-120.00

1 2 3 4 5 6 7 8 9 10 11 12 13 14 15 16 17 18 19 20

Year

Fig. 10.1 A simple spreadsheet for a single project, lacking interest charges. The graph responds instantly to the change of any parameter so the sensitivity of the date of zero cumulative cash flow can be investigated

which is the list of items up to year 5, will be very much more detailed, and it is better to lay out the sheet with a column for each year, entering the cash items in rows.

Figure 10.1 is a simple example made up for a 20 MWe project. It is (deliberately) deficient in that the cost of capital is zero, and there is no interest on accrued earnings. The cost data were real estimates at 2007, taken from a report commissioned by the New Zealand Geothermal Association [2009]. Standard accounting practice for recording transactions—see, for example, Marsh [1980]—is not used; instead, income is shown positive and expenditure negative. The vertical rows represent time in years. The activities during the construction period of 5 years are itemised and costs assembled based on the following:
Gross output of station = 20 MWe
Cost of station at $2.2 million/MWe = $44 million
Net output is 94 % of gross = 18.8 MWe
Total wells required = 8 (5 production and 3 injection)
Cost of drilling = $5 million/well
Steamfield cost = $18 million

The operating and maintenance costs during the life of the station are assumed to be $2.2 million per annum.

In the first project year, the costs are for exploration and planning, assumed to be $3.5 million. To the year 3 drilling costs, $2 million has been added for preliminary design and permitting. The major capital items of steamfield and power station itself have been assumed to be paid for in two parts, 30 % in year 4 and the balance in year 5.

The station is assumed to operate at a capacity factor of 95 %, that is, it operates at full output for 95 % of every year. However, in the first year of operation, a pessimistic capacity factor of 50 % has been assumed to allow for testing and early operating problems. If the selling price of electricity is P c/kWh, the annual revenue is calculated as follows:

$$\begin{aligned}\text{Annual revenue} &= \text{net output rate (kW)} \times \text{capacity factor} \times 8760\ (\text{hrs/year}) \\ &\quad \times \text{P}(\text{c/kWh}) \\ &= 18.8 \times 0.95 \times 8760 \times \text{P}/100\ (\text{dollars})\end{aligned}$$

The operating and maintenance costs have been deducted from the revenue for each year, and the remainder has been paid to reduce debt.

In reality, interest will be charged on loans. Funds will not simply be paid to sit in bank accounts until the cost arises; instead, detailed arrangements for making them available will be required, and yearly accounting is unlikely to be adequate. However, the purpose of this simulation is simply to show the items making up the cash flow. A graph of cumulative cash flow has been added, and using Excel©, it can be made to respond immediately to changes made to the chart. For example, the selling price of electricity can be entered into a single cell which is picked up in the calculation of revenue for every year. The spreadsheet can thus be used to see the effects of various changes, to either the parameters needed in the calculation which are linked to the value in a single cell under the heading of "data" on the sheet or changes to the cash flow items column. Changes worth examining, for example, might be:

- A delay of 1 year after year 4, in which nothing is done
- The need to drill one replacement well in year 10 and another in year 16
- A decrease in the annual load factor from 95 % to 80 %
- A decrease in the full load net output to say 15 MWe (as a result of steam supply shortages)

The cash flow pattern shown in the graph of Fig. 10.1 has the characteristic shape of all large projects; the debt builds up gradually at first and then steeply (even more steeply when interest is included), reaching a maximum (minimum on the graph) which coincides with the time at which income begins to be generated. Income from the following years reduces the debt gradually until it is paid off, when the cumulative cash flow curve crosses the zero axis. The time from first expenditure to this point is the payback period, and at this time, the project can be said to have broken even. After this, the net cash flow is positive and profit is being made for the remainder of the project life. Figure 10.1 coupled with Table 10.1 shows a break-even point 16 years after the first year of production if the electricity is sold at 8 c/kWh.

Having set up the programme, it can be used to investigate the economic effect of any changes to the project plan, but it must include the proper valuation of money, which is addressed next.

10.3 Economic Evaluation of Projects

All major power projects require significant starting investment over a construction period of a few years and might take 15 years or more to recover that investment. It might seem odd that the economic performance of such projects is still mainly assessed by reducing the considerable cash flows over the project life to a single parameter and then comparing the values of this parameter for all the projects being considered. According to Leung and Durning [1978], the economic comparison of projects in the electricity supply industry as a whole was rather empirical until the 1950s, after which the methods described here came into use. It must be remembered that the prediction of events and cash flows tens of years into the future is accompanied by uncertainties which even detailed analysis cannot eliminate. Economic decisions call for judgement and do not rely only on the result of a formal analysis. A committee responsible for selecting alternatives would equip itself with several of the single parameters but then exercise its collective wisdom.

10.3.1 Discounting the Cash Flow

The single parameter methods which involve the timing of costs and revenues can be described as discounted cash flow methods. Figure 10.1 shows the cost of the first item (exploration and planning) as $3.5 million. If interest was charged on this loan at 5 % pa, then without any further expenditure, the debt in successive years becomes \$3.68 million after the first year and then 3.86, 4.05, 4.25 and so on. The value in any year is the initial expenditure times $(1 + r)^t$ where r is the interest rate on the debt and t is the number of years since the debt was incurred. Suppose that it was decided to put money aside now to spend on this work 3 years into the future, what sum would be sufficient? The answer is just enough to grow to exactly \$3.5 million at 5 % pa, i.e. by an annual multiplying factor of 1.05; the sum is \$3.02 million. This number is the discounted value of the expenditure or its present value, "present" meaning at the time the money was set aside. The concept is that of interest, but the terminology is different because sums of money are being brought backwards in time rather than forwards, and they shrink rather than grow. Thus the present value, PV, of a future amount Q is

$$PV = Q/(1 + r)^t$$

The factor $1/(1 + r)^t$ is called the discount factor, and every cost or revenue throughout the project can be brought backwards in time to a single figure valued at year 0 (or any other year) by multiplying it by the appropriate discount factor. What was called the interest rate in working forward in time is called the discount rate in working backwards, and they may not have the same value. Money is not necessarily borrowed from a bank, it might be raised as bonds by a company who has many uses for it apart from building a power station and its value is represented by the rate at which profit could be made if it was used for other things. This might be higher than a bank lending rate. The discount rate is often called the "cost of capital".

10.3.2 Net Present Value

A new spreadsheet incorporating discounting is shown as Fig. 10.2. The discount factor for each year is shown in column H, calculated from

$$\text{Discount factor} = 1/(1 + i)^t$$

where i is the discount rate and t the year number, given in column A. The formula in the spreadsheet for year 8 is

$$\text{Discount factor} = 1/(1 + 0.01^{*}\$C\$33)^{\wedge}A8$$

so a change in discount rate stated as a % in cell C33 transfers through the whole spreadsheet.

The annual net cash flow is the sum of costs and revenues. The costs and revenues are as defined for Fig. 10.1 but are entered in this table as end of year figures, without any interest. All transactions are assumed to occur at the end of each year, and no interest is incurred during that year. The annual net cash flow is the sum of costs and revenues, and by applying the discount factor for that year, a

	A	B	C	D	E	F	G	H	I	J	K	L
1	Year	Cash Flow items	Drilling costs	Total costs	Revenues	Annual net	Cumulative	Discount	NPV of annual	Cumulative		
2						cash flow	cash flow	factor	net cash flow	NPV		
3	0											
4	1	exploration and planning		3.50	0	-3.50	-3.85	0.9091	-3.18	-3.18		
5	2	drill 3 exploration wells	15	15.00	0	-15.00	-18.85	0.8264	-12.40	-15.58		
6	3	drill 3 wells, preliminary design, permitting	15	17.00	0	-17.00	-35.85	0.7513	-12.77	-28.35		
7	4	drill 2 wells, pay 30% pipeline and power plant costs	10	28.60	0	-28.60	-64.45	0.6830	-19.53	-47.89		
8	5	pay balance of construction costs		43.40	0	-43.40	-107.85	0.6209	-26.95	-74.83		
9	6	first year only half capacity output		5.00	10.95	5.95	-101.90	0.5645	3.36	-71.47		
10	7			2.20	21.90	19.70	-82.19	0.5132	10.11	-61.36		
11	8			2.20	21.90	19.70	-62.49	0.4665	9.19	-52.17		
12	9			2.20	21.90	19.70	-42.79	0.4241	8.36	-43.81		
13	10			2.20	21.90	19.70	-23.08	0.3855	7.60	-36.22		
14	11			2.20	21.90	19.70	-3.38	0.3505	6.91	-29.31		
15	12			2.20	21.90	19.70	16.32	0.3186	6.28	-23.03		
16	13			2.20	21.90	19.70	36.03	0.2897	5.71	-17.33		
17	14			2.20	21.90	19.70	55.73	0.2633	5.19	-12.14		
18	15			2.20	21.90	19.70	75.43	0.2394	4.72	-7.42		
19	16			2.20	21.90	19.70	95.14	0.2176	4.29	-3.13		
20	17			2.20	21.90	19.70	114.84	0.1978	3.90	0.77		
21	18			2.20	21.90	19.70	134.54	0.1799	3.54	4.31		
22	19			2.20	21.90	19.70	154.25	0.1635	3.22	7.53		
23	20			2.20	21.90	19.70	173.95	0.1486	2.93	10.46		
24												
25												
26		**Data**										
27		Installed capacity (MWe)	20									
28		Full load net output (MWe)	18.8									
29		Annual load factor	0.95									
30		Operations and maintenance costs million$/yr	2.2									
31		Total annual output (kWh)	1.56E+08									
32		Electricity selling price c/kWh	14.0									
33		Cost of capital %	10									
34		Cost per well $million	5.0									
35												
36												
37												
38												
39												
40												

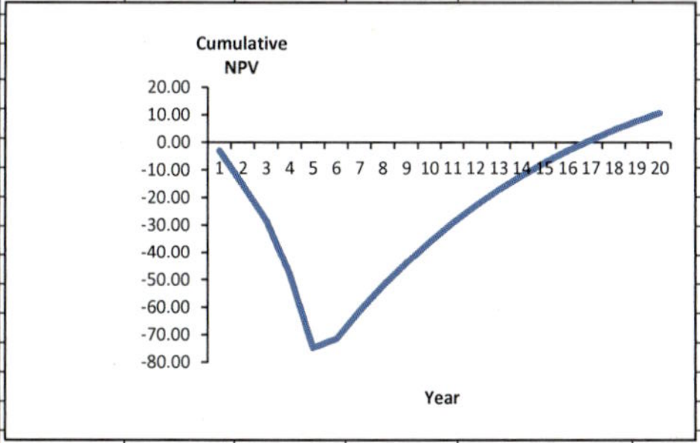

Fig. 10.2 The single project of Fig. 10.1 incorporating interest and discount factors

column of NPV values for every year is obtained—these are the values at the present time (year 0) of the net cash flow in any future year. These values are accumulated in column J, from which it can be seen that all costs have been recovered by year 20 with a cost of capital of 10 % and an electricity selling price of 12.5c. This would probably not be an attractive proposition, and a higher selling price and lower cost of capital would produce a higher cumulative NPV, which is a $ value representation of the project, one idea of the profit.

If the selling price of electricity is regulated, directly or indirectly, the cumulative NPV at the end of the project is not very helpful for choosing between alternatives—all NPV's are likely to be similar. Comparing NPV's would be helpful for the producer of unregulated price goods, because the option with the biggest profit would be most attractive. For the electricity industry, a method known as the "present worth of revenues" was introduced in the USA (Marsh [1980]); it represents a change in viewpoint rather than a significant change in calculation procedures. An acceptable return on investment is chosen and applied to all of the options being considered, and annual payment of this return becomes a cost to the project. Instead of the net annual income being the difference between revenues calculated for the expected electricity price and costs, the net annual income is decided upon first and added to the costs, which leaves a figure for the revenues to be collected from the sale of electricity. The option which has the lowest required revenues is the preferred one.

10.3.3 Payback Period

The simplest of the single parameters by which to judge the economic performance is payback period, this being the time after the start of expenditure at which the project has earned enough revenue to pay back the money invested—it was defined with reference to Fig. 10.1, although its calculation there did not involve discounting. For the conditions shown in Fig. 10.2, the payback period is 20 years.

In comparing options, the alternative with the shortest payback period is supposedly preferred. Making payback period the sole criterion would effectively make cancellation of the debt the absolute priority. There could be legitimate reasons for this, perhaps a need to have the capital available for another project as soon as possible or because the project is risky and the shortest possible exposure to risk is called for. The revenues earned by the project up to this time are included in the calculation, but later revenues are not, so using payback period as the single criterion makes no attempt to estimate the benefits of the project as a whole.

10.3.4 Internal Rate of Return

There is a discount rate at which the project of Fig. 10.2 only just breaks even if the life is 20 years and the cost of electricity is fixed at 14 c/kWh. This can be found by adjusting cell 31 and watching the final cumulative NPV, and it turns out to be about 12 %. This discount rate is called the internal rate of return (IRR) or the discounted cash flow rate of return. It has been calculated here from the net annual cash flow stream, but it can also be explained as the discount rate at which the present value of the cash flow stream of costs exactly equals the present value of the revenue stream over the project life. The order of the arithmetic operations, finding the net and then discounting, makes no difference. It marks the discount rate at which the project changes from one that makes an economic profit to one that makes an economic loss. The project might still be justifiable by providing non-economic benefits.

The IRR is informative but limited in the information it provides. The project life must be defined for this calculation, from which it follows that the IRR is an improvement over payback period as a representation of the project. If the cost of capital were to reach the IRR, the project would only just break even. The more the cost of capital falls below the IRR, the greater will be the NPV at the end of the project, so as great a margin as possible would normally be sought. However, as stated, the cost of electricity is usually regulated. For electricity generation, it might be said that an IRR value well above the cost of capital would provide, not a high profit but a high margin of security against the risk of unexpected problems and hence costs.

In summary then, the IRR is used for comparison with the cost of capital; its calculation does not consider net revenues so cannot arrive at a $ valuation of the project.

10.3.5 Levelised Cost of Electricity

It was demonstrated above that by adjusting the discount rate by trial and error, the discount rate at which the project would just break even could be found. In the same way, the cost of electricity could be varied to find the value of discount rate at which the project just breaks even. This value is called the levelised cost of electricity, and it is defined as the sum, over the project life, of discounted costs divided by discounted sales in kWh:

$$\text{Levelised cost of electricity} = \frac{\sum_{i=1}^{n} \frac{(\cos ts)_i}{(1+r)^i}}{\sum_{i=1}^{n} \frac{(sales)_i}{(1+r)^i}}$$

The levelised cost of electricity is commonly used to compare alternative methods of generation; like IRR, it provides some guidance as to possible profitability. The lower the levelised cost compared to the price at which the electricity can actually be sold, the greater the profit margin.

10.4 Factors Affecting Project Life

Project life is important in calculating project NPV and IRR. The life of a geothermal project depends primarily on the life of the geothermal resource, equivalent to the fuel for a fossil-fuelled station. Unlike fossil fuels, geothermal energy is not transportable over long distances, and the station must be built near to the resource. If the supply runs out altogether, there is usually no alternative but to sell the power station equipment. More likely, the supply of geothermal fluid or its temperature will decrease over the life of the project. The Wairakei station, NZ, was constructed with three stages of turbine, high-, intermediate- and low-pressure machines. Several years after commissioning, the delivery pressure of the steam from the wells decreased so the high-pressure turbines were removed and used at another resource.

Heavy mechanical engineering equipment is often regarded as having a 25-year life, although this is a rule of thumb rather than an accurate determination. The critical factors are wear, corrosion and mechanical stress (fatigue), and the level of detailed calculation of these factors that has gone into the design depends on the particular industry. In today's aircraft industry, weight is critically important from the point of view of performance, and if structural weight is to be minimised, the life of components at given stress levels has to be exceptionally predictable to ensure passenger safety. In comparison, the Wairakei power station turbines were commissioned in 1956 and are still operating in 2012 after 56 years, because the manufacturer had to allow a margin to ensure the performance and safety of his equipment, and it was not possible to calculate this margin accurately at that time.

10.5 Other Economic Aspects: Sensitivity Analysis and Risk

The suggested use of a spreadsheet to examine variations in the proposed project is in fact a sensitivity analysis, in that it can be made to show the effect on the cash flow curve of any changes from the planned project or variations in the cost of services and items. Sensitivity analyses also need to be carried out at the project selection stage when single parameter representations of the project are being considered. There are several modern methods of illustrating risk—and hence sensitivity—visually, such as "tornado diagrams" and "spider plots", but the basic aspects are as covered here.

As an example, consider the sensitivity of the cumulative NPV at year 20 to variations in the cost of drilling a well and the annual load factor of the completed station. The spreadsheet of Fig. 10.2 allows the cost per well to be called from a single cell (C34) whenever it is required to produce an annual cost. The cost has been varied in increments of 10 %, both positive and negative, applied to the datum case value of $5 million per well, and for each value, the NPV has been recorded. The results are shown in Fig. 10.3.

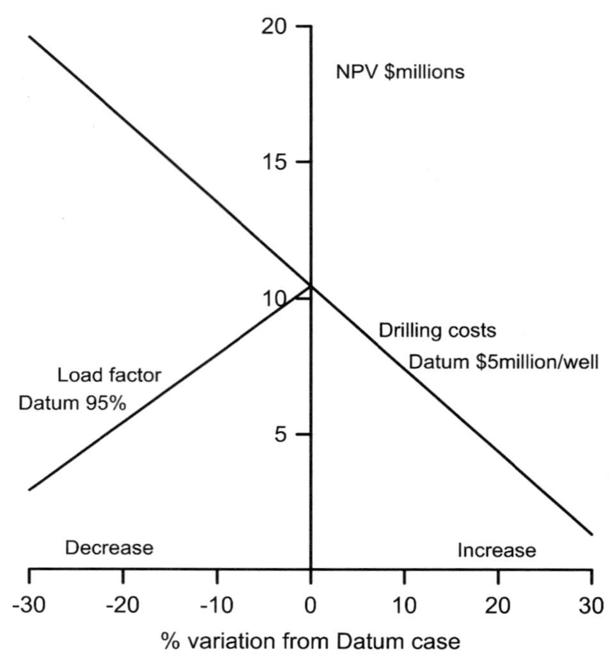

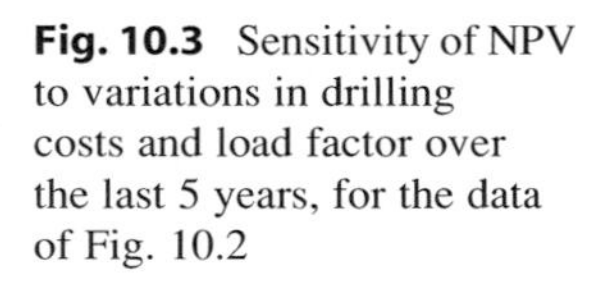

Fig. 10.3 Sensitivity of NPV to variations in drilling costs and load factor over the last 5 years, for the data of Fig. 10.2

The cumulative NPV has a value of $10.46 million after 20 years in the datum case of Fig. 10.2 (14.0 c/kWh selling price and 10 % cost of capital). It decreases as well cost increases and vice versa, and the variation in NPV is linear. The graph also shows a test of the sensitivity of NPV to load factor, for example, due to reduced resource output. This has been simulated by reducing the load factor over the last 5 years; Fig. 10.3 shows that a severe load factor reduction of 40 % in the 95 % value taken as the datum case over this relatively short period would reduce the NPV to almost zero, at which the project would only just break even—in purely financial terms, it removes the profit from the entire project.

Sensitivity analysis is discussed in more detail by Allen [1991]. More recently, Sanyal [2005] considered the sensitivity of geothermal electricity generation costs to various factors and also addressed the minimisation of levelised generation costs using enhanced geothermal systems (Sanyal [2010]).

10.6 Consideration of a Geothermal Station in a Network

A network is the group of interconnected power stations available to an organisation to supply electricity. The demand for electricity, referred to as the electrical load, is likely to follow a regular pattern which must be determined, particularly in bringing electricity into a new rural area. In Western countries, the domestic electrical load typically increases very rapidly from a low level overnight to a morning peak as users begin their day. The load decreases after breakfast, increases to a high level for a short period at lunchtime then decreases in the afternoon, peaks at evening mealtime and returns to its low level overnight. Industrial loads will be virtually continuous for some activities such as mining and ore processing and variable for commerce and light industry, peaking between breakfast and lunch and then again in the afternoon until work and business ceases for the day. The precise load at any time is unpredictable, being subject to random influences. The single electricity generating company considered here might have several power stations, some fossil fuelled (coal, gas or diesel), one geothermal and one hydro-station. It is not possible to store electricity per se, so the total output of the stations in operation must be controlled to exactly match the total load. Electricity can be converted to some other form of energy that can be stored—the spinning turbines of all the power stations in a network provide some storage like this, and flywheels have been proposed but are not used. Pumped storage is an option, in which excess electricity drives pumps to move water from a low-level lake to a high-level lake so that it can be reconverted to electricity by running the water back down again through a water turbine. From a purely engineering point of view, this is attractive because the pumping of water and generation of electricity are both mechanically efficient processes, so not much energy is lost. Few have been built presumably because their capital cost is unjustifiable.

The total load may be unsteady, but a large proportion of it may be constant. A load–duration curve is produced for the system, which shows how many MWe of the total load is always present, and this is referred to as the base load. The available power stations are ranked according to the cost of electricity produced and their ability to respond quickly to load changes, so each is appropriate to some part of the load–duration curve.

10.6.1 Marginal Cost and Load Following

The faster a car goes, the greater the fuel consumption per kilometre travelled; in economic terms, the speed increase has an associated cost increase which is called a marginal cost increase. When several stations are available to supply an increase in load, the one with the least marginal cost should be used first. This is not purely an economic factor, however, but depends on the rate at which any of the available stations can increase its load—axial steam turbines are limited in the rate of rise of temperature that can be applied without causing mechanical problems due to thermal expansion.

10.6.2 Base Load Service

In general, power stations are an integral part of an electrical transmission network distributing electricity to various loads. Such a system may be owned and operated by the government of the country or by commercial companies.

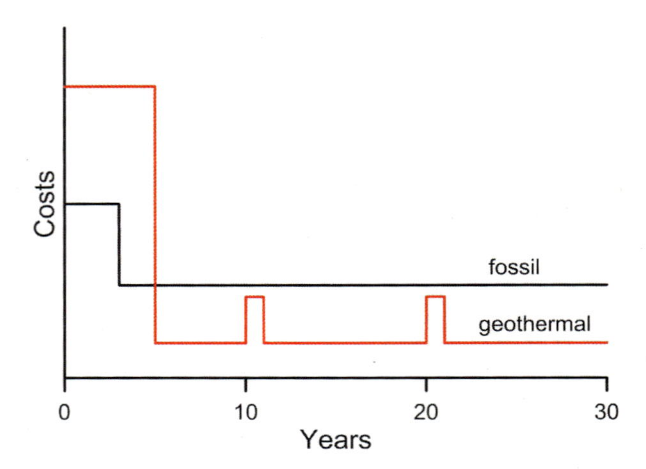

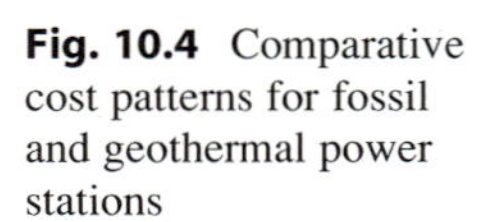
Fig. 10.4 Comparative cost patterns for fossil and geothermal power stations

Figure 10.4 shows the difference in the pattern of capital and operating costs between a fossil-fuelled station and a geothermal station.

The fossil-fuelled station was built over a period of 3 years and requires a supply of fuel throughout its life, which has been assumed to be 25 years. There is a significant cash flow throughout the station life due to the purchase of fuel. The geothermal station was built at a higher cost over a period of 6 years; this includes the exploration and drilling to confirm that the geothermal resource can supply enough geothermal fluid to power the station fully over the 25-year period. The geothermal station is "capital intensive" in that the bulk of the 25-year costs is invested before the station begins its output. The cost has been incurred whether the station is used or not, whereas the fossil-fuelled station avoids the fuel cost if it is not used. Geothermal stations should be operated continuously, to provide for that part of the load that is always present, which is referred to as base load. Geothermal stations have this capital intensiveness in common with nuclear power stations, which are also used to meet the base load, and for different reasons, neither type is capable of the fastest load following performance, which further favours their base load use. Hydro-stations are also capital intensive and, in principle, have zero fuel cost, like geothermal, so would-be candidates for meeting base load. However they are capable of faster load following, and their fuel supply is seasonal and unpredictable from year to year so must be conserved; both of these factors count against their automatic use to meet base load.

10.7 Energy Analysis

In New Zealand at the present time, it is possible to buy small cast iron water pipe fittings at a cost which would not pay for the electricity to melt the metal they are made of, let alone its machining and transportation to market. Viewed in isolation, the selling price is uneconomic. The reason may be that having set up the foundry and embarked on castings requiring a large amount of metal to be processed, the marginal cost of producing small items is low enough to make it profitable. Large projects require the same scrutiny as to value for money, but an assessment is not possible intuitively. This issue arose in the 1970s in connection with the UK nuclear electricity generating programme, when it was raised by Chapman [1974]. At the time, oil supplies were restricted worldwide, and it was proposed to build nuclear stations to provide alternative sources of electricity. This required the construction of a series of nuclear reactors fuelled by refined uranium and moderated by heavy water (D_2O), both the product of energy intensive processing and thus requiring a supply of electricity from the existing fossil-fuelled power stations. Chapman's message (see also the response by Wright and Syrett [1975]) was that although one of these reactors could produce more energy than was used in its construction, building many of them at a high rate did not produce the required result of decreasing the national dependence on fossil fuels. The programme, it was argued, absorbed all of the generated output and more before it was completed, which would have been many years into the future.

This issue is better understood today, for example, it has recently been examined by Frick et al. [2010] in an analysis which they refer to as "life cycle assessment". Their reasoning parallels that of Chapman, namely, that political efforts to reduce greenhouse gas emissions and the consumption of finite energy resources have led to proposals to extract low-temperature geothermal water for electricity generation via binary power plants. The conversion efficiency is low, governed by the second law of thermodynamics, while high-grade (high-temperature) energy is required for well drilling, pipelines and manufacture of the power plant. The auxiliary power consumption of binary cycle plant is relatively high, and low-temperature geothermal fluid sources require fluid to be produced by pumping. All of these factors extend the project payback period. Inevitably, since the problem is so complex by involving, in principle, every energy consuming activity which makes a contribution to the power station construction, the analysis by Frick et al. draws only general conclusions. The value of their contribution lies in continuing to draw attention to the issue and demonstrating how to examine it.

10.8 The Steam Sales Contract

There are a few geothermal power projects in which the wells and pipelines are owned by one party and the power station and transmission lines by another. This arrangement splits an otherwise seamless engineering undertaking into two separate parts which are required to cooperate very closely. The split is not physical but organisational – Fig. 10.5. Each party has its own staff and management structure, and their interaction is defined in the Steam Sales Contract. Several physical points are defined at which the responsibility changes from one party to the other, for example, a point on the main steam pipeline and a point on the pipeline returning condensed steam to the field for disposal. It is convenient to consider that the power station has a "fence" around it, in terms of the boundary of responsibilities.

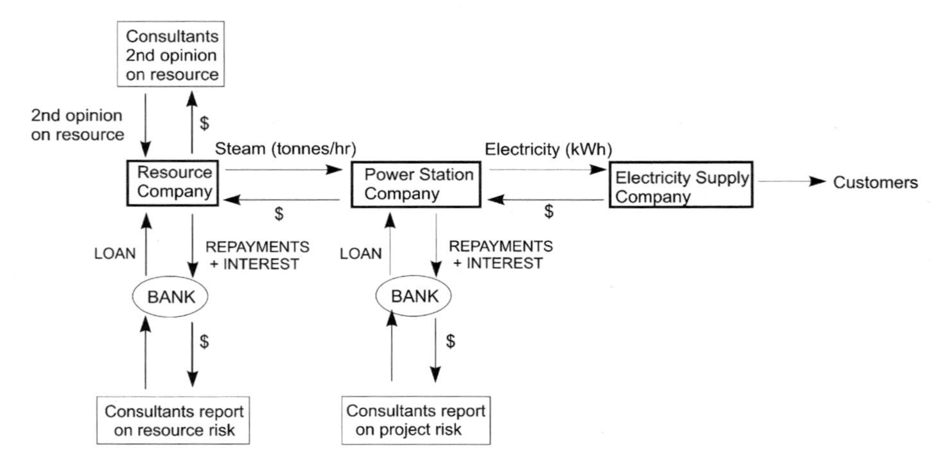

Fig. 10.5 Organisation chart for parties served by a Steam Sales contract

Referring to the parties as the steam supplier (Resource Company) and the station operator (Power Station Company), the purpose of the contract is to arrange for a reliable supply of steam sufficient to run the power station at full capacity. Assuming base load operation is intended, the contract will state the intended output in MWe and the steam conditions and flow rate required to produce this. The station operator pays the steam supplier for the steam actually supplied at a defined quality, so measurement equipment must be installed at an agreed point near the (hypothetical) fence. The measured parameters are likely to be flow rate, pressure and dryness fraction, assuming the steam to be saturated, and the flow rate and chemical species of any liquid carried with it. The steam will include gas, which generates power as it passes through the turbine but requires power to remove it from the condenser. If the gas flow rate turns out to be variable in the long term, then the contract may have to be written to account for it. The measurements and all related data, instrument specifications, etc. must be available to both parties, and the water properties (steam tables) to be used for calculations related to the contract should be specified together with the calculation procedure. The instrumentation must be maintained to keep the measurements within the manufacturer's tolerance. The power station output will need to be measured at a defined point because the voltage is usually stepped up in stages downstream of the alternator, and each transformer introduces losses; the input terminals of the first transformer after the alternator might be chosen. Since the purpose of the measurements is to define how much the station operator pays to the steam supplier, a target amount per month or quarter is likely to be specified, with variations in payment if the actual amount supplied differs from the target. Both parties require an incentive to keep their performance to the contracted expectations, and any shortfall should therefore result in a penalty. Output in excess of the target might be attractive to the station operator, depending on arrangements for the sale of the electricity, in which case the payment formulae will include this possibility.

The arrangements at the end of the contract period must be defined. The contract may be renewable, but if not, then ownership of the assets constructed by the Resource Company must be addressed in the contract.

There are secondary matters to be contracted for in addition to the supply of steam. Electricity is required to operate the wells, pipelines, instrumentation, water supply and disposal pumps and staff facilities, and since geothermal stations are often at remote locations, the power station usually provides for this load. The load must be defined in the contract since the value of the electricity supplied must be accounted for in the transaction.

Depending on the location of the fence and the arrangements for separating water from the steam supply, separated water might have to be pumped out to the field from the station. Steam condensate will certainly have to be disposed off from within the fence.

Increases in the cost of drilling, casing and labour, for example, required for the execution of the contract, can be accommodated for by reference to national economic indices, to avoid the contractor building in contingency arbitrarily.

The details of the contract just described are relatively straightforward to define for an existing reservoir development and power station, where the resource is understood and the plant already in existence. Sometimes a steam sales contract is foreseen from the outset, perhaps due to an inability of one of the parties to raise enough capital for the entire project. Design aspects of the facilities will then need to be discussed. For example, steam jet ejectors use steam but have a lower capital cost than mechanical gas extractors, which consume electricity. There is an incentive for the Power Station Company to reduce its capital outlay and its parasitic electrical load by using ejectors, but this would increase the steam demand, to the detriment of the Resource Company. The parties involved would reach agreement on such issues and on the form of the contract before any related expenditure is committed to. The length of time for capital recovery is long, and each party needs to be certain that the other can perform as required.

Few steam sales contracts are published, but Puente and Andaluz [2001] reviewed the performance of a steam sales contract at the Cerro Prieto resource between the Mexican public electricity company, CFE, and a private contractor. The contract was for a period of 10 years beginning in 1991 and was for the supply of 800 tonnes/h of steam; it appears that CFE already had its own wells and steam supply system for the power station. It was awarded following a call for tenders and functioned well for several years until Mexico suffered an economic crisis as a result of which the contractor's supply fell below the contracted rate, incurring financial penalties which made its position worse. CFE's own steam supply declined due to lack of investment, causing CFE and the contractor to reach agreement with a second (US) contractor in the form of a steam supply contract for 1,600 tonnes/h, twice the original contracted flow rate. At the time of writing of the paper, the contract term had expired, and a new contract was being sought.

References

Allen D (1991) Economic evaluation of projects. Institution of Chemical Engineers, UK
Chapman P (1974) The ins and outs of nuclear power. New Scientist, 19 December 1974
Frick S, Kaltschmitt M, Schroder G (2010) Life cycle assessment of geothermal binary power plants using enhanced low-temperature reservoirs. Energy 35:2281–2294
Leung P, Durning RF (1978) Power system economics: on selection of engineering alternatives. J Eng Power 100:333–343
Marsh WD (1980) Economics of electric utility power generation. Oxford Engineering Science Series, Oxford
New Zealand Geothermal Association (2009) http://www.nzgeothermal.org.nz
Ogryzlo CT, Randle JB (2005) Financing the San Jacinto-Tizate geothermal project, Nicaragua. World Geothermal Congress
Puente HG, Andaluz, JI (2001) Steam purchase contract: a singular Mexican experience Geoth Resources Council Trans 25:33–36
Sanyal SK (2005) The cost of geothermal power and factors that affect it. Keynote Lecture, World Geothermal Congress
Sanyal SK (2010) On minimizing the levelised cost of electric power from enhanced geothermal systems. In: Proceedings of World Geothermal Congress
Turvey R (1968) Optimal pricing and investment in electricity supply. Allen and Unwin, Australia
Turvey R, Anderson D (1977) Electricity economics: essays and case studies. Johns Hopkins University Press, California, US, for World Bank
Wright J, Syrett J (1975) Energy analysis of nuclear power. New Scientist, 9 January 1975

The Power Station 11

The prime mover for bulk electricity generation worldwide is the turbine, coupled to a condenser and alternator. A working fluid is taken through a cycle of changes of temperature and pressure and through the prime mover to produce torque, which involves heat supply and rejection to the surrounding environment. The dominant choice of working fluid remains as water, and the steam cycle has undergone over a century of development. Geothermal steam is not supplied as part of a closed cycle, but turbines for conventional (fossil-fuelled) power stations have required only relatively minor modifications in design to make use of it. However, the low temperature of most geothermal heat supplies compared to fossil-fuelled boilers has led to the choice of organic fluids as the working fluid, for which the design of the power station equipment, including the turbine and condenser, must be altered quite significantly.

This chapter discusses historical trends in steam cycle power stations and then explains the development of the steam Rankine cycle and the modifications introduced to make it approach the Carnot cycle more closely. As well as providing the background for geothermal steam plant, it provides a starting point for considering designs for alternative working fluids. Details of turbines, condensers and ancillary equipment for geothermal steam power stations are introduced and organic Rankine cycle and trilateral flash cycle plant are discussed.

11.1 Introduction

Electricity is generated by the rotation of a coil in a magnetic field and the power station houses the machinery to do this. Since about 1900 most power stations have used steam turbines as the drivers (prime movers) and fossil fuels as the heat source. The use of geothermal steam to drive a turbine is a departure from fossil-fuelled practice in that the steam enters the station from the wells and leaves it as condensate—the turbine is a machine placed in a once-through steam flow. In fossil-fuelled stations a boiler generates the steam which passes through the turbine into a condenser and back to the boiler. This is convenient because the water can be

A. Watson, *Geothermal Engineering: Fundamentals and Applications*,
DOI 10.1007/978-1-4614-8569-8_11, © Springer Science+Business Media New York 2013

maintained at a high level of purity, reducing corrosion so as to allow metals to be used at high temperature over a long working life. But it is not merely a convenience; it complies with the concept of a heat engine as originated by Carnot. A heat engine is defined as a machine which takes a fixed mass of fluid—the working fluid—through a cycle of processes during which heat is transferred to and from the fluid and mechanical work is produced. Thus a steam turbine using geothermal steam is not strictly a heat engine, as it does not operate continuously on the same mass of steam. This is a theoretical rather than a practical matter, and geothermal steam turbines are virtually identical to those used in fossil-fuelled stations, except for design differences necessitated by the relatively low steam temperature and pressure used compared with modern fossil-fuelled plant. The turbine is essentially an item of machinery within which the momentum of the steam flow is transferred to the rotating shaft; it is not integral to the steam cycle and in fact very little of turbine blade design involves thermodynamics. The same issue regarding the cycle arises in internal combustion engines, which continuously refresh the working fluid, yet their performance is calculated as if they were true heat engines.

Despite this, there is merit in studying the Rankine cycle as applied to fossil-fuelled generation for consideration of geothermal turbines. It is the main cycle for steam plant using water as working fluid and has direct application to geothermal turbines using working fluids other than water—organic fluids—which must be conserved for cost and environmental reasons, so are used in a true heat engine cycle. The Carnot cycle is the ideal, regardless of working fluid, and the Rankine cycle departs from it but is the closest practical alternative. Various modifications to the cycle have been adopted over the last century to bring it closer to the ideal, and these have led to modifications to plant design. The fossil-fuelled steam Rankine cycle is a datum, and once understood, the variations introduced for geothermal application follow very easily. It will be found that the design of the power plant involves a good deal of optimisation. In common with most heavy engineering equipment, the actual design details are commercially sensitive, and a purchaser must understand the general operation of the equipment at a fundamental level (or engage a consultant) for a critical examination of it.

It is not proposed to deal with electricity per se in this book. As noted, Faraday established the principle of generation in 1831, but a generator was not built until the 1870s; a brief history is given by van Riemsdijk and Brown [1980]. For large-scale generation, three-phase alternating current has been usual since the early days, and the generator is called an alternator. The voltage frequency is either 50 or 60 Hz, produced by rotating the alternator at 3,000 or 3,600 rpm, respectively. The alternator is driven by a prime mover, a steam turbine for the largest electrical output, and the turbine is usually designed to run at the same speed as the alternator. Turbine and alternator are usually coupled directly, with their axes in line, the pair being known as a turbo-generator or more specifically a turbo-alternator. Turbines are designed to rotate at "synchronous speed", i.e. to give 50 or 60 Hz electrical output, and precise speed control is essential. Gear boxes to alter speeds are sometimes used but only for smaller power outputs, since gearboxes are expensive and involve greater wear and tear than turbines or alternators; turbines using organic working fluids do not suit such high-speed rotation as steam turbines and may use gearboxes. A transmission grid is required to distribute the electricity produced, usually operating at 440 kV or more, much higher than the generation voltage of typically 11 kV, and turbo-alternators feed into transformers and a switchyard for connection to the grid (see Fig. 1.2). There are a few electrical equipment design issues specific to geothermal power stations. One of these is the abnormally high concentration of H_2S in the atmosphere around geothermal power stations, due partly to the natural surface activity to be found in many geothermal resource areas but partly also to the discharge to atmosphere of the non-condensable gas extracted from the condenser. The gas causes corrosion of the copper used in exposed switches and related conductors. Another issue is the earthing mat, an extensive mesh of copper conductors buried beneath the power station foundations to provide a good earth connection as a datum voltage level. Geothermally altered soils provide a very acidic, corrosive environment for buried copper conductors.

11.2 Historical Trends in Thermal Power Stations

The dominant feature in distributed electricity supply of all types, not just geothermal, over the 120 years of its existence has been the growth in consumption and hence installed capacity, and this is expected to continue. It has been estimated by the IAEA [2012] that world electricity supply will double in the period 2000–2030, requiring an investment equivalent to US$550 billion per annum for new and replacement plant. Approximately half of the investment goes to generation and half to distribution equipment. Steam has been and will continue to be the dominant working fluid used to drive turbo-alternators, typical single machine outputs of which have increased from 1 MWe in 1900 to 25 MWe in 1912, 128 MWe in 1928, 600 MWe in 1960, and so on, with the largest single unit being in the order of 1,000 MWe at the present time.

Efficiency has been an important consideration throughout. In fact it has been an important consideration since the first days of steam engines—those used for draining British coal mines burned so much coal that where possible the water was discharged upstream of a watermill to recover some of the value. Improvements in efficiency over the period to about 1970 were made as the result of continuous increase in steam temperatures and pressures; by the 1960s steam temperatures had reached 550 °C or thereabouts, a level not much exceeded because of blade material problems (creep). Geothermal steam conditions were far below those of contemporary fossil-fuelled stations when geothermal resources came into use in the period 1950–1960 and represented a step backwards in the efficiency of conversion of heat to electricity; the same was true of nuclear-generated steam at that time.

Pressures in modern fossil-fuelled steam plant are now often supercritical. The "oil shock" of the 1970s stimulated further research and development; there was concern about the supply of petroleum fuels worldwide and a consequent increase in their price. Combined cycle plant became common, gas turbines with their exhaust providing the heating to boilers now termed steam generators, with thermal efficiencies well above 40 %. Attention was also paid to working fluids other than water in the search for ways to make better use of lower temperature heat sources. Alternative working fluids had been considered much earlier, mercury in 1923 and diphenyl-oxide in 1934. Both of these have a very high critical temperature and moderate critical pressure, advantageous for use in the high-temperature part of the cycle, whereas organic fluids were aimed at application in the low-temperature part of the cycle, as waste heat recovery units. Small unit sizes of the latter were developed, for example, by Ormat Ltd, which could stand alone to recover heat from suitable sources, and the general type became known as organic Rankine cycle (ORC) plants. This began a trend to smaller unit sizes, which suited application to geothermal energy, and a departure from the practice of housing all generating plant in a building. Free-standing units exposed to the weather became possible, and the task of bringing into use resources in remote locations was made easier and more economic. The ultimate capacity of a newly explored geothermal resource and the capital cost of the exploration make it appropriate to build power stations in stages, which the availability of smaller units also helps.

In the present era, global warming and general concern for the environment are additional driving forces for the development of geothermal generation and use of low-temperature fluids, and a second round of the search for alternative working fluids is evident.

11.3 Generation of Electricity Using Steam as the Working Fluid in a Cycle

11.3.1 The Rankine Cycle as a Representation of the Carnot Cycle

The Carnot cycle was introduced in Chap. 3 as a means of defining entropy; a fixed mass of fluid was taken through four processes which changed its thermodynamic state but brought it back to the starting point. The fluid did not flow; it had no velocity at any stage but merely changed its thermodynamic state as it was presented to four different thermal boundary conditions in turn. Now the focus is on the steady flow of H_2O around the circuit shown in Fig. 11.1, a boiler supplying steam to a turbine and condenser, with the liquid water pumped back to the boiler.

A large mass of fluid flows continuously around a circuit, passing through items of equipment that change its thermodynamic state. Heat is added to the loop and work is extracted but all rates of change are zero—any parameter measured at any fixed point in the loop is invariant with time. In thermodynamic terms this is described as a closed cycle plant. The fluid now has velocity, and every particle of it passes through a cycle just as did the fixed mass in Chap. 3.

The Carnot cycle is the ideal and can only be approximated in practice. Figure 11.2 shows both the Carnot cycle, GCDFG, and the Rankine cycle,

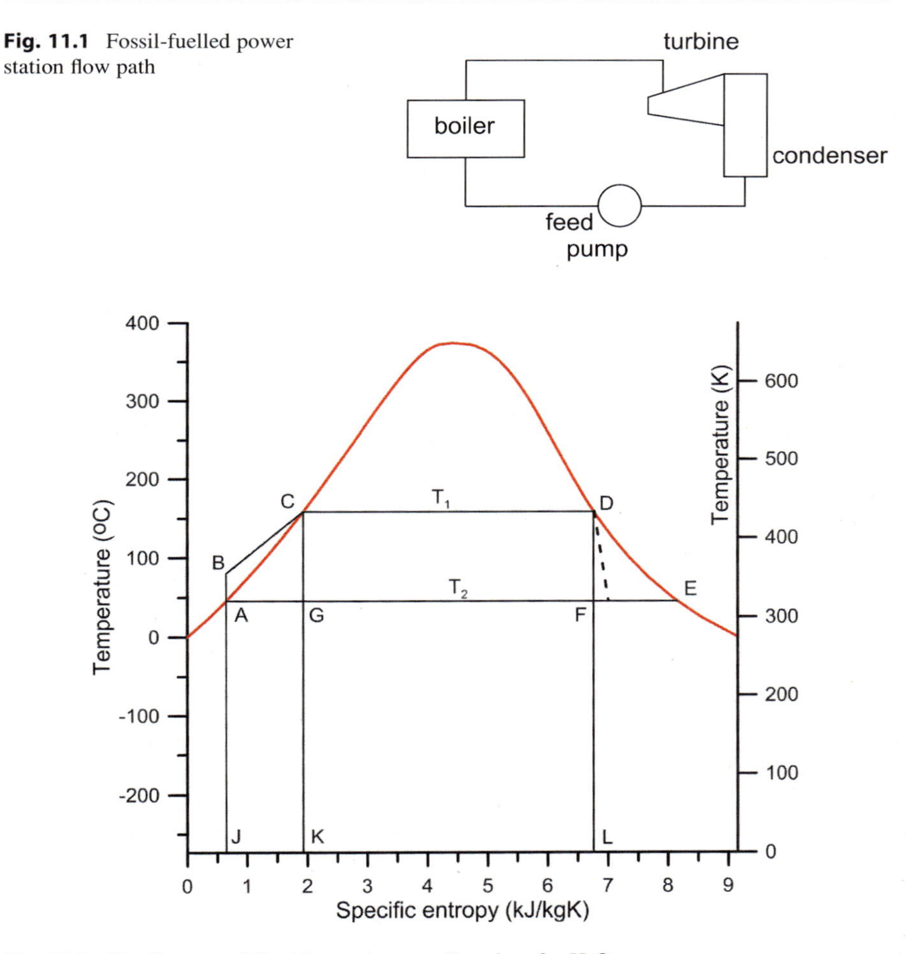

Fig. 11.1 Fossil-fuelled power station flow path

Fig. 11.2 The Carnot and Rankine cycles on a T–s chart for H_2O

ABCDFA, on a T–s diagram for water. The envelope of saturation conditions can be plotted from the steam tables; for other fluids it has a different shape which has significant implications to be discussed later. It will be recalled that the horizontal lines across the envelope represent the path taken by fluid changing from saturated water at C to saturated steam at D, and in reverse from E to A. The fluid is two-phase everywhere inside the envelope. The Carnot cycle has two isothermals, CD at temperature T_1 and FG at T_2, and two isentropic stages, GC and DF. It was not conceived with water and steam in mind as the working fluid, but with a single-phase gas. An isothermal supply and rejection of heat to/from the working fluid are easily achieved because condensation and evaporation are isothermal processes. However, halting the condensation from F exactly at G would be difficult to arrange economically, even today, and the isentropic expansion and compression processes are theoretical ideals. Rankine presented an achievable cycle in 1854 as the best

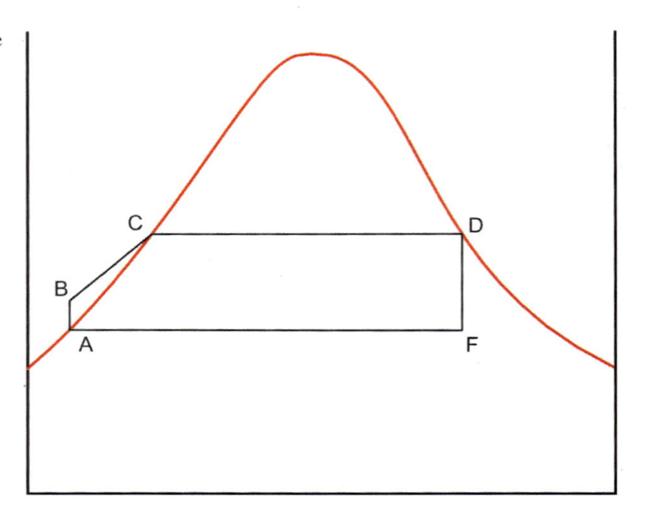

Fig. 11.3 The Rankine cycle on a T–s chart for H_2O

that can be done to approximate the Carnot cycle, and it is named after him. It applies to both reciprocating engines and turbines, but a turbine will be assumed here (Fig. 11.3).

Choosing point A as the beginning of the Rankine cycle, where the working fluid is saturated liquid at the lower temperature, the liquid pressure is raised until it reaches the boiler pressure at B, after which it is heated in the boiler at constant pressure, following the line BC. It begins boiling at C and at low to moderate upper temperatures, T_1, the evaporative heat addition over CD is much greater than the heat addition over BC, so the cycle approximates the Carnot cycle by receiving most of its heat at constant temperature. Work is done on the liquid over path AB, but since water is very incompressible, this is small compared to the work extraction when the saturated steam at D is expanded through the turbine, path DF—steam is very compressible. The expansion DF is not truly isentropic due to friction in the turbine flow passages—the real path is shown dotted in Fig. 11.2; the friction converts mechanical work into an extra source of heat, thus departing from Carnot's ideal circumstances. At the end of expansion, F, the fluid enters the condenser where it is completely condensed to saturated liquid at A. Heat is removed isothermally over path FA, and the process departs from the Carnot cycle only in not stopping at G.

The parameters at various points around the cycle can be calculated in an "informal" way because the mass flow rate is constant throughout—only the steady flow energy equation in its simplest form need be used, together with the definitions of specific entropy, specific enthalpy and dryness fraction. By informal is meant that though the equations of continuity of mass and energy are being solved at each point, it is not necessary to write both down formally. For the Carnot cycle, suppose the upper temperature in Fig. 11.2 is the saturation temperature for a pressure of 6 bars abs, which is 158.8 °C, and the condenser operates at 0.1 bar abs for

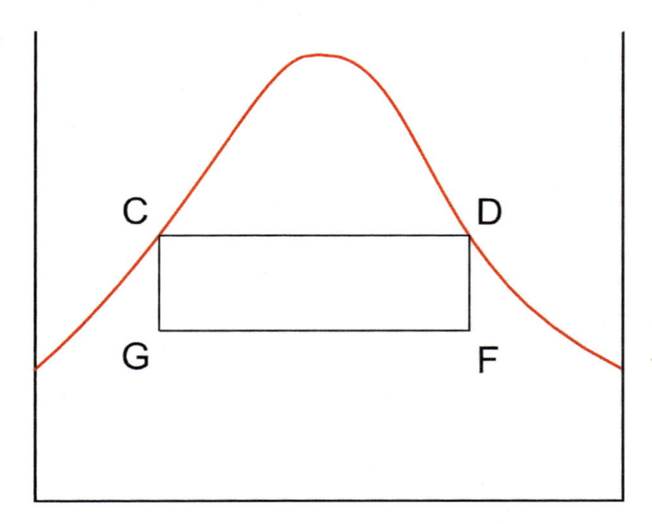

Fig. 11.4 The Carnot cycle on a T–s chart for H_2O

which the saturation temperature is 45.8 °C. The properties required for the calculation are as follows:

Ps (bar abs)	T_s (°C)	S_f	S_{fg}	S_g	h_f	h_{fg}	h_g
6.0	158.83	1.931	4.828	6.759	670.5	2085.6	2756.1
0.1	45.81	0.649	7.500	8.149	191.8	2392.1	2583.9

The units of specific entropy s are kJ/kgK and of specific enthalpy are kJ/kg. With Fig. 11.4 as reference, first find X_G, the dryness fraction at G:

$$1.931 = 0.649 + X_G^* 7.500 \text{kJ/kgK}$$

giving $X_G = 0.1709$ and thus $h_G = 191.8 + 0.1709 * 2392.1 = 600.69$ kJ/kg

Now find X_F:

$$6.759 = 0.649 + X_F^* 7.500 \text{ kJ/kgK}$$

giving $X_F = 0.8147$ and thus $h_F = 191.8 + 0.8147 * 2392.1$ kJ/kg $= 2140.56$kJ/kg

Using the steady flow energy equation, Eq. (3.6)

$$Q - W = (h_2 - h_1) + \frac{1}{2}\left(u_2^2 - u_1^2\right) + g(z_2 - z_1) \tag{3.6}$$

neglecting the kinetic and potential energy terms and assuming no heat loss, the work output from the turbine and work input in raising the pressure back from condenser pressure to 6 bars abs are, respectively,

$$h_D - h_F = 2756.1 - 2140.56 = 615.54 \text{ kJ/kg}$$

$$\text{and } h_C - h_G = 670.5 - 600.69 = 69.81 \text{ kJ/kg}$$

Thus the net work output from the cycle is 615.54–69.81 = 545.73 kJ/kg.

Alternatively, the net work output can be calculated as the difference between the heat added and the heat rejected, which on Fig. 11.2 is area KCDL—area KGFL which is

$$(158.83 - 45.81)^* s_{fg6\ bar\ abs} = (158.83 - 45.81)^* 4.828 = 545.66 \text{ kJ/kg}$$

These two differ only by rounding errors.

The efficiency of the Carnot cycle, η_C, is the net work output divided by the heat supplied, which is

$$\begin{aligned} \eta_C &= \text{net work output}/T_{S6\ bar\ abs}{}^* s_{fg6\ bar\ abs} \\ &= 545.73/\big((158.83 + 273.15)^* 4.828 \\ &= 0.2617 \text{ or } 26.17\%. \end{aligned}$$

Alternatively by definition

$$\eta_C = (T_1 - T_2)/T_1 = (158.83 - 45.81)/(158.83 + 273.15) = 0.2616$$

Two arithmetic points should be made here, the dryness fraction should be calculated to four significant figures for steam table work, and the arithmetic was simplified by calculating the temperature difference in °C rather than convert both to degrees K, a dangerous shortcut practice best avoided.

The calculation of efficiency amounts to calculating the ratio of heat supplied to heat rejected in the condenser and is the same for any cycle. It is very simple for the Carnot cycle but less so for the Rankine. The heat rejected is the area of the rectangle JAFL and presents no problem but the heat supplied is the area JABCDL. If the feed pump work is neglected, and the route is assumed to be JACDL, this involves finding the area between the envelope and the temperature axis such as the area shown shaded in Fig. 11.5. This is the Gibbs function for saturated water, which used to be listed in early steam tables for the benefit of steam plant designers.

Since the focus here is on geothermal power plant, full Rankine cycle calculations are not needed, but there are three cycle modifications of significance. The first is the use of feed heating to bring the cycle shape closer to the Carnot cycle; from Fig. 11.2 it can be seen that the main departure from it is the result of being unable to stop the condensation process at G. The significance of a departure can be judged by the change in area from that of the Carnot cycle which it causes, and the change here is JBCK, which is certainly significant compared to KCDL. It represents additional heat discharged to the condenser, which must be made physically larger and more expensive as a result. This departure far outweighs the area modification due to the departure from isentropic of the expansion in

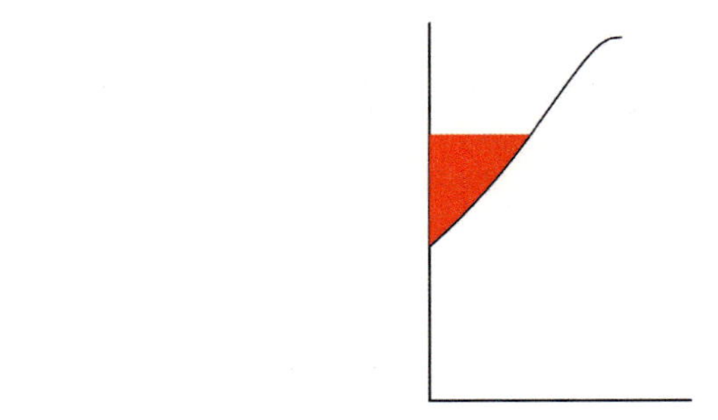

Fig. 11.5 The *shaded area* is the Gibbs function for saturated water

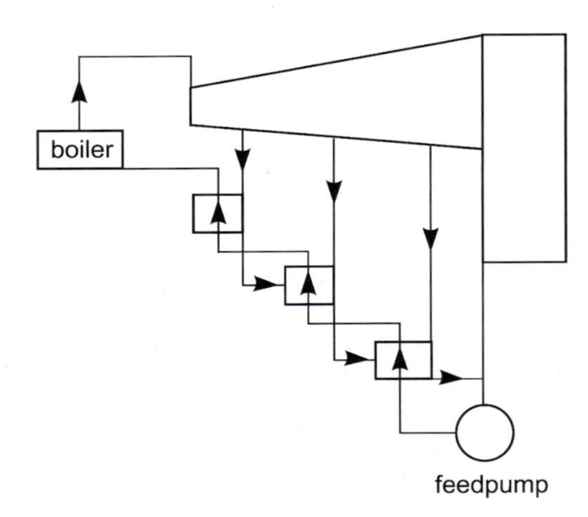

Fig. 11.6 Three stages of feed heating

the turbine. The heating of the condensate takes place at constant pressure, and it was realised early in the twentieth century that taking a small amount of steam from the turbine at various stages through the expansion would provide heat which could be transferred to the condensate, thus bypassing the condenser. Imagine that steam at a temperature halfway between T_1 and T_2 in the expansion on Fig. 11.2 was transferred to the condensate at almost the same temperature. This is possible because the condensation of steam provides heat transfer with only a small temperature difference. The result is a loss of power in the turbine, but a smaller rejection of heat in the condenser, and a consequent saving in condenser cost. This practice is known as regenerative feed heating, and in large fossil-fuelled plant, it is carried out by extracting steam at typically three pressures (temperatures) along the expansion. As shown in Fig. 11.6, the condensate of the highest temperature steam extraction after it has given up its h_{fg} in the first heater is flashed into the second to add to the next steam extraction, and so on. The optimum number of stages is the result of an engineering–economic analysis. When the idea of feed heating was first thought of,

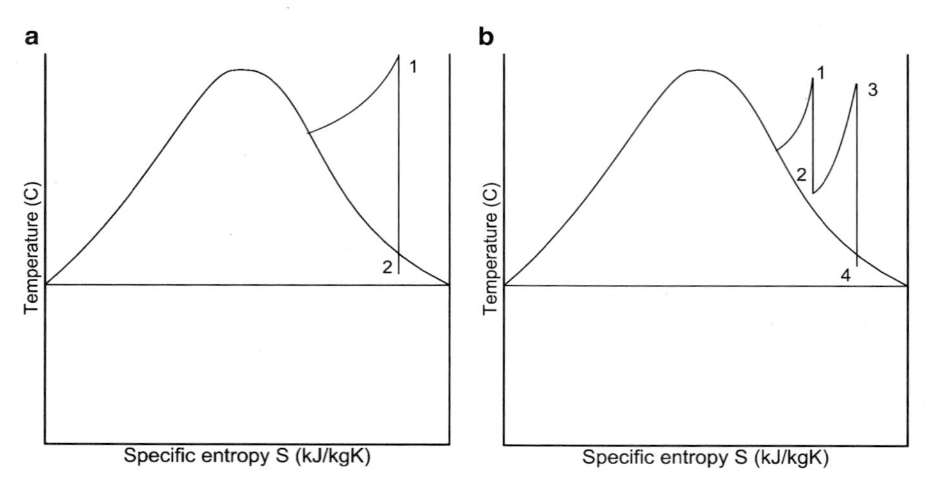

Fig. 11.7 (**a**) Superheating, with expansion from 1 to 2 and (**b**) reheating, with expansion from 1 to 2, then reheating to 3 for expansion to 4

low-temperature boiler exhaust was used to provide the heat, but economic optimisation led to the use of extracted (bled) steam.

The second and third departures from the Rankine cycle involve superheating the steam; what has been referred to as the Rankine cycle so far might be more accurately called the saturated steam Rankine cycle. The saturation envelope has a temperature maximum at the critical point, and the heat added during the evaporation part of the saturated steam cycle becomes a smaller proportion of the total heat added in the cycle as the top temperature T_l is increased, that is, with increase in boiler pressure. The departure from the Carnot cycle is then greater. Examination of the steam tables reveals that h_{fg} decreases with increasing saturation pressure. The bottom temperature is nominally fixed by atmospheric conditions which represent the lowest practicable heat rejection temperature of perhaps 25–30 °C. Increasingly higher top temperatures were sought to increase the thermal efficiency; however, saturated steam expanded from high pressure becomes very wet in the turbine before it has reached the condenser; water droplets damage the turbine blades by eroding them if the wetness is greater than about 10 %. Figure 11.7a shows one remedy, referred to as superheating, in which the saturated steam emerging from the evaporation part of the boiler is superheated to point 1 by being exposed to higher temperatures in the boiler. This point is at higher entropy than for the original saturated steam and the shape of the saturation envelope is such that moving point 1 to the right makes the steam less wet after it expanded to condenser temperature at point 2.

Figure 11.7b shows the cycle modification referred to as reheat. The steam supply to the turbine is superheated to point 1 and expanded to a temperature in the region of the saturation envelope, point 2, at which it is extracted from the turbine and passed back through the boiler to be reheated at a lower pressure to point 3 and expanded to condenser conditions at point 4. In practice the turbine can be made up of two separate machines, coupled on the same axis with the alternator, one passing only the high-pressure steam and the other the reheated steam. Optimising the maximum superheat temperature and the end of the first expansion is again a complex engineering—economic problem—and in the context of this book is a problem that would need to be re-examined for every different working fluid.

11.3.2 Steam Turbines in Practice

High-speed reciprocating steam engines were developed in the early part of the twentieth century but were outstripped by turbines. Apart from a device ascribed to Hero of Alexandria in about 50 A.D. which is often used as a way of explaining how a reaction turbine works, the first description of a steam turbine was that by Giovanni Branca (1629) in Italy. Turbines were built and developed in Britain by Parsons (1884), in France by de Laval (1889) and Rateau (1898), in the USA by Curtis (1896) and in Sweden by Ljungstrom (1910)—many were working on the idea at the same time.

There is nothing in the Rankine cycle itself which implies benefit from using a turbine. van Riemsdijk and Brown [1980] quote from a lecture given by Parsons explaining his reasoning, which was that since water turbines have a high mechanical efficiency, 70–80 %, which he put down to water having a small compressibility, it should be possible to make a steam turbine with so many stages that the pressure drop across each would be small enough for the expansion of the steam in that stage to be small also. By analogy, it would be reasonable to expect the same high stage efficiency as for a water turbine, and the mechanical efficiency of the whole multistaged turbine should thus be much higher than that of a reciprocating steam engine, although the thermodynamic efficiency would still be governed by the second law of thermodynamics via the Rankine cycle. It was the mechanical efficiency of the machinery which Parsons was considering. The turbine can be almost perfectly balanced and extracts energy from flow right up to entry to the condenser, whereas in a reciprocating engine, the moving mass of the piston changes direction at the end of each stroke and uses some of the expansion of the steam to clear the cylinder. The turbine was thus ideal for electricity generation, mechanically efficient, requiring low maintenance and having a high rotational speed.

Pictures of turbines are available on the websites of turbine manufacturers such as Mitsubishi Heavy Industries [2012] and others. Large machines might have different turbines mounted on the same shaft, so that the whole expansion of the steam takes place in separate stages of machinery. The separation into stages may be associated with the cycle or for mechanical reasons. Some of the important aspects of turbine design can be illustrated by considering different types of machine.

A condensing turbine exhausts the steam at the lowest practicable temperature for performing work, which will be the temperature for which the condenser was designed. A back-pressure turbine exhausts steam at a pressure that allows more work or heat to be extracted. The reason may be to allow the steam to be used for some other purpose, such as a chemical processing plant, or to pass the steam back into the boiler for superheating or reheating. A high inlet pressure machine may be split into stages simply for mechanical design reasons, such as a benefit from changing the mean diameter at which the blades rotate. Alternatively, there may be two sources of steam available, the high-pressure steam at inlet and a lower-pressure supply. In this case the turbine exhaust steam and the other supply will be added together, requiring a turbine with a bigger cross-sectional area for flow.

The passage of the steam through the turbine blades produces a rotation which drives the alternator but also produces an axial thrust, which on a large turbine may require a very large bearing surface to carry it. Sometimes turbines are designed to be back to back so the steam flow enters at the centre and divides to flow in opposite directions. The blades must be designed differently, to produce the same rotation, but the axial thrust is cancelled—again an optimisation study is required by the manufacturer.

Figure 11.8 shows a mixed pressure turbine for a nuclear power station, chosen to illustrate the points above. The high-pressure steam enters at the top on the drawing and moves to the left through the high-pressure turbine, which has four stages, each a row of blades preceded by a row of fixed blades known as nozzles to redirect the steam flow. On leaving the high-pressure turbine, the flow direction is changed through 180°, lower-pressure steam is added and the total flow passes through a second turbine with four stages before leaving to be reheated. The steam passage through the second turbine is at a greater radius, providing a bigger area of cross section for the flow, which has increased in mass flow rate by the extra steam added. In accordance with Parsons' ideas, the pressure drop over each stage is kept small. The axial thrust of each of the eight stages is balanced as much as possible, to keep the net axial thrust small; thrust bearings can be incorporated to resist the net thrust but invoke frictional torque and a power loss. The turbine rotational speed is dictated by the alternator output frequency, but several variables remain to manipulate and minimise the net axial thrust.

It can be seen from the figure that the steam can flow through the blades as intended but it can also bypass the blade because there must be a gap between the rotating parts (the shaft and blades) and the stationary parts (the nozzles and their attachment to the fixed outer casing). This bypass route is made as small as possible, as any steam flowing through it is wasted, and this is the reason for the complicated structure in this area. Elaborate sealing methods have been developed.

All axial flow steam turbines look superficially the same. At the high-pressure end, turbine blades are short relative to their diameter, but as the condenser is approached, they must become longer because of the increased cross-sectional area required by the increasing volumetric flow rate, and their active length is at a larger radius. The cross-sectional shape of the passages relative to the turbine axis is indicated by Fig. 11.9 in which the first nozzle is drawn as an actual nozzle, which they sometimes were in older, smaller designs, and the other sets of nozzles as fixed blades.

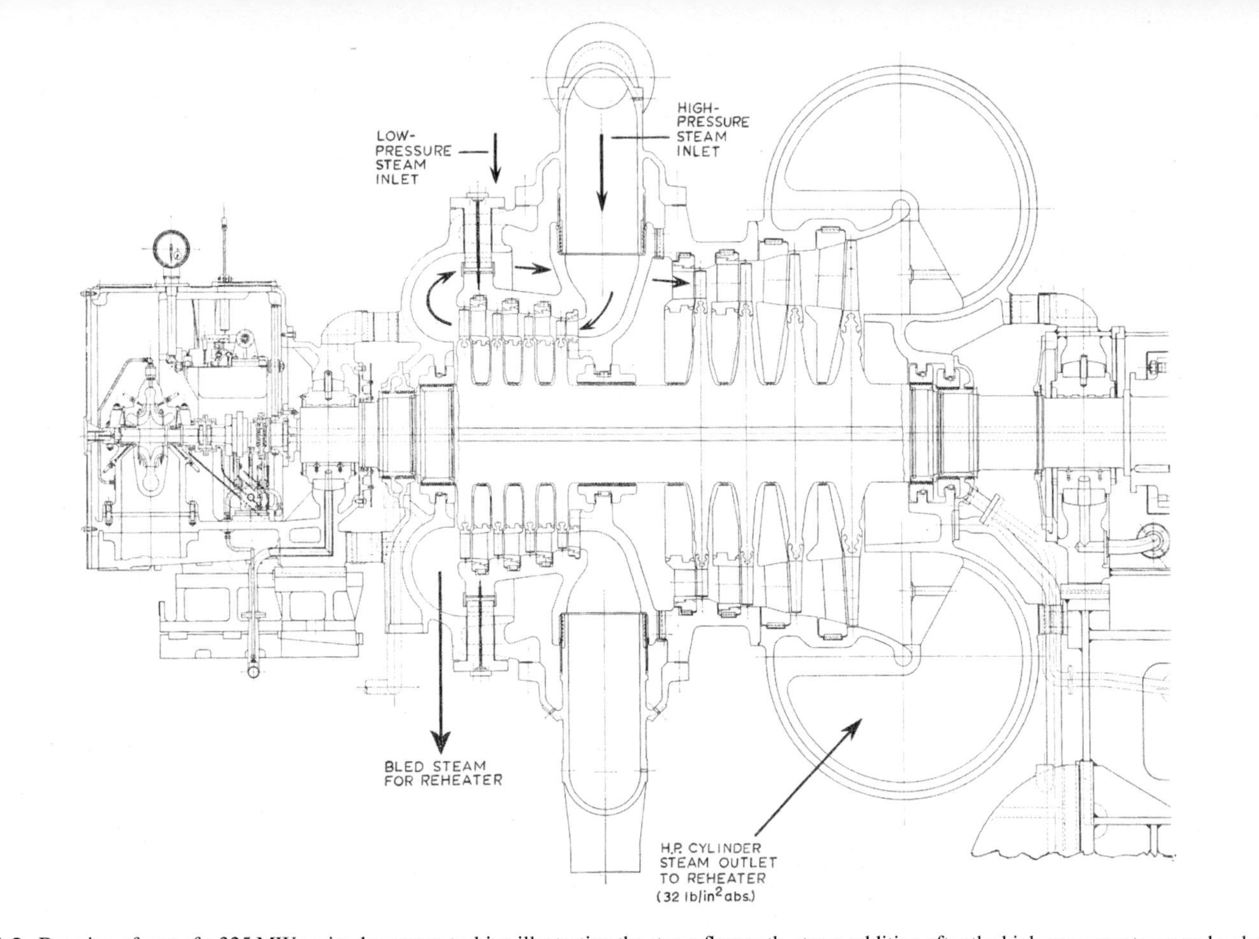

Fig. 11.8 Drawing of part of a 325 MWe mixed pressure turbine illustrating the steam flow path, steam addition after the high-pressure stages and a change in flow direction and passage diameter (reproduced from Worley, Proc (1963–64) by permission of SAGE Publications Ltd.)

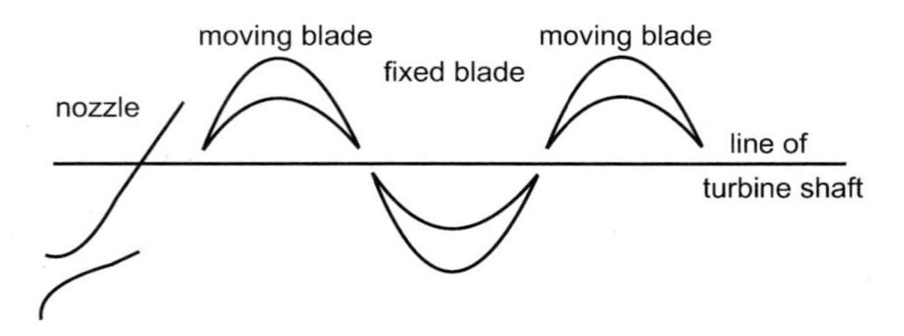

Fig. 11.9 Looking radially inwards towards the turbine axis of rotation, showing a single nozzle and moving and fixed blades. The fixed blades are also referred to as nozzles. The flow passes from left to right

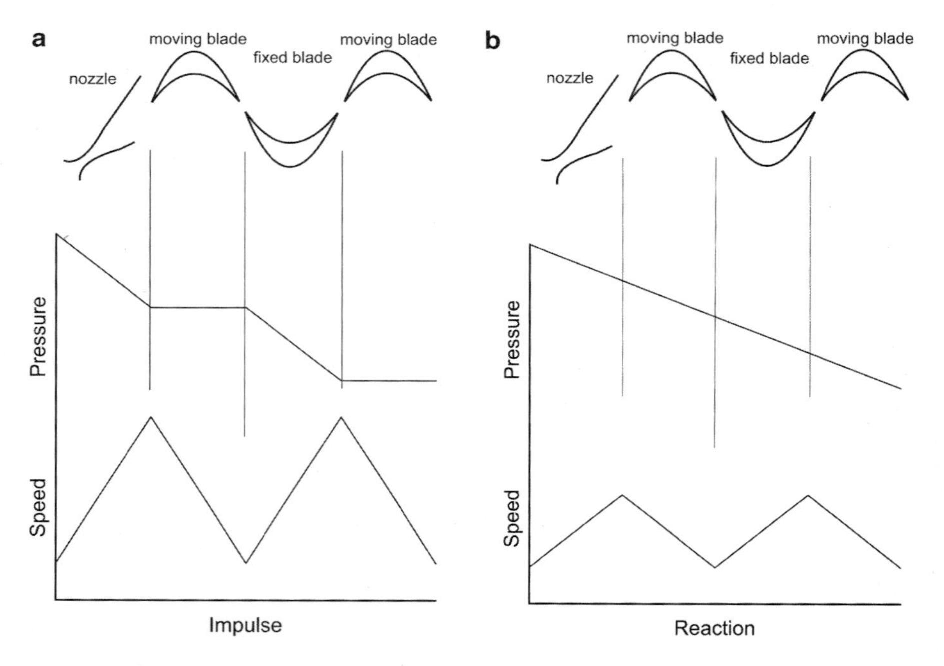

Fig. 11.10 Comparison of the two basic turbine types (**a**) an impulse turbine and (**b**) a reaction turbine

The blades are close together, both rotating and fixed, and the flow passage is a narrow curved slot of sinuous shape.

Each stage of nozzle and blade pair can be designed as either an impulse or reaction type; these have different passage shapes and represent the limiting case, and in practice some turbines are designed with passages representing a mixture of different proportions of impulse and reaction effect. The two types are compared in Fig. 11.10.

In an impulse turbine, Fig. 11.10a, the steam is accelerated in the nozzles by making the pressure fall, according to Bernoulli's theorem, but the blade passage is

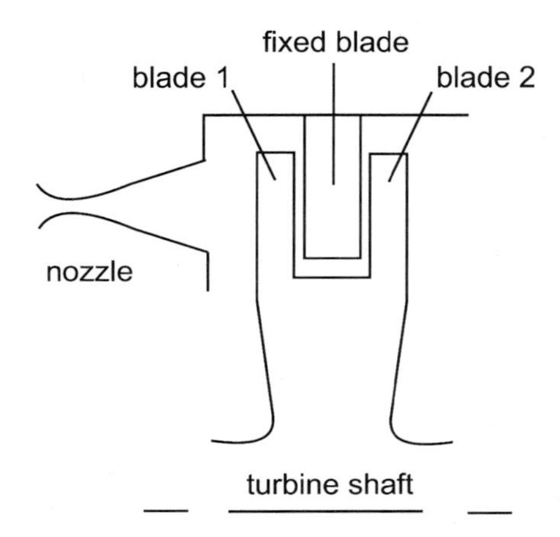

Fig. 11.11 A Curtis velocity-compounded stage

designed so that there is no pressure reduction there. This suggests a uniform cross-sectional area for flow, but to accommodate pressure reduction due to friction, the cross-sectional area is increased slightly. There is a speed reduction in the rotating blade slot, however, because the flow imparts its momentum to the rotating wheel (note, speed not velocity, the flow direction is not being considered here). The next nozzle set accelerates the flow again in exchange for a further pressure reduction, and so on. The other type is the reaction turbine, Fig. 11.10b, in which the slot dimensions are chosen so that the pressure falls uniformly through both nozzles and blades. The speed increases in the nozzles and falls in the blades as before, but speeds are lower than in the impulse turbine because the designer's aim is to make the force on the blade the result of change of momentum plus a force due directly to pressure drop.

That the pressure remains constant through an impulse turbine blade simplifies construction because there is no pressure gradient to cause the steam to leak past the blades around their ends, and labyrinth-like seals are unnecessary. Clearances in the radial direction between moving blades and the casing and between diaphragm and rotor can be relaxed from their usual very small values.

In about 1897, Curtis in the USA adopted what is referred to as the velocity-compounded impulse stage. It is shown diagrammatically in Fig. 11.11 and is usually the first stage of the turbine. The steam leaves the nozzle at a speed twice that shown in Fig. 11.10a. Its momentum is transferred in two increments; half of the speed is lost in the first set of moving blades, and the remaining half in the second set, to bring the steam speed down to a low level again, usually in a manifold. The second manifold is useful in allowing the steam pressure to become circumferentially more uniform as it may have entered the turbine through a partially complete ring of nozzles, a common arrangement for the steam entry. In consequence, both of the Curtis stage blades may be carried on a single disc between entry and exit manifolds—Fig. 11.11.

The first stages of either type of turbine lose efficiency because the blade height is short, so that a large proportion of the surface area in the steam passage is that of the outer ring and the inner rotor surfaces; these provide no contribution to work but contribute to frictional losses. The nozzles turn the steam through an angle without doing work as the nozzles do not move, but at the cost of a further frictional loss. In addition, if the nozzle ring is incomplete, at any particular time only a portion of the blade ring is being supplied with steam and is able to produce work. The friction losses result in heating of the steam and a recovery of some of the specific enthalpy drop, an effect referred to as a "reheat factor" (Wrangham [1948]—in the author's opinion, older textbooks are a good source of information on turbine design; see also Kearton [1945]). The mean circumferential velocity of the blades is a design factor and may be adjusted by the designer in the way the increase in annular area per stage is provided for—an increase is necessary because of the increasing specific volume of the steam as it passes through the machine. The stage increase can be provided by adding length to the blades at their inner or outer radii, or at both ends, or by increasing the root radius.

Finally, it remains to demonstrate what is involved in the design of the steam passages. The problem is one of fluid mechanics, not thermodynamics, and is based on the steady flow energy equation in the Bernoulli form (Sect. 4.2.4). The early days of turbine design predate computers and calculators, and the design equations bear a resemblance to the approach to flow through an orifice plate as set out in Sect. 8.3.1, with factors introduced to allow for frictional losses, etc. An orifice plate in a pipe is, after all, a variable cross-sectional area flow passage. The area of the passage between turbine blades is rectangular and must match that between the nozzles, but each is varied along the flow to produce the speed or pressure change required and curved to produce the change in direction, thus giving a change in velocity that can be analysed with a velocity diagram as shown in Fig. 11.12.

Figure 11.12a, the absolute velocity of the steam leaving the nozzles is v_1. The nozzles are stationary, but the blades are moving to the left, and the relative velocity of v_1 to the blades is u_1. The fluid leaves the blade slot with a velocity relative to the blade surfaces, but this has an absolute velocity v_2 when the movement of the blades is accounted for. The velocity diagram is shown as Fig. 11.12b and the component of most interest is the change in velocity that the steam undergoes through the blade passage, which is Δv. If the mass flow rate is $\dot{m}$, then the force on the blade per blade slot is $\dot{m}\ \Delta v$ (kgm/s^2 or Newton). This total force is at an angle to the direction of motion and the actual force driving the blade is $\dot{m}$(DE), the component normal to the axis of rotation, which produces the torque to drive the alternator. The change in momentum associated with the velocity CE causes the end thrust on the rotor; this velocity component has not been drawn in the figure, for clarity. The blade speed varies with the radius at which the blades are set on the rotor disc, and it is easy to see that the nozzle and blade angles must vary with radius. Very long turbine blades, near the exhaust into the condenser, for example, must be twisted as a result.

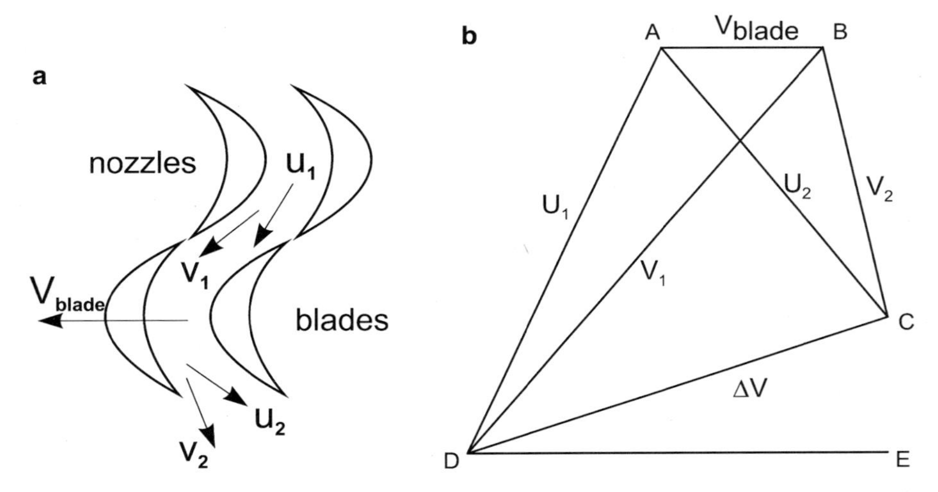

Fig. 11.12 Showing as (**a**) the flow passages in a nozzle and turbine blade pair and the absolute and relative steam velocities, and (**b**) the velocity diagram plotted to arrive at the torque and axial thrust of the pair. The axis of rotation (not shown) is vertical

11.3.3 Transient Performance of Steam Turbines

The response of the turbine to changes in load is important in power stations of any type, and the characteristics are quantified by the turbine manufacturer. In a recent study Kosman and Rosin [2001] report the effect of start-up and load variation on the service lifetime of turbine components. The speed of rotation must be kept constant in order to produce the correct frequency, 50 or 60 Hz, and elaborate speed governors are used. In addition, the power output must balance the load presented to the alternator, so the steam governor must be capable of providing both small and large variations in the flow rate. The required variation is achieved by restricting the flow area through which steam can enter the turbine, thus reducing the pressure, which results in no significant loss of specific enthalpy (throttling takes place) although there is a loss of opportunity to produce work. On machines with a large steam flow, the entry to the nozzles is sometimes through several ports, each with a valve fitted, and the number of valves passing steam is increased as the demand rises, to avoid the difficulty of adjusting a single large valve by a small amount.

Figure 11.13 shows the Willans line, the variation between steam consumption and power output, which was found experimentally by Willans in 1888 to be a linear relationship (Wrangham [1948]) if the turbine is governed by throttling. The line has a slope B and is described by

$$\dot{m}_{turb} = B.(L_e) + \dot{m}_0 \tag{11.1}$$

where

$\dot{m}_{turb}$ is the total steam flow rate to the turbine (kg/s)
L_e is the electrical load (MWe)

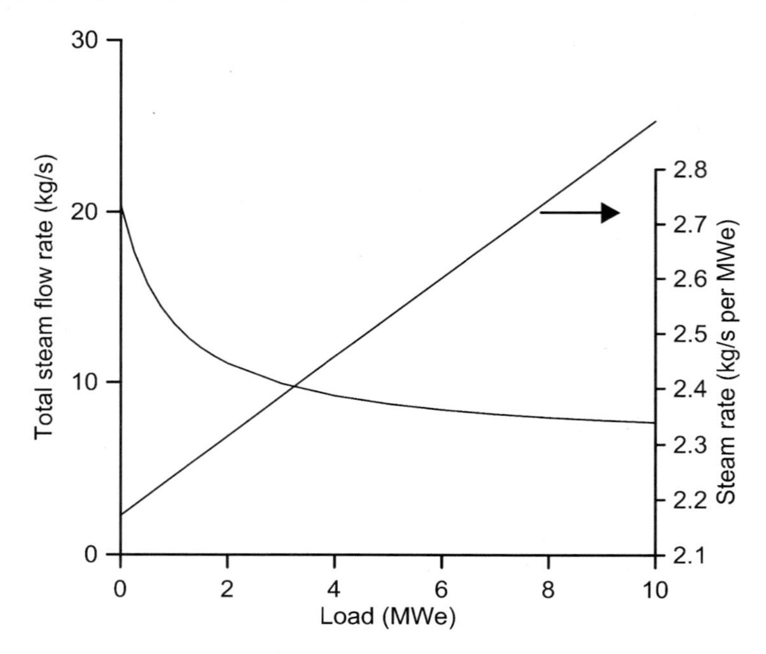

Fig. 11.13 The Willans line for a throttle governed turbine

$\dot{m}_0$ is the minimum steam mass flow rate required to keep the turbine at synchronous speed as load tends to zero (kg/s). Kearton [1945] suggests that $\dot{m}_0$ is 10–14 % of the full load mass flow rate.

An inverse curve is shown, representing $\dot{m}_{turb}/L_e$ in kg/s per MWe, which becomes asymptotic to the Load axis at a value of steam flow per MWe (kg/s per MWe) which is useful to characterise the turbine performance and is sometimes called the steam rate. The rate increases as the load decreases because the steam is throttled and the blades, although rotating at the same speed, are not working at their design condition. For a geothermal condensing steam turbine operating at six bars abs inlet pressure, the steam rate is typically in the range 2.2–2.4 kg/s per MWe.

Geothermal steam is low pressure, high specific volume, which means that the physical size of the governing valves is large for the power output of the turbine as compared to fossil-fuelled plant, and the steam inlet pipe carrying the governor valve is of large diameter. Together with the presence of silica in the steam, this prompted Mitsubishi Heavy Industries Ltd [1998] to develop a butterfly governing valve.

A further issue in relation to transient steam turbine performance is non-base load use. As explained in Chap. 10, their cost structure makes base load use of geothermal power stations very desirable, as availabilities of around 95 % can be achieved. A practical reason making base load use preferable arises because of the very small clearances necessary between the moving and fixed parts of the turbine to restrict steam leakage between stages. Small differential temperatures between fixed and moving parts are required if the clearances are not to be reduced to zero, when the parts would touch. It is necessary to increase the temperature of the turbine slowly, and manufacturers specify a maximum rate of temperature increase for particular designs. In the past it was the practice in some power stations to pass hot air through the turbine before steam was admitted. Kearton [1945] states that the Ljungstrom radial flow steam turbine warms up much more quickly than an axial flow turbine; they were built in capacities up to 50 MWe and have been examined recently by Marcuccilli and Thiolet [2010] with a view to application to geothermal organic Rankine cycle plant.

11.4 Using Geothermal Steam

11.4.1 Flashing the Well Discharge at Several Pressures

The discharge characteristics vary from well to well, and wells are grouped and their output flashed at a predetermined pressure. There is merit in keeping the turbine inlet pressure and hence the inlet temperature as high as possible, because this increases the conversion efficiency. The higher the flash pressure, however, the more energy is left in the separated water, but it can be flashed at a lower pressure and more steam produced. This could be added to the same turbine fed by the first flash partway through its length, which would then be called a pass-in turbine or mixed pressure turbine—an example has already been shown in Fig. 11.8. In liquid-dominated geothermal resource use, two or three flash stages only are used, as illustrated in Fig. 11.14. However, as will become clear in Chap. 12, the choice of flash pressure and the number of stages depends on geochemical considerations and not just on turbine performance.

Wairakei, which is an old design made complicated by unusual circumstances explained in Chap. 14, has a large number of relatively small turbines. The steam

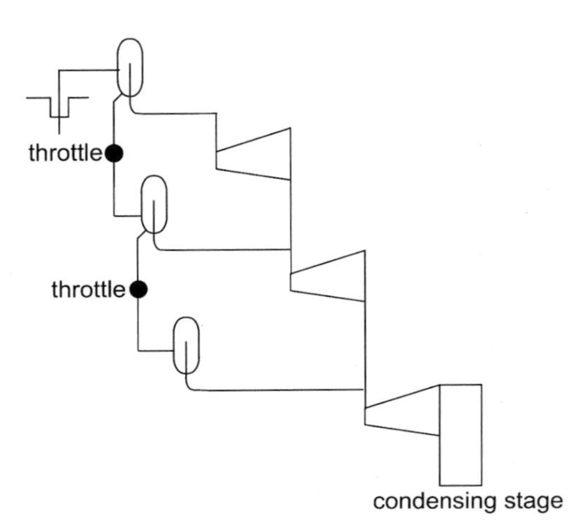

Fig. 11.14 Well discharge flashed in three stages

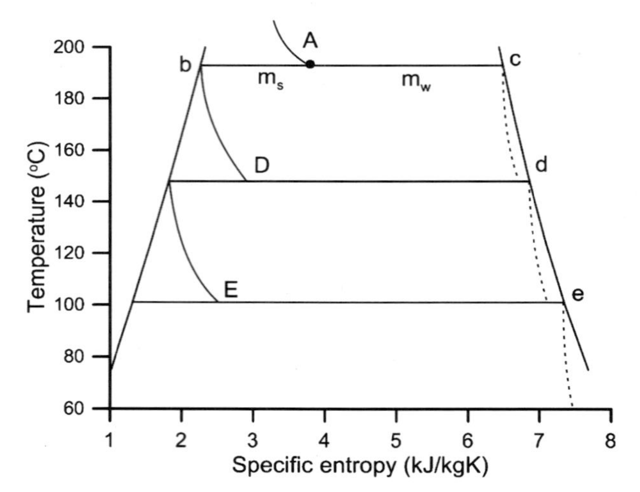

Fig. 11.15 T–s diagram for a triple flash well discharge

was supplied at 13.4, 4.45 and 1.0345 bars abs, with the exhaust of the higher pressure machine combined with new steam flashed from separated water, and the last machine exhausting into a condenser—see Thain and Carey [2009] for details of the arrangement.

The optimum wellhead pressure for a given well discharge characteristic using either single flash or double flash can be calculated, but even adjacent wells may have different characteristics and the choice of type and number of turbine units is the result of a wider optimisation, involving manufacturing costs as well as thermodynamic performance, within the restrictions imposed by resource geochemistry. The theoretical optimum considering only the discharge and flow separation may have a small influence on the adopted design.

The steam in a geothermal power station does not follow a cycle in the plant itself, but the entire combination of resource, well and steamfield equipment and power station can properly be viewed as a cycle. To compare different systems, an analysis based on the second law of thermodynamics must be carried out, usually called "exergy analysis". This is not dealt with in this book, but has been defined by ASTM [2006] and is set out in detail by DiPippo [2012] and other thermodynamics texts. Exergy analysis is increasingly being applied to larger entities than power stations, e.g. country regions (see Sciubba et al. [2008]), and is worthy of a more extended treatment than is possible here.

To calculate the power output for a proposed number of flash stages, a T–s diagram such as that of Fig. 11.15 is required. This figure has been drawn to represent the Wairakei stage pressures mentioned above. Local atmospheric conditions dictate the temperature at which heat is rejected so there is no merit in continuing the T–s diagram to absolute zero. It is usually assumed that specific enthalpy is conserved in a flash expansion, and if the fluid in the resource is a high-temperature liquid, the isenthalpic line enters from above left and ends at the first flash pressure, with saturation temperature of 193 °C, at point A. The mixture is then separated into saturated water and steam at opposite sides of the T–s envelope, points b and c, respectively. The simple arithmetic rules governing the properties of a two-phase mixture have been explained in Sect. 3.4.2 and already used above to analyse the Carnot cycle. The mass proportion of steam and water are as depicted in the figure, calculated using

$$h = h_f + Xh_{fg} \tag{11.2}$$

where h is the specific enthalpy of the resource fluid and the property values and dryness fraction are for the appropriate temperature, in this case 193 °C. The dryness fraction is

$$X = \frac{\dot{m}_g}{(\dot{m}_g + \dot{m}_f)} \tag{11.3}$$

so the mass flow rates are proportional to the length of the lines bA and Ac. The separated water from the first flash arrives at point D after the second flash, and likewise at point E after the third. The steam produced is at 148 °C and 101 °C and is represented by points d and e. The exhaust from the high-pressure turbine must be added to the calculated mass flow rate entering at e from the second separated water flash, and the same when considering the final turbine inlet. Since the steam entering any turbine stage is saturated, it will be wet on leaving, and interstage separators may need to be used to ensure that the ongoing steam is dry saturated at inlet to the turbines.

11.4.2 Geothermal Steam Turbines

Geothermal steam carries with it water, dissolved solids and noncondensable gases. The steam supply from a liquid-dominated geothermal resource is likely to be carrying some water containing dissolved silica and calcium carbonate. Solid deposits on turbine blades are the result, and Kubiak and Urquiza-Betran [2002] considered the effect on performance of a reduction in flow area and shape of the passage between blades and nozzles. At some point through a condensing turbine, the steam crosses the envelope of the T–s diagram and becomes wet. The water droplets do not behave like steam in the flow passages, as their momentum is not passed on efficiently to the blades, and they damage the blades and nozzles by their impact. The steam mass flow rate is gradually reduced as the steam becomes wetter. Circumferential water drains are fitted, designed to catch water droplets and remove them from the flow. Non-condensable gases actually produce work in flowing through the turbine, but this and more is required to remove them from the condenser, as discussed below. Turbine maintenance necessitated by the dissolved solids carried in the steam has been addressed by Sakai et al. [2000] and by Matsuda [2006], with discussion of stress corrosion cracking and surface damage by impact and corrosion.

11.4.3 The Possibility of Superheating Using an External Heat Source

Using an external heat source would in principle allow the geothermal steam to be superheated and reheated, similar to Fig. 11.7, which would result in a higher electricity output. Several studies have been carried out; see Kestin et al. [1980]. However geothermal steam carries dissolved solids which could give deposition and corrosion problems in superheaters. Furthermore, there is no obvious source of heat at a high enough temperature that could not be used more effectively in a non-geothermal power station; however, particular circumstances might exist that would make the idea economically attractive.

11.4.4 Heat Rejection at Geothermal Steam Power Stations: Condensers and Related Equipment

The problems of heat rejection from geothermal steam plant are greater than those for fossil-fuelled plant in that the geothermal steam carries with it the non-condensable gases mentioned above. However condenser construction is simplified by being able to mix the steam to be condensed with the cooling water because the steam is not part of a high-purity water cycle, allowing direct-contact or jet condensers to be used, in which steam emerging from the turbine enters a vessel into which jets of cold water are sprayed. Once mixed, the cooling water and steam condensate cannot be separated. The alternative is to keep the steam and cooling water separate by using a tubular condenser with cooling water on the inside of a bank or matrix of tubes exposed to steam on the outside. These have a long history of use in fossil plant, but slight differences are required for geothermal plant, such as gaps through the tube matrix to allow the passage of non-condensable gases, a provision which reduces the pumping power required to maintain the vacuum. Tubular condensers are not unknown in geothermal stations. Modes of condensation and non-condensable gas blanketing of condenser surfaces have already been discussed in Sect. 7.5.

Figure 11.16 shows the idea of a barometric direct-contact condenser, named because of its resemblance to a barometer (Fig. 3.1). The vessel contains the spray equipment and sits on a large diameter vertical tube approximately 10 m long, either in a pit or above ground. Imagine a valve on the bottom of the tube, opened when the vessel and tube are filled with water. The vessel would be evacuated as the water ran out and would remain so in use as long as the water level was maintained in the tube and provided the non-condensable gases were continuously removed. The condensate collects in the tube and the water level and hence vacuum are maintained by allowing it to flow out of the tube end. If this is in a pit, the water may have to be pumped out, with further consumption of electricity.

In fossil-fuelled power plant, the turbine is often supported immediately above the condenser, which is physically larger, so that the steam passes through the minimum length of large diameter duct—large diameter because the specific

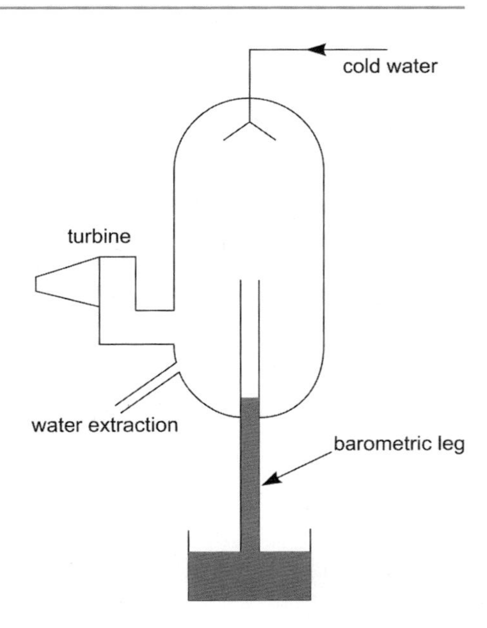

Fig. 11.16 Diagrammatic arrangement of a barometric leg direct-contact condenser

volume of the steam at low pressure is very large and the flow rate from the turbine exit is slowed to minimise the kinetic energy loss.

Kestin et al.'s [1980] study contains a detailed discussion of condensers for geothermal use, including materials of construction, which is still very relevant.

Although geothermal steam power stations produce more condensate than is required to top up wet cooling towers, the decreasing availability of cheap, clean water for cooling fossil and nuclear power plant has led to a greater use of air-cooled condensers. Very large ones have been built in Europe, including natural draft concrete cooling towers of the configuration usually used for wet cooling. More often finned tubes are arranged in two banks set at an angle, referred to as A frame.

Condenser vacuum is often maintained by extracting the non-condensable gases using steam-jet ejectors in stages. The ejector is a device based on Bernoulli's equation, in which steam, an available, compressible, high-pressure fluid, is accelerated through a nozzle to reach a minimum pressure at its throat—less than the required condenser pressure—after which the flow area is increased, the flow decelerates and the pressure rises. The throat is connected to the condenser where the non-condensable gases collect, and the combined output passes into a small jet condenser, the vacuum for which is produced by a second-stage ejector. The pressure of the fluid leaving the second-stage ejector is above atmospheric so requires no more power consumption to be discharged to atmosphere, although the concentration of H_2S requires it to be discharged at high level or dispersed in some way. The steam consumption of the ejectors is significant, and alternative ways of producing vacuum are used, as reviewed by Ozkan and Gokcen [2010].

Multistage radial flow gas compressors are common, illustrated by Mitsubishi Heavy Industries Ltd [1998], and in the case of high gas-output resources can be physically large items of equipment. Liquid ring vacuum pumps are also in use, the principles of operation of which are shown by one manufacturer, Nash [2012].

The final stage of heat rejection is to transfer the heat collected from the condenser to the environment. Thermal power stations of all types are built adjacent to rivers or estuaries where possible, to reject heat to them (in other words to use them as a source of cooling water). This is the case at Wairakei, where the river water is clean enough to be used in jet condensers; issues relating to water contamination are mentioned in Chap. 14. Alternatively, heat can be rejected to the air by means of cooling towers. On exit from the condenser, the cooling water is sprayed into a rising column of air created by fans on top of an open slatted structure, often with wooden slats; the fans are powered using the generated electricity. The cooled water falls into a pond beneath, from where it is returned to the condenser. The performance of the cooling towers is dependent on the atmospheric temperature. In fossil-fuelled stations natural draft cooling towers are common. The warm condenser cooling water is carried to some height within the tower and sprayed downwards over the cross-sectional area. Atmospheric air enters the bottom of the tower, which is open for a height of only a few metres, unlike the forced draft towers, because the upper part of the tower is shaped like a jet ejector, accelerating the warm air upwards and drawing in fresh air. The towers are made of reinforced concrete. The natural draft cooling tower at Ohaaki, New Zealand, designed for 120 MWe, appears to be the only one of its type used for a geothermal power station. The H_2S extracted from the condenser is released into the rising column of air in the tower, above the water sprays.

11.5 Plant Using Working Fluids Other Than Water

Steam is a suitable working fluid for converting heat into work over the temperature range 160–550 °C at the turbine inlet. Efforts to convert a greater proportion of the energy from fossil fuels led to trials of other working fluids either singly or in combination with steam, but they have found major use only for geothermal generation in the last few decades—in fact they are probably in the majority, numerically. DiPippo [2012] presents a detailed cycle analysis of organic Rankine cycle geothermal plant (and geothermal steam plant) of which there are many in use in different arrangements.

11.5.1 Organic Rankine Cycles

In comparing the Carnot cycle with the fossil-fuelled Rankine cycle in Sect. 11.3.1, the departures of the latter from the ideal were identified. These departures reduced the efficiency of conversion of heat to work and they were to some extent exacerbated by the shape of the saturation envelope for water. Alternative working fluids were sought at an early stage in power plant development, driven by improvements in the strength of materials at high temperature—Wood [1970] stated that "*claims that some other fluid is preferable to water are a recurring phenomenon*" but went on to note their suitability for low-temperature heat sources. Materials to use high source temperatures were available, temperatures much higher than the critical for water. To retain the advantage of evaporation, a working fluid that was still two-phase at temperatures of 500 °C or more was needed. Rogers and Mayhew [1967] present a numerical example of a "binary cycle", one with two working fluids, water and mercury, operating in two separate but combined cycles with separate turbines. Mercury had some of the required theoretical features but was abandoned in the face of practical difficulties.

Many modern geothermal plants for resources producing water or two-phase well discharges operate with binary cycles, steam coupled with an organic working fluid. The latter necessarily operates in a closed cycle as it is expensive and unsuitable for discharge to the environment. The aim is to convert more of the heat to work than can be achieved using steam plant. Angelino and Colonna di Paliano [2000] identify the same issue in respect of molten carbonate fuel cells—a large proportion of the heat released cannot be used in the primary conversion device. They provide a review of heat recovery to various organic working fluids which are then used with a "conventional" ORC plant. The choice of working fluid for any application is governed by the degree of hazard in its use from the point of view of toxicity, potential for atmospheric pollution, fire hazard, ease of transport, cost and availability, etc. Many different working fluids are currently being considered for use with enhanced geothermal resources; see, for example, Kalra et al. [2012]. In the meantime, using "conventional" geothermal resources, a significant amount of installed geothermal ORC plant is of Ormat Technologies [2012] design and manufacture, using isopentane or *n*-pentane as working fluid. The critical point for isopentane is 187 °C and 33.8 bars abs and geothermal separated water is available at similar temperatures or less.

Although not shown in this book, the h–s chart rather than the T–s chart is the choice of steam turbine designers, and the P–h chart is the choice of ORC designers. For the present purposes, the T–s chart best displays the points to be made. Figure 11.17 shows the general shape of the T–s saturation envelope for isopentane and that for water; several other potential working fluids also have the important characteristic of sloping backwards on the gas side of the envelope, so that s decreases as T decreases. The envelopes of Fig. 11.17 have been superimposed to compare the shapes, and the actual property values are quite different; plotted correctly the two envelopes would be widely separated and of different sizes. Curve A is a proper representation of the T–s envelope for water, but curve B has been drawn by eye and is not meant to represent any particular organic fluid, but to have the general characteristics that have a bearing on its thermodynamic performance as a working fluid.

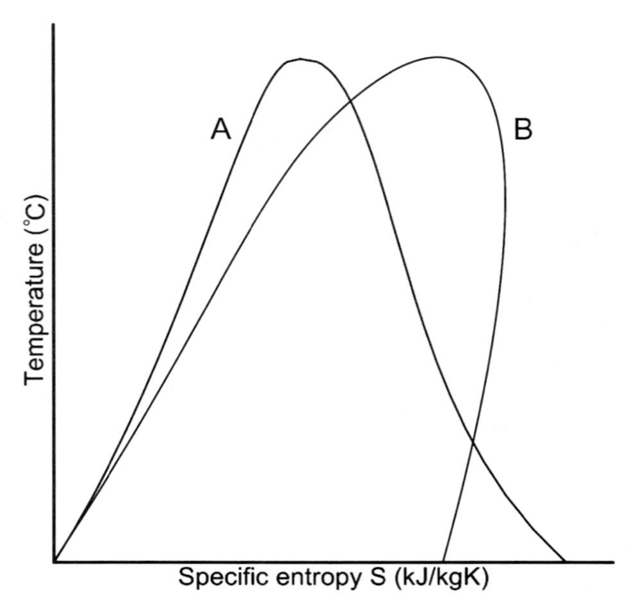

Fig. 11.17 Comparison of the general shapes of T–s envelopes: (**A**) for water and (**B**) a sketch representing some organic working fluids, including isopentane. The envelopes are not drawn to scale or placed in their correct relative positions on the axes

The important differences in respect of Rankine cycle plant design that follow from the envelope shapes are as follows:

- An isentropic expansion of saturated vapour results in a wet mixture at turbine exit for H_2O and a superheated vapour for working fluid B, provided that the expansion starts at a suitable temperature. This removes concern about blade erosion by droplets.
- The slope of the liquid side of the envelope is less steep for B than for water so the departure from the Carnot cycle is greater for the organic fluid, which, taken in isolation, would tend to make regenerative heating more attractive than for steam Rankine cycles.
- The saturation envelope for B is fairly wide near its summit, which is beneficial in maximising the proportion of heat added during evaporation and helping to reduce the effect of departure from the Carnot cycle resulting from heating in the liquid state (see Fig. 11.2).
- At 30 °C, a suitable temperature for rejection of heat to the environment and thus the lowest temperature potentially required in the cycle, the saturation pressure of isopentane, in particular, is approximately 1 bar abs. This is mechanically convenient as the condenser leakage potential is minimised.
- The density of gaseous organic fluid is high, which leads to a comparatively smaller turbine for a given output—the rate of transfer of momentum in the blades increases with density.
- Choosing an upper temperature of 220 °C for separated geothermal water, and a heat rejection temperature of 30 °C, the specific volume ratio for saturated

vapour, V_{220}/V_{30}, is 382 for water and approximately an order of magnitude less for isopentane and other organic fluids. As for density, this too permits many fewer stages for an ORC turbine than for a steam turbine working between the same temperatures (recall Parsons' original reasoning in Sect. 11.3.2).

- The velocity of sound in organic fluids is lower than in steam, and turbine blade tip speeds must be kept subsonic. This might require a gearbox between turbine and alternator, to increase the alternator revs/min to synchronous speed.

The current range of plant using isopentane typically has a two-stage turbine (Legmann and Sullivan [2003]) in line with the observations above. As regards regenerative feed heating, the small number of stages in the turbine reduces the opportunity to bleed working fluid—the flow is large enough but the temperature differentiation is probably too small and some heat would be transferred using larger than necessary temperature differences, thus defeating the plan to approach the Carnot cycle requirements. However the advantages of regenerative heating were its effect in reducing the condenser capacity as well as improving cycle efficiency, and there could be benefit from this aspect. A large proportion of the heat supply is rejected in the condenser because of the low source temperature, which makes the condenser heat load large. The typically air-cooled condensers consume electricity and are physically large. Purely in terms of cycle efficiency, regenerative feed heating would be more beneficial to organic working fluid cycles than to steam cycles in view of the slope of the T–s envelope on the liquid side. A low-temperature heat source is available in the separated geothermal water flow.

As regards superheating or reheating, the slope of the T–s envelope on the gas side makes it unnecessary to adopt superheating or reheating for the reasons it was adopted in steam cycles. The maximum temperature of the heating fluid must be taken advantage of, and deviation from the Carnot cycle is not a reason for avoiding superheat, which many organic Rankine cycles employ in their optimised configuration. Sohel et al. [2010] report on a dynamic model of the performance of a 5.4 MWe organic Rankine cycle power plant of unspecified design. They provide a T–s diagram showing that the working fluid is superheated, but leaves the turbine before reaching condenser conditions, entering a heat exchanger (termed a recuperator) which transfers heat back into the working fluid heating part of the cycle, perhaps exchanging work from the turbine for a reduced condenser heat load.

There is a great deal of current literature concerned with selecting a working fluid from the large range of organic fluids available. For example, Franco and Villani [2009] provide a detailed discussion of the methodology and factors for design optimisation of this type of plant for geothermal fluid temperatures in the range 110–160 °C. Such is the range of fluids on offer that their focus is on methodology rather than engineering factors.

Perhaps the most noticeable features of the majority of ORC plant, of Ormat Inc design and manufacture, are that they can stand in the open, and the prime mover is small, very much smaller in bulk than the heat exchangers which are shell and tube type. The condensers are air cooled with the working fluid contained in almost horizontal banks of finned tube with cooling fans—the tubes are set at an angle to the horizontal to allow the liquid to drain for collection. The condensers cover a significant area, but can be elevated above the rest of the plant. These features will have an effect on the engineering–economic optimisation of the design, which thus cannot be analysed from a purely engineering performance standpoint.

Throughout the development history of the fossil-fuelled Rankine cycle reviewed above, the search for optimum economic performance by manipulating the cycle appears to have been the primary focus. It must be remembered however that the full commercial activity of designing and supplying the main power station plant involves research and development, design and then manufacturing which includes consideration of the capital assets of foundry and machine shop, then manpower, finance, transport to site, maintenance, reliability and overall environmental impact and energy analysis of the type discussed in Sect. 10.7—no small task. Thermal efficiency of the plant is important but is by no means the dominant factor in making a selection and cannot be viewed in isolation. In common with all commercially developed power industry equipment, what is marketed is the result of private manufacturing industry optimisation.

11.5.2 An Example of Dual Working Fluid Plant

Legmann and Sullivan [2003] describe the Rotokawa 1 development, New Zealand, which is an example of using steam from the wells directly in a back-pressure steam turbine and using its exhaust together with the separated water to supply organic Rankine cycle generators. The station was designed and manufactured by Ormat Inc. The total designed output of the plant is 30 MWe of which 14 MWe is produced by a multistage reaction steam turbine and the balance by three 2-stage organic Rankine cycle (ORC) turbines. The steam turbine rotates at 3,000 rpm and the ORC machines at 1,500 rpm. A reconstructed process flow diagram from details in the paper is shown as Fig. 11.18, from which it can be seen that two of the ORC units receive their heat supply from the steam turbine exhaust.

The other two operate as a pair in tandem, using heat from the separated water. The process is at its simplest, and the paper reports that the power station is very reliable with an availability of 98 % and output 10 % higher than the design figure.

Ormat supplied a second-stage power station at Rotokawa, this time with a simpler process flow arrangement, shown by Sohel et al. [2009]. The steam flow drives a 35 MWe steam turbine which exhausts to a steam to isopentane heat exchanger, thus supplying the heat for a 7.5 MWe ORC unit. The separated water provides the heat source for a second ORC unit of the same output. Sohel et al. demonstrated that an improvement in efficiency had been achieved for summer operation by increasing the heat rejection capacity in the air-cooled condenser by using water. The gain in power output on the hottest day amounted to 6.8 % but averaged over the year was only 1 %; it is a reflection of the variation of weather conditions and provides no basis for judging the optimum cycle configuration adopted by the plant designers.

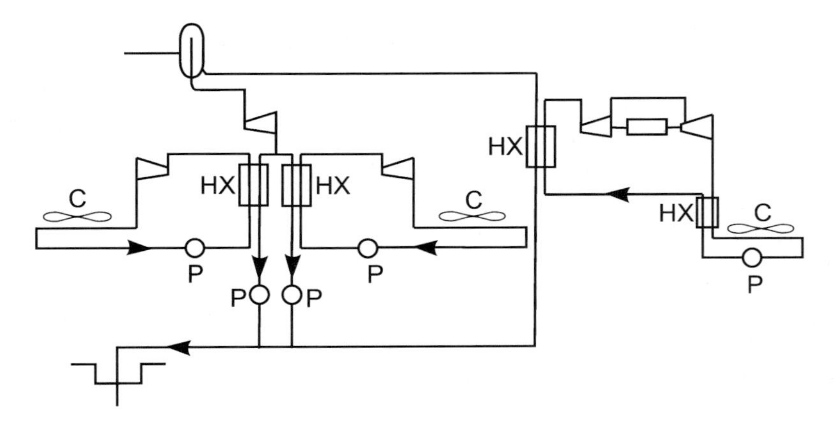

Fig. 11.18 Interpretation of the combined steam and ORC plant described by Legmann and Sullivan [2003]. The closed loops contain isopentane, HX indicates heat exchangers, P pumps and C air-cooled condensers

11.5.3 The Kalina Cycle

The Kalina cycle first seems to have entered the geothermal literature via the paper by Kalina and Liebowitz [1989] which considered the application of a cycle using an ammonia–water mixture as working fluid.

It will be recalled from Sect. 3.3 that Carnot's ideal circumstances for converting heat to work required that any temperature reduction must produce work; otherwise, the heat flow would be wasted. It was convenient for the Carnot cycle that evaporating liquid remained at constant temperature, because it provided the ideal constant temperature heat source. Now consider transferring the heat from a geothermal liquid, say separated water, in a long heat exchanger. The heating liquid remains single phase and its temperature gradually falls between inlet and exit as it transfers its heat to another fluid. Figure 11.19 represents the situation, with the separated water heating fluid entering from the right and experiencing falling temperature as it transfers its heat, line A; consider temporarily that the horizontal axis represents the distance through the heat exchanger. Travelling in the opposite direction is the working fluid of an ORC, entering from the left at lower temperature, line B. The working fluid heats up to saturation temperature at "a" on the graph and remains isothermal until it is totally evaporated at "b", after which it shows an increase of temperature with position as before. If heat can be transferred at a suitable rate at point "a", then beyond that, the temperature difference is unnecessarily large, and the temperature of the heating fluid is being reduced wastefully. To clarify the details, the horizontal axis represents the amount of heat transferred, which is linearly related to heating fluid temperature change. This graph occurs in heat exchangers from many industries, chemical processing and nuclear, fossil and geothermal power plant, including ORC plant. The benefit of the Kalina cycle is that the working fluid is an ammonia–water mixture, and as the mixture increases in

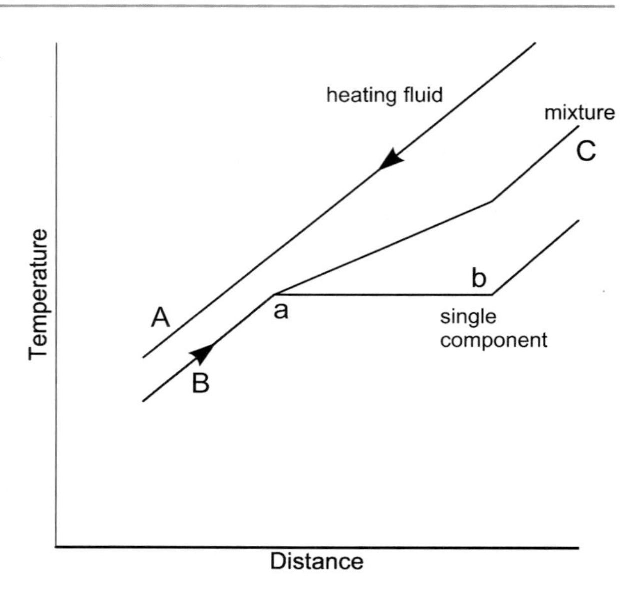

Fig. 11.19 The fundamental idea behind the Kalina cycle is to reduce the change of temperature difference between the heating and evaporating fluid

temperature, the ammonia comes out of solution and the saturation temperature of the mixture rises. It can be made to vary as the line marked "mixture", line C on Fig. 11.15, which on average shows a lower temperature drop between the heating fluid and the evaporating mixture. A heat engine which must take its heat supply from a flowing stream of single-phase liquid (i.e. almost constant specific heat) will have a higher efficiency if this line can be followed instead of the single component line.

Potentially, the Kalina cycle has a higher efficiency than the ORC cycle, and a 1997 study by Lu (reported as Lu et al. [2009]) showed that the proposed Kalina cycle design referred to as KCS11 was expected to have a higher efficiency than an existing ORC plant. Welch and Boyle [2010] report on the design and construction of two Kalina cycle plants. DiPippo [2004] carried out a comparative performance analysis of several ORC plants and the only operating geothermal Kalina cycle plant and concluded that the Kalina plant may produce a 3 % higher output for the same operating conditions. The plants examined had different local conditions; the Kalina plant is in Iceland where heat rejection is easier than for the ORC plant in the USA because ambient temperatures are lower. The Kalina cycle plant is more complicated than ORC plant and has not attracted significant capital investment yet. Its proper place in the hierarchy is yet to be established.

11.5.4 The Trilateral Flash Cycle and Two-Phase Prime Movers

The trilateral flash cycle is shown on the T–s diagram of Fig. 11.20, which is a modification of the Rankine and Carnot cycle diagrams of Fig. 11.2 for water as the working fluid; the cycle is proposed for organic working fluids also. Saturated water

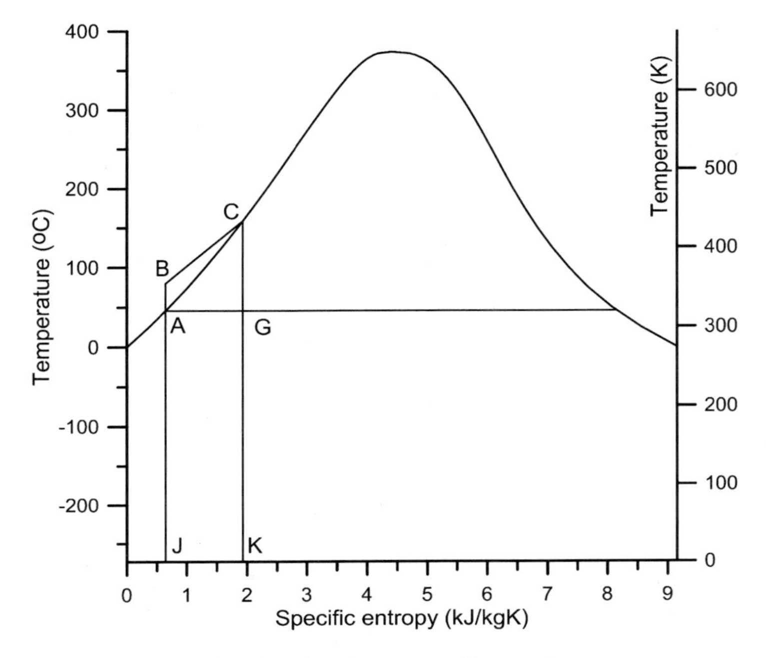

Fig. 11.20 The trilateral flash cycle, plotted on a water T–s envelope

arrives at the condenser at A, is pumped up to maximum pressure at B from where it is heated in the compressed liquid phase to C. From C the water is expanded to G from where it is condensed back to A. The cycle has a triangular shape. Isentropic expansion CG has been shown in the figure, but in reality the fluid state will be slightly right of G as specific entropy increases due to frictional heating in the expanding fluid. This increase in specific entropy may be greater than that in a steam turbine, as the working fluid is saturated liquid at C and thereafter is a disorganised two-phase mixture. The figure has been drawn for a maximum pressure of 6 bars abs and a condenser pressure of 0.1 bar abs, thus working between 159 and 46 °C, and the Carnot efficiency is the ratio of the areas ABCG to JBCK—compare the equivalent areas for the Carnot and Rankine cycles of Fig. 11.2, which uses the same temperatures. The average temperature at which heat is transferred to the fluid is less than in Fig. 11.2, so the theoretical efficiency must be less than that of the Rankine cycle, but the equipment is much simpler.

The cycle as shown here is described as a wet vapour cycle. Smith [1993] notes that Ruths [1924] took out a patent based on his idea that heat stored in an accumulator could be used to advantage by discharging the whole amount through a reciprocating steam engine or turbine. The geothermal equivalent of Ruths' proposal is to discharge a well producing two-phase flow directly into a prime mover. Smith et al. [1995] provide a review of suitable prime movers developed for the geothermal and chemical industries, including the bi-phase turbine and screw expander, and he later focused on geothermal use—Smith et al. [2005]. The screw expander has continued to attract attention but few have been built; an early one built in the USA has recently been replaced by a standard ORC plant (Buchanan and Nickerson [2011]) but that is barely significant given the small amount of development effort applied to their mechanical design. Welch and Boyle [2010] provide information on a recently developed type.

The fluid in a two-phase mixture does not travel at a uniform speed and, intuitively, should give a lower conversion efficiency than the well-ordered flow through a turbine. However Brown and Mines [1998] compared an ORC plant and a trilateral flash plant and concluded that the prime mover for a trilateral flash cycle system would require a machine efficiency of only 76 % to give it the same overall conversion efficiency of heat to work as an ORC plant with a typical turbine of 85 % mechanical efficiency. As before, the thermodynamic cycle issues are easy to address in considering the use of the trilateral flash cycle and two-phase prime movers, and the real problem is a lifetime energy (exergy) optimisation—is more thermodynamic work expended in manufacturing the equipment than is generated by it, and are the total environmental effects worthwhile? Smith et al. [2005] presented reasons for incorporating screw expanders into ORC plants, providing evidence of increased output and decreased cost.

References

Angelino G and Colonna di Paliano P (2000) Organic Rankine cycles (ORC) for energy recovery from molten carbonate fuel cells AIAA 2000-3052, 35th Intersociety Energy Conversion Engineering, Nevada, USA

ASTM American Society for Testing and Materials (2006) Standard guide for specifying thermal performance of geothermal power plants, ASTM E974-00

Brown BW, Mines GL (1998) Flowsheet simulation of the tri-lateral cycle. Geoth Resour Counc Trans 22:105

Buchanan T and Nickerson, L (2011) Expansion and repowering of mammoth geothermal resource: selection of generating cycle and expander technology. Geothermal Resources Council Annual Meeting, San Diego, USA

DiPippo R (2004) Second law assessment of geothermal binary plants generating power from low temperature geothermal fluids. Geothermics 33:565

DiPippo R (2012) Geothermal power plants: principles, applications, case studies and environmental impact. Butterworth-Heineman, Oxford

Franco A, Villani M (2009) Optimal design of binary cycle power plants for water dominated medium temperature geothermal fields. Geothermics 38(4):379–391

IAEA (2012) World energy supply, Bulletin 46/1 http://www.iaea.org

Kalina AI, Liebowitz HM (1989) Application of the Kalina cycle technology to geothermal power generation. Geother Resour Counc Trans 13:605

Kalra C, Bequin G, Jackson J, Laursen AL, Chen H, Myers K, Hardy HK, Zia J (2012) High potential working fluids and cycle concepts for next generation binary organic Rankine cycles for enhanced geothermal systems. In: Proceedings of 37th workshop on geothermal reservoir engineering, Stanford, USA

Kearton WJ (1945) Steam turbine theory and practice. Pitman, London

Kestin J, DiPippo R, Khalifa HE, Ryley DJ (eds) (1980) Sourcebook on the production of electricity from geothermal energy. US DoE, Brown University, Rhode Island

Kosman G, Rosin A (2001) The influence of startups and cyclic loads of steam turbines conducted according to European standards on the component's life. Energy 26:1083

Kubiak JA, Urquiza-Betran G (2002) Simulation of the effect of scale deposition on a geothermal turbine. Geothermics 31:545–562

Legmann H, Sullivan P (2003) The 30MWe Rotokawa I geothermal power project; 5 years of operation. Iceland Geothermal Conference

Lu X, Watson A, and Deans J (2009) Analysis of the thermodynamic performance of Kalina cycle system 11 (KCS11) for geothermal power plant—comparison with Kawerau Ormat binary plant. In: Proceedings of ASME Energy Sustainability Conference, San Francisco, USA

Marcuccilli F, Thiolet D (2010) Optimising binary cycles thanks to radial inflow turbines. In: Proceedings of World Geothermal Conference, Bali, Indonesia

Matsuda H (2006) Maintenance for reliable geothermal turbines. GRC Transact 30:755

Mitsubishi Heavy Industries Ltd (1998) Geothermal power generation, a company booklet, revised edition

Mitsubishi Heavy Industries Ltd (2012) http://www.mhi.co.jp

Nash (2012) http://www.gdnash.com/e-library.aspx#howitworks_

Ormat Technologies Inc (2012) http://www.ormat.com

Ozkan NY, Gokcen G (2010) Performance analysis of single flash geothermal power plants: gas removal system point of view. World Geothermal Congress

Rogers GFC, Mayhew YR (1967) Engineering thermodynamics, work and heat transfer. Longman, London

Ruths J (1924) Method and means of discharging heat storage chambers containing hot liquid used in steam power and heating plants. UK Patent 217,952

Sakai Y, Yamashita M, Sakata M (2000) Geothermal steam turbines with high efficiency, high reliability blades. GRC Transact 24:521–526

Sciubba E, Bastianoni S, Tiezzi E (2008) Exergy and extended exergy accounting of very large complex systems with an application to the province of Siena, Italy. J Environ Manage 86:372

Smith IK (1993) Development of the trilateral flash cycle system part 1: fundamental considerations. Proc Inst Mech Eng 207:179–194

Smith IK, Stosic N, Aldis C (1995) Trilateral flash cycle system – a high efficiency power plant for liquid resources. WGC, China

Smith IK, Stosic N, Kavacevic A (2005) Screw expanders increase output and decrease the cost of geothermal binary power systems. Geother Resour Counc Trans 29:787–794

Sohel MI, Sellier M, Brackney LJ, Krumdieck S (2009) Efficiency improvement for geothermal power generation to meet summer peak demand. Energy Pol 37:3370

Sohel MI, Krumdieck S, Sellier M, Brackney LJ (2010) Dynamic modeling of an organic Rankine cycle unit of a geothermal power plant. In: Proceedings of World Geothermal Congress

Thain IA, Carey B (2009) 50 years of power generation at Wairakei. Geothermics 38(1):407–413

van Riemsdijk JT, Brown K (1980) The pictorial history of steam power. Octopus Books, London

Welch P, Boyle P, Sells M, Murillo I (2010) Performance of new turbines for geothermal power plants. GRC Transact 34:1091–1096

Wood B (1970) Alternative fluids for power generation. Proc I Mech E 184(1):1969–1970

Worley NG (1963) Steam cycles for advanced Magnox gas-cooled nuclear power reactors. Proc Inst Mech Eng 178(Part 1, No 22):1963–1964

Wrangham DA (1948) The theory and practice of heat engines. Cambridge University Press, Cambridge

The Steamfield

12

The steamfield delivers the two-phase discharge from production wells to separators, from which steam is piped to the power station and separated water to the injection wells; there is little difference in principle between steamfields for steam and ORC power plant. A bird's-eye view would show a very spindly arrangement of pipes following a zigzag route, some blocks of equipment, and some holding ponds and access roads. The pipes may be several kilometres long. Given the level of attention paid to thermodynamic efficiency in using a relatively low-temperature heat source, it might be thought that thermal insulation and minimising pressure drop in the pipelines are the primary concern in steamfield design. They are important, but so also is the avoidance of scale deposition, slug flow in the two-phase lines, entry of condensate into the turbines and flashing of saturated water in injection lines, all of which have the potential to cause the power station to shut down, resulting in loss of income and expensive repairs. The control of wells in response to changes of electrical load and other disturbances must be designed for. Flash steam power plant can be selected according to the turbine inlet pressure(s), but it is only when attachment of all the wells to the power station is considered that the selection can be finalised. This chapter outlines the design considerations.

12.1 Overall Design Considerations

12.1.1 Silica Deposition

In Chap. 11 only the thermodynamic issues of flash steam plant were discussed, but the silica concentration in the separated water is an important parameter. Of the species dissolved in geothermal fluids, silica most often influences design because it can deposit to form a hard scale on equipment, in amounts capable of completely blocking a pipe. A recent paper by Brown [2011] gives the background and includes discussion of chemical anti-scaling additives. Some of these were originally developed for multistage flash evaporator desalination plants, where their use was standard, but they are used only for particular circumstances in geothermal plant.

A. Watson, *Geothermal Engineering: Fundamentals and Applications*,
DOI 10.1007/978-1-4614-8569-8_12, © Springer Science+Business Media New York 2013

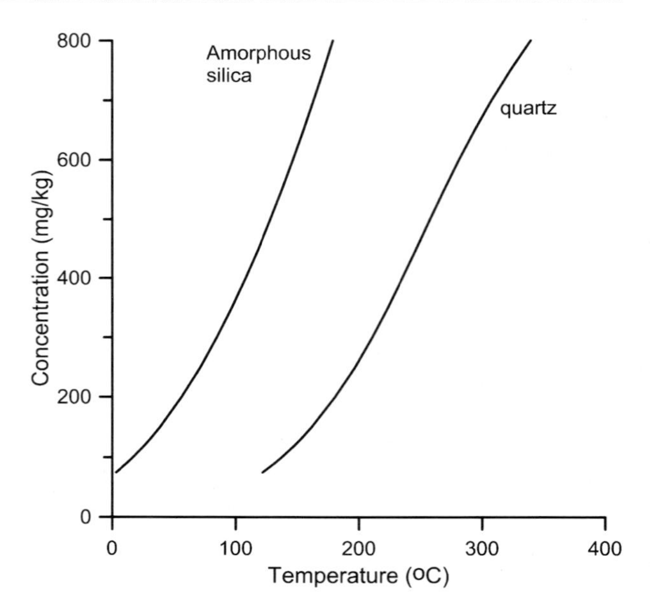

Fig. 12.1 Variation of quartz and amorphous silica solubility with temperature (°C), according to equations given by Fournier [1977, 1986]

Silica is dissolved as quartz in the undisturbed water in the resource, but if it deposits at the surface, it does so as amorphous silica. The quantity of silica in solution is determined using the solubility of quartz and the tendency to deposit using the solubility of amorphous silica. The solubility of both forms reduces strongly with temperature and is higher for amorphous silica than for quartz, as shown in Fig. 12.1. By the time high-temperature resource water reaches the surface, its temperature is reduced, and it is the amorphous silica saturation level that is important. Although this is greater than that of quartz at the same temperature, part of this advantage is taken up by an increase in concentration as a result of flashing in the well and in the separators, which increases the concentration by removing water. At the pressures and temperatures involved, the solubility of silica in steam is negligible for the present purposes. The margin between the solubilities of quartz and amorphous silica can sometimes all be taken up, leaving the separated water supersaturated with respect to amorphous silica. It will then deposit, but possibly slowly and according to several secondary effects such as pH, as explained by Brown [2011]. The solubilities are compared in Fig. 12.1.

Fournier [1977] provided simple equations for the solubility of both forms, providing Eq. (12.1) for amorphous silica. Later he produced an equation quoted by Brown [2011] and shown here as Eq. (12.2), explicit in T presumably because of its importance in geothermometry.

$$\text{Amorphous} \quad \log_{10}\mathrm{C} = -731/\mathrm{T} + 4.52 \qquad (12.1)$$

valid over the temperature range $0 < \mathrm{T} < 250\ ^\circ\mathrm{C}$

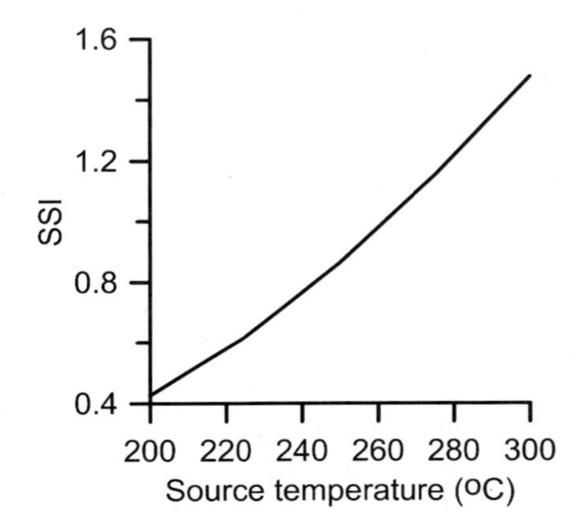

Fig. 12.2 Showing silica saturation index (SSI) as a function of source temperature for a separation pressure of 6 bars abs

$$T = -42.196 + 0.28831^{*}C - 3.6685E{-}4^{*}C^{2} + 3.1665E{-}7^{*}C^{3} + 77.034^{*}\log C + 273.15 \tag{12.2}$$

valid over the temperature range 20 < T < 330 °C

where C = concentration in mg/kg

T = temperature in K

Equation (8.23) gave the increase in concentration of a dissolved species in saturated water when it is flashed.

$$C_f = \left[\frac{\dot{m}_D}{\dot{m}_f}\right].C_D = \frac{C_D}{(1-X)} \tag{8.23}$$

where C_D and C_f are the concentrations in the total discharge and separated water, respectively, and X the dryness fraction. The dryness fraction from Eq. (3.18) is

$$X = \frac{h - h_f}{h_{fg}} \tag{12.3}$$

and h_f and h_{fg} are functions of temperature, like the solubilities, so variations with T are important. The parameter used to judge the potential for silica to be deposited, taking into account all the relevant processes, is the silica saturation index (SSI), which Brown [2011] defines as the ratio of the silica concentration in the liquid divided by the equilibrium amorphous silica solubility at the conditions prevailing. Figure 12.2 illustrates the effect of source temperature on silica saturation index if the source is liquid saturated with quartz and the separation pressure is 6 bars abs.

The quartz solubility is high at temperatures in the region of 300 °C and flashing this source fluid at 6 bars abs results in a value of silica saturation index of SSI > 1, indicating that scaling will occur. If the well produces from a formation containing two-phase fluid, the concentration of quartz in solution depends on the history of the two-phase mixture; the origins of excess enthalpy were discussed in Sect. 8.1.2.

The steamfield design will have begun before drilling all of the production wells. There are economies to be gained by drilling several deviated wells from a single site, and perhaps at least one from every site will have been drilled and tested so the chemistry of the fluids in that part of the resource will have been established, and there is a reasonable chance that wells yet to be drilled will have the same chemical characteristics. The discharge characteristics, including chemical characteristics, will have been established sufficiently to allow the fluid processing to be decided upon, single or double flash steam and/or ORC plant, and non-condensable gas content of the total discharge will have been estimated in connection with power station condenser design. A silica saturation index of 1.0 may be a nominal design criterion in selecting separation pressures, but it can sometimes be exceeded safely. This is a resource-specific matter calling for geochemical expertise, bearing in mind the secondary geochemical effects that have an influence.

12.1.2 Overall Steamfield Layout

Many of the tasks involved in the engineering design of the steamfield are repetitive and time-consuming, and although they rely on publicly available research material, the expertise tends to lie with developers and consulting companies rather than with academics. It is common in process engineering to start out with one design then extrapolate it by a margin when a new design is called for, a process of gradual improvement the results of which are commercially confidential and not released.

The wells have different discharge characteristics and discharge chemistry, but there are economies to be had by standardising pipelines and separators, so in a new field, production wells may be grouped to produce a reasonably similar total flow rate into each separator station. This is quite different from the original design of Wairakei, for example, which used wellhead separators and now has a mixture of design approaches—see Thain and Carey [2009]. Figure 12.3 shows the elements of a hypothetical steamfield set out as a process diagram. The steamfield has three separators each supplied by three production wells and discharging the separated water to three injection wells. Separation at wellhead was the usual practice in early developments of liquid-dominated resources, including Wairakei, to avoid piping two-phase fluid, but this meant that second-stage flash steam would require steam pipelines equal in length to those of the primary steam, which would be uneconomic. James [1967] made the case for transporting the entire two-phase well discharge at Wairakei to separators near the power station. In the same year the Otake (Japan) 12.5 MWe geothermal power station began operation, adopting the same solution. The results of two-phase pipeline experiments were reported by Takahashi et al. [1970]. Since then two-phase pipelines have become more common and provide some freedom in siting the separator stations. Whether a single separator is used depends on the power output of the wells and the proven separator design available, and the single separators shown in Fig. 12.3 might be separator stations (groups of separators). Separator design is empirical, and a separator is unlikely to perform to specification over the full range of flow from zero to

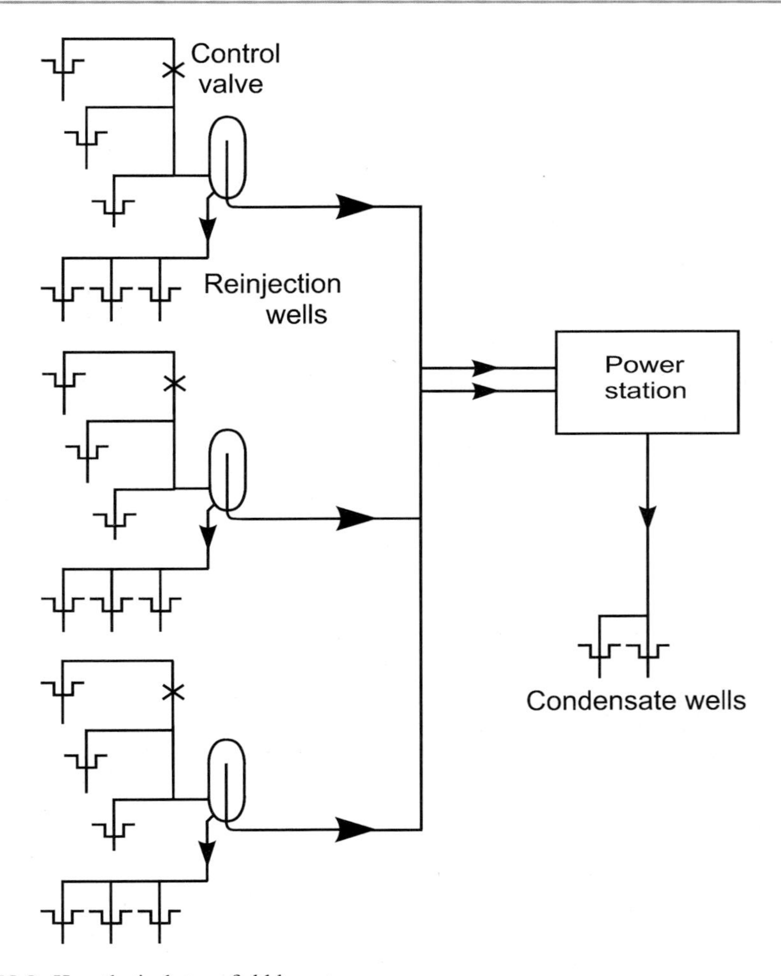

Fig. 12.3 Hypothetical steamfield layout

maximum input, so valves can be fitted to production well output pipelines to control the flow. Variations between production wells may occur and some individual control or setting of wellhead pressure is likely to be needed. Because the steamfield covers long distances, it is appropriate to motorise some valves so they can be operated remotely and quickly from the power station. This requires a steamfield control system that can be manually operated for some changes, but has some automatic features to deal with unexpected changes such as failures. The main components of the arrangement are drawn in a "Process and Instrumentation" diagram, an example of which is given by Ciurli and Barelli [2009] for the Monteverdi field in Italy, a dry steam resource, so there is no separated water. A very extensive steam pipeline network for a liquid-dominated resource is that of Cerro Prieto, shown by Garcia-Guttierez et al. [2012], but like Wairakei, it is an old development.

Some standby production wells are required, and these can be used as an extra means of control of flow to the separator station. Standby injection capacity is also necessary. The figure shows duplicate steam pipelines to the power station, as a reminder that "all eggs should not be put into a single basket" and to allow for maintenance outages, and also provides spare separated water and condensate injection capacity. Condensate is disposed of to separate condensation wells; it is not mixed with the separated water as it is usually very acidic as a result of the dissolved gases.

A resource assessment will have helped decide on the power station capacity to be installed, and the well discharge characteristics and chemistry will have guided the selection of turbine inlet pressures (assume a single flash steam turbine arrangement for this discussion). The steam line diameter can be chosen based on the range of diameters of pipe available and the number of pipelines proposed, assuming a suitable mean velocity of steam in the pipes. The number of wells required at full output can be estimated from the steam consumption of the turbine(s) indicated by the potential suppliers, and some standby wells will be required also. The consequences of a separator in Fig. 12.3 being out of action may be that the steam supply from the remaining connected wells can be increased. This might require more wells to be discharged, each with a higher wellhead pressure than normal to allow a higher pressure drop through some of the steam lines, which will have to transmit a greater flow rate than normal to make up for the separator station that is out of commission. Consideration must also be given to the possibility of wells running down in the longer term, or their discharge characteristics changing—in general, a "what if" examination must be carried out, based on a list of possible occurrences.

12.2 Pipeline Design

12.2.1 Steam Pipeline Design

There are both structural and process issues in the design of cross-country steam pipelines, which may be kilometres in length and have diameter of 1.0 m or more. The forces on the pipeline are internal pressure, which produces hoop and axial stresses as discussed for casing in Sect. 5.4.1, forces due to the change in momentum as the fluid flows around a bend, analogous to those already discussed in relation to turbine blade design, and axial stresses due to thermal expansion. The linear thermal expansion coefficient for steel is typically 12.0E-6/°C, so filling a cold, empty steam pipeline at atmospheric conditions with steam at a pressure of 6 bars abs will cause an increase in length of 1.6 %, or 1.6 m per 100 m. This is absorbed by building flexibility into the pipeline, either by making it take a deviated route (Fig. 12.4a, b) or by introducing bellows (Fig. 12.4c). The flexibility in the zigzag arrangement may be sufficient with the pipe rigidly constructed. Alternatively, longer, less flexible, straight lengths may be used with extra flexibility

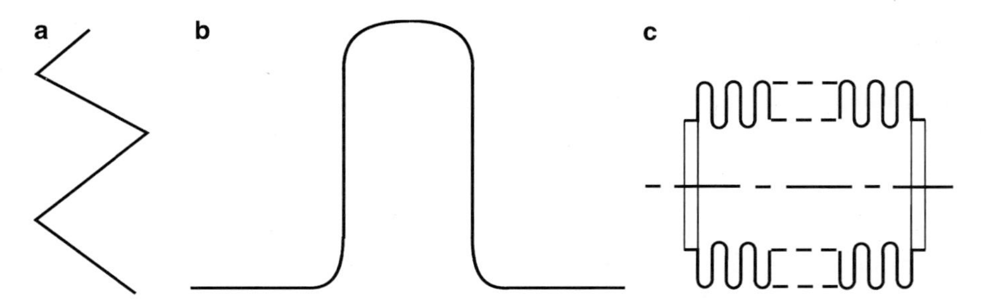

Fig. 12.4 Showing pipeline design features: (**a**) plan view of a horizontal zigzag layout, (**b**) a vertical expansion loop and (**c**) expansion bellows detail

provided by bellows at every corner, bellows being a short length of pipe with a cross section as shown in figure. The zigzag pattern can follow the lie of the land provided that the supports are suitably designed. The expansion loop, Fig. 12.4b, which does not need bellows, can be either vertical or horizontal—vehicular access to both sides of the pipeline is facilitated if it is vertical, avoiding the need for bridges or tunnels, but greater structural support is needed. The pipeline in general must be held above ground on stable supports that restrain it but permit axial and rotational movements as designed to provide the flexibility. Pipeline routes are always walked by the designers and surveyed, with geotechnical assessment carried out to ensure firm foundations for the supports.

Stress analysis for geothermal pipelines is usually carried out to comply with ANSI code B31.1 [2001], which gives formulae for wall thickness, details of flanges and pipe fittings, etc. Measures to protect the pipeline against overpressurisation are required, such as bursting discs.

As a result of frictional pressure drop, the steam flow in the pipe becomes wet, and heat loss through the insulation adds to the wetness. Although the processes are thermodynamically disadvantageous, the condensate produced is beneficial in providing a means of removing silica from the steam, a process referred to as "scrubbing". The steam flow leaving a steam well or separator carries a small amount of liquid spray which dries out, leaving amorphous silica droplets. These are aggregated into the water droplets, which also dissolve part of the noncondensable gas in the steam flow. The resulting acidic solution can be drained from the bottom of the pipe, where it collects. Freeston [1981], who reported that the steam in the Wairakei steam pipelines was typically up to 0.5 % wet, carried out experiments to improve the drainage of water from the bottom of the pipe and developed a design of "drainpot", to be placed at intervals along the bottom of the pipeline. Sulaiman et al. [1995] give details of an investigation of silica scaling in steam pipelines at the Kamojang geothermal field, Indonesia.

The calculation of pressure drop and that of heat loss through the insulation are important for the reasons explained. Steam is a single-phase fluid—the small degree of wetness can be ignored so far as pressure drop is concerned—and the approach to frictional pressure drop was set out in Sect. 4.3.4. Equation (4.50) is the relevant one, with the definition of Fanning friction factor in Eq. (4.51). The cross-sectional flow area A can be introduced and the equation is then

$$\frac{1}{2}\rho\bar{u}^2.f.\pi d = A\frac{dP}{dx} \tag{12.4}$$

The mass flow rate can be introduced as

$$\dot{m} = \rho\bar{u}A \tag{12.5}$$

and some manipulation leaves a useful form for the calculation of pressure gradient:

$$\frac{dP}{dx} = \frac{2f}{\rho d}\left(\frac{\dot{m}}{A}\right)^2 \tag{12.6}$$

Steam velocity is usually in the range 30–50 m/s, and a pipe of 600 mm diameter, for example, flows at a Reynolds number of almost 4.0E6, assuming saturated steam at 6 bars abs. This is outside the range of the smooth pipe friction factor given as Eq. (4.55), a common problem in geothermal pipelines for both steam and two-phase flows, as most industrial applications are for flows at higher pressure in smaller diameter pipes. McAdams [1954] gives a version with a range to Re = 3.0E6 as follows:

$$f = 0.00140 + \frac{0.125}{\text{Re}^{0.32}} \tag{12.7}$$

Both laminar Eq. (4.53) and turbulent flow Eqs. (4.55) and (12.7) are plotted in Fig. 12.5; the turbulent flow ones are coincident. The laminar flow curve is not continuous with the turbulent flow curves because the flow is unsteady at the laminar–turbulent transition region, a Reynolds number of about 2,100. The fact that in laminar flow the shear stress is a greater proportion of the kinetic head of the flow than in turbulent flow is counter-intuitive. Both of the turbulent flow curves are correlations of experimental results for smooth pipes. A family of curves for a range of pipe roughness can be found in many fluid mechanics texts.

The pressure drop over valves and fittings must be taken into account. The flow is disturbed and has increased losses, so the pressure drop is greater than pro rata to the length of flow passage involved. The additional pressure drop is given for each type of fitting as the equivalent length of straight pipe, which makes calculation easier.

Thermal insulation usually consists of fibrous material of low thermal conductivity which traps air and keeps it motionless—since it also has low thermal conductivity, the combined effect is large. The insulation must be kept dry and it is normal practice to contain it around the pipe by means of aluminium sheet. An expression

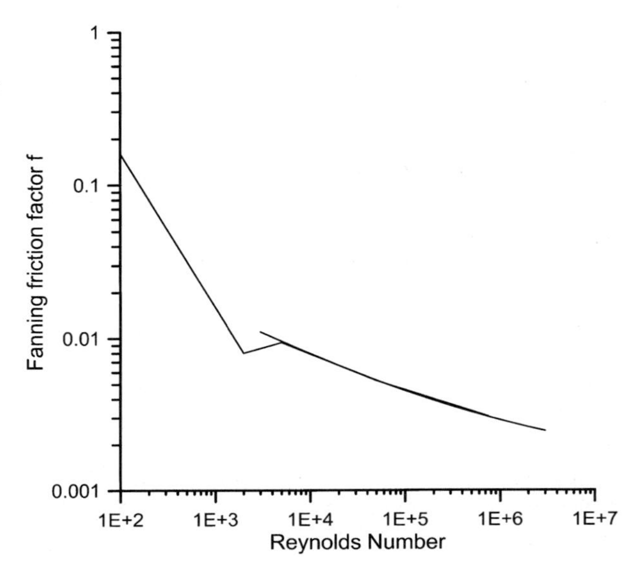

Fig. 12.5 Showing the variation of Fanning friction factor with Reynolds number for smooth circular pipes

for the heat loss through any thickness of insulation on a pipe can be calculated using the steady-state, zero internal heat generation form of Eq. (4.65), which is

$$\frac{d}{dr}\left(r\frac{dT}{dr}\right) = 0 \tag{12.8}$$

The pipe radius can be taken as the inner radius of the insulation, and integrating gives

$$r\frac{dT}{dr} = B_1 \tag{12.9}$$

where B1 is a constant. If the total heat loss per metre of pipe is Q_{loss}, it can be written as

$$Q_{loss} = \lambda.2\pi a.\left(\frac{dT}{dr}\right)_{r=a} \tag{12.10}$$

which is equivalent to a heat flux at the inner surface of the insulation of

$$\dot{q}_{r=a} = \frac{Q_{loss}}{2\pi a} \tag{12.11}$$

so

$$B_1 = \frac{a.\dot{q}_{r=a}}{\lambda} \tag{12.12}$$

A second integration of Eq. (12.9) results in an expression for temperature as a function of radius through the insulation

$$T = B_1 \ln r + B_2 \tag{12.13}$$

and introducing the outer boundary condition of ambient temperature, T_{amb}, at the outer radius b, an expression for the heat loss as a heat flux at the pipe wall can be found as follows:

$$\dot{q}_{r=a} = \frac{\lambda \Delta T}{a \ln\left(\frac{a}{b}\right)} \tag{12.14}$$

where ΔT is the temperature difference over the insulation.

The outer boundary condition should strictly be a convective one, since the pipe is exposed to the wind and rain, and even in still air will have a natural convection plume, but it is sufficient to assume uniform temperature at the local ambient, T_{amb}. There is a maximum thickness beyond which the outer surface area is increasing the heat loss faster than the extra thickness is decreasing it, although this is something of an academic distraction. Ovando-Castelar et al. [2012] have estimated that the heat loss through Cerro-Prieto steam pipelines is equivalent to 2.5 % of the installed capacity of the power station.

Various methods of simulating the performance of steam pipeline networks have been developed and are mentioned in the literature. Perhaps the first was that of Marconcini and Neri [1978]. The fundamental problem is that a network of interconnecting pipes supplying steam to a power station may be supplied by a large number of wells. The steam mass flow rate provided by a well is a non-linear function of wellhead pressure, similar to that of Fig. 8.1, but with constant specific enthalpy. Every well has a unique characteristic. A steady-state solution only is required. Ciurli and Barelli [2009] demonstrated the usefulness of the approach by applying it to the problem of reconfiguring the network of the Monteverdi geothermal field, Italy.

12.2.2 Two-Phase Pipeline Design

There are many similarities in the design of steam pipelines and two-phase pipelines—the same structural issues arise and the approach to thermal insulation is the same. There is no scrubbing to be incorporated. Yet the design of pipelines for two-phase flow is much more difficult than for steam, and the calculation methods used are not given much publicity.

The empirical nature of two-phase flow predictions was emphasised in Sect. 7.7. Equation (7.14) was formulated to show that the pressure gradient in a two-phase flow could be considered to be made up of three components, due to friction, acceleration and gravity. The results of predicting steady-state well discharge characteristics were demonstrated in Sect. 8.5; the calculation methods are generally acceptable provided some parameters are left to be adjusted by trial and error.

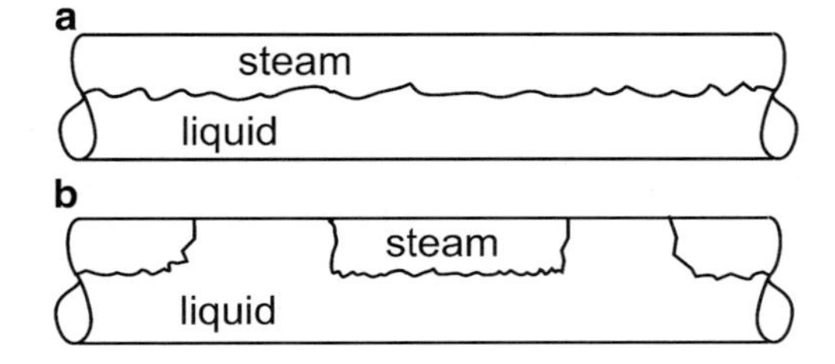

Fig. 12.6 Sketch of two possible flow regimes in horizontal, adiabatic two-phase flow (**a**) stratified flow (**b**) slug flow

The acceleration component was shown to be small, and it was remarked that if the flow had been heated, acceleration would have been greater due to thermal expansion of the flow; the flow in a well that has been discharging for some time is adiabatic, and the flow through an insulated two-phase pipeline is virtually the same. A compilation of data by the Engineering Sciences Data Unit [2002] to assist the calculation of frictional pressure gradients in two-phase horizontal adiabatic flow notes that acceleration contributed only 1 % to the total pressure drop over the many experiments considered.

The gravitational component is at its largest in the flow in a vertical well, and Fig. 8.13 showed that it ranged from 20 to 80 % of the total pressure drop. Two-phase pipelines must follow the lie of the land, so will have uphill and downhill sections, but usually at modest slopes, so the gravitational component will be less, but not negligible. The slope could be kept to a certain value, given some guidance.

The empirical nature of the pressure drop calculation, plus the nature of two-phase flow itself, gives rise to uncertainty in the results of pressure drop predictions. The problem is exacerbated by geothermal pipelines needing to be larger in diameter than in other industries and thus larger than those used to formulate the empirical calculation methods. Of the 6,453 measurement data points in the ESDU compilation, only 11 came from a set of experiments with a pipe diameter of 0.3048 m (12″), the rest were much smaller. Two-phase pipeline diameters of 0.75 m or more are used at some fields. In New Zealand, the method of calculation produced by Harrison [1974] still appears to be used; it is based on a series of experiments carried out at Wairakei, but using pipe diameters smaller than required in modern practice.

The main design difficulty arises because of the potentially damaging effect of slug flow, and an inability to accurately predict when it will occur. Analogous to the flow regimes in vertical two-phase flow shown in Fig. 7.6, Fig. 12.6 shows two of several recognised regimes of horizontal flow.

At a low-enough flow rate, the liquid flows along the bottom of the pipe, but given the right circumstances, the liquid surface can develop waves that rise and completely fill the tube, trapping a steam pocket between them. These piston-like slugs of water have considerable momentum and can produce large forces at bends. Both uphill and downhill sections of pipe can be imagined to produce slugs as the water in the upper part of the wave is influenced by gravity. Experiments in which recognisable flow regimes are created at measured superficial velocities of steam and water have been carried out, sufficient that boundaries between the regimes can be drawn on the plotted datum points—regime charts. Ghiaasiaan [2008] offers the opinion that a chart with axes of superficial gas and liquid velocities produced by Mandhane et al. [1974] (reproduced in Foong et al [2010]) is probably the most widely accepted map for horizontal flows. Superficial velocities were defined in Sect. 7.7.2.

As an example of the difficulty of dealing with this issue, Foong et al. [2010] explain how they strengthened the pipe and supports to withstand slug flow loads rather than rely on choosing the operating parameters to avoid the regime using a flow map, and this approach has merit. Apart from the reduction in power station output that would follow a pipeline failure, the environmental damage to flora from a spill of separated water is significant, leaving a patch of ground that may take many years to become revegetated to its previous condition.

12.2.3 Design of Pipes Carrying Water, Including Injection Wells

Injection pipelines to deliver separated water at saturation conditions and with a high-silica content to injection wells are designed based on steady-state operation as for the other steamfield pipelines. Provision must be made for them to be emptied into holding ponds to avoid scaling, in the event of an emergency, and the holding ponds must be kept empty of rainwater in preparation. Pipeline drains need to be provided at locations of low elevation. Water velocities up to 3 m/s are typical and the same method of pressure drop calculation applies as in Sect. 12.2.1 above. At some geothermal fields, separated water must be pumped to injection wells, which in the past has led to problems of silica deposition in the pumps. Kotaka et al. [2010] report that two-stage injection pumps are used at two New Zealand geothermal projects, to increase injection capacity, with the implication that this approach may have resulted from an optimisation study involving the cost of injection wells, the availability of resource area for injection and formation permeability.

The diameter of injection wells is a parameter that can be selected, rather than simply accept the standard production well diameter of 9 5/8″, although the permeability of the receiving formations and the standing water level in the well need to be known first, so a standard diameter well may be the choice for the first one. The total flow rate of separated water to be injected can be estimated from the discharge well characteristics, the number of flash stages proposed and their pressures. The available pressure head to drive the water into the formation, in the absence of pumps, is equivalent to the difference in water levels between the well standing shut and the water level in the injection pipeline when the well is receiving fluid—this could be the separator water level, at maximum. At a given flow rate, some of this head is lost in frictional pressure drop through the pipeline and casing down to the formation, and the balance is available as a pressure difference above the undisturbed formation pressure. Using the Theis solution and the frictional pressure drop calculation approach outlined for steam pipeline pressure drop above, the effect of selecting the next larger sizes of production casing for a proposed well can be examined. The maximum use should be made of the formation capacity to accept fluid, consistent with value for expenditure—casing costs and associated drilling costs increase with diameter (casing costs are of the order of 20–25 % of the total well cost), and the total increase can be compared with the cost of drilling standard wells.

Where a tubular condenser is used, the condensate will be separate from the cooling water. It will be acidic as a result of the dissolved gases and may be disposed of to a dedicated condensate injection well. Otherwise, with direct contact condensers, the condensate is mixed with the cooling water, which is fresh water to which biocides are added to avoid the growth of organisms in the cooling towers and pond (basin). A continuous discharge and replacement with fresh condensate is usually arranged, and the "overflow" is injected.

12.3 Steamfield Equipment Other Than Pipes, Including Separators and Silencers

12.3.1 Separators

The separators used in liquid-dominated geothermal developments are cyclone separators, in which a high-speed two-phase flow is injected tangentially into a tall vertical cylinder at about mid-height. The water moves to the wall of the cylinder under centrifugal action and flows downwards as it rotates, forming a pool in the base of the cylinder, from which it is drained. The steam is displaced upwards to the cylinder axis where it enters an axial pipe and flows downwards and out to the steam pipeline. Cyclone separators are used in many industries, although mainly smaller in size than geothermal separators. The separation efficiency for geothermal service is often quoted as 99.95 % or higher, and the literature suggests that measurements of efficiency within 0.05 % are possible. The literature on the general topic is very large, but most applies to the separation of solids from a solid–gas two-phase flow. Liquid two-phase flows are separated in the petroleum industry, but the conditions are sufficiently different for a different approach to have been adopted in many cases. For geothermal two-phase flows, a standard arrangement has been adopted, for which the design review by Bangma [1961] set the scene. The basic design is shown in Fig. 12.7 and is based on separators originally designed for Wairakei—see Thain and Carey [2009]. The steam exits the vessel through its base. This is not the only design proposed for geothermal use, Cerini [1978] reported on a rotary separator which was a combination of separator and two-phase turbine, and Schilling [1981] reported on a design with a swirling entry through a vertically upwards divergent nozzle into a large diameter vessel from which dry steam left at the top. The Bangma pattern has persisted, but increased in size. Computerised fluid dynamics modelling has been applied recently by Pointon et al. [2009] and by Purnanto et al. [2012]; however, separators are developed rather than designed from scratch, so far as the process is concerned. Their structures are designed according to standard pressure vessel codes. Part of

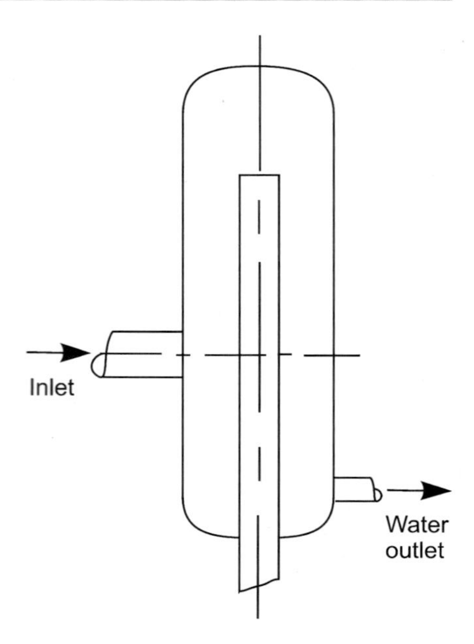

Fig. 12.7 Vertical axis separator as investigated by Bangma [1961]

Pointon et al.'s study was concerned with a structural failure investigation, vibration of the central steam exit pipe having led to its failure.

Bangma [1961] described a range of tests, including, for example, spiral inlet instead of tangential, for what has continued to be the basic design, illustrated by Fig. 12.7.

Bangma recommended the proportions for the design, with the cylinder internal diameter three times the steam exit pipe diameter, and the two-phase entry seven steam pipe diameters below the top of the vessel. In his tests the vessel diameter was 0.38 m and the maximum two-phase flow rate that was used was 126 kg/s. In comparison, the separator examined by Pointon et al. [2009] had a diameter of 3.3 m and the two-phase inlet flow rate was 520 kg/s; the diameter had increased by 8 times for an increase of flow rate of 4. Between these two contributions, Lazalde-Crabtree [1984] reviewed the design approach and produced design recommendations for separators and steam driers.

In the majority of designs, the separated water does not exit the main separation vessel and enter the injection or next-stage flash pipeline directly. Instead it enters a holding tank which is conveniently built into the base of the main vessel. The water level in this tank is a matter of some importance to transient behaviour of the system arising from faults.

12.3.2 Silencers

Silencers were introduced in Chap. 8, essentially separators as described in the last section, but open at the top so that the separation occurs at atmospheric pressure. They were used at Wairakei and elsewhere and are illustrated by Thain and Carey [2009]. They were commonly used for flow measurement, when they were usually constructed of sheet metal, perhaps with wooden slatted cylinders; however, they are not required for the chemical tracer methods described in Sect. 8.4.4, which is the main advantage of the methods. On a completed development, permanent silencers are justified and are usually made entirely of concrete, providing long-term arrangements for well discharge and testing. Being unpressurised, the structural design of these is less demanding than for separators and pipelines.

12.3.3 Steam Discharge Silencers

Silencers for single-phase fluids consist of perforated tubes surrounded by permeable material that diffuses and attenuates the pressure waves that have entered through the perforations. They have been used for power plant steam discharge silencing, for example, with a design for geothermal plant by Nishiwaki et al. [1970], and for fossil-fuelled stations by Clayton and Cramer [1979] for very large flow rates of high-pressure steam. Injecting water into a steam flow would provide some silencing and might be arranged at a geothermal resource using a standard discharge silencer and recirculating the injected water.

For high-mass flow rates over a long period of time, a different method is normally used, called a rock muffler; this device may also be used for two-phase flows. It is made by terminating the pipeline in several short branches, each with a blocked end and perforated with many holes, not unlike slotted liner but usually circular holes. The pipeline is directed into a pit, where the branches are covered with rocks—hence its name. The design criteria are set out by Zein et al. [2010] who considered a portable version. The criteria included allowing a large enough area of holes to allow subsonic flow through them, which would reduce the noise level at source.

12.4 Transient Performance of the System of Wells, Steamfield and Power Station

The entire system is usually designed to operate as base load plant, but faults do arise and the system must be designed to accommodate them. A control system is required in addition to last ditch equipment such as safety valves and bursting discs.

12.4.1 Water Hammer

Water hammer is a potential problem in all piping of liquids and two-phase mixtures, from domestic to industrial scale, including large water ducts for hydro-electric power stations. The sudden closure of a valve or some other type of flow disturbance causes a compression wave (a shock wave) to travel along the pipe, rebounding off closed ended branches, etc., and increasing the stress on the pipeline and its fittings. It occurs in pipelines carrying pure liquids at uniform temperature well below saturation temperature, conditions for which the phenomenon is reasonably predictable so that pipelines can be suitably reinforced. It also occurs where the liquid contains dissolved gas or is close to saturation so that "column separation" can occur, i.e. the formation of a gas pocket or a bubbly mixture in the pipeline (cavitation), and methods of predicting column separation and their effects are poor.

According to Bergant et al. [2006], the physics of water hammer was first recognised in France in 1858 and the sudden closure of a downstream valve was the first system to be studied, in Russia by Joukowski (1897). They cite the example of a pipe supplying a hydroelectric power station in Japan failing—the concrete manifold to the turbines split open as a result of the high-pressure wave and the flow in the pipeline cavitated as a result of the low-pressure wave and caused the pipe to collapse inwards. An accidental sudden downstream valve closure initiated the events.

Leaving aside column separation, the equations for a water hammer analysis that sufficiently represent a straight pipe are a continuity equation and an axial momentum equation. For a compressible fluid flowing in an axisymmetric pipe, the continuity equation is Eq. (4.6) written in axial coordinates

$$\frac{\partial \rho}{\partial t}+\frac{\partial(\rho u)}{\partial x}+\frac{1}{r}\frac{\partial(\rho r v)}{\partial r}=0 \tag{12.15}$$

and the corresponding momentum equation is Eq. (4.8). For a shock wave the effects of viscosity are neglected, leaving the equation as

$$\rho\frac{Du}{Dt}=\rho\left(\frac{\partial u}{\partial t}+u\frac{\partial u}{\partial x}+v\frac{\partial u}{\partial r}\right)=-\frac{\partial P}{\partial x} \tag{12.16}$$

Radial velocities and variations can be neglected, leaving a one-dimensional problem. Assuming that the fluid has a small compressibility, Eq. (4.81) used in the development of porous media flow

$$\frac{\partial \rho}{\partial t}=\rho c\frac{\partial P}{\partial t} \tag{4.81}$$

can be employed to further simplify Eq. (12.15) from which ρ can be cancelled to give

$$\rho c\frac{\partial P}{\partial t}+\frac{\partial u}{\partial x}=0 \tag{12.17}$$

to be solved with

$$\rho\frac{\partial u}{\partial t}=-\frac{\partial P}{\partial x} \tag{12.18}$$

Note that the momentum convection term in the x direction has been discarded as negligible. Streeter and Wylie [1967] solved this pair of equations numerically, including a term added to the momentum equation to account for wall friction and incorporating a friction factor; several software packages for water hammer are commercially available and are reviewed by Bergant et al. [2006].

The simplicity of this pair of equations, Eqs. (12.17) and (12.18), might well lead to doubts that they could predict events in a real pipeline several kilometres in length with bends, valves and fittings. In the introduction to their book, Sharp and Sharp [1995] state that the basic theory is well developed for single-phase fluid but that significant radial variations do occur and that friction cannot be adequately represented as indicated above. They appear critical of the simplicity adopted for numerical solutions. For designers the remedy seems to be to carry out a water hammer analysis and be guided by the results, but soften the response of valves and mechanical moving parts and protect the system with bursting discs.

12.4.2 The Potential for Flashing in Separated Water Pipelines as a Result of Fault Conditions

Water at saturation temperature can flash almost explosively, because the specific volume of the steam is orders of magnitude higher than the water; short pipelines carrying it attracted design attention in connection with fossil-fuelled power stations, in particular, the boiler feed water pump inlet piping. The condensate is delivered from the condenser to a storage tank at the elevation of the top of the boiler. It then flows down to the boiler feed pumps through an essentially vertical pipe, but with bends and valves; the storage tank height is chosen to provide enough pressure at the pump to exceed its suction head. Dartnell [1985] shows the arrangement for a typical 500 MWe power plant. The tank acts as a "de-aerator" by being supplied with steam to keep the water at saturation temperature—the tank is maintained at a pressure of up to 10 bars abs and is a sealed pressure vessel, with a large volume of saturated steam over the saturated water. If the turbine trips or moves away from its designed steady operating condition, the steam supply to the tank may be reduced, the water in the tank may cool and the subsequent condensation of the steam above the water may cause the pressure in the tank to fall. The suction of the feed pumps may then reduce the water pressure in the downcoming flow to a level at which the water flashes, and the steam bubble formed eventually collapses, with the effects discussed in Sect. 7.5. Wilkinson and Dartnell [1980] list nine catastrophic failures in such pipes and valves due to steam bubble collapse in UK power stations over a 20-year period.

The relevance of this is that water at saturation temperature may be in transit along a pipe when some upstream event causes the supply pressure to fall. A pressure change travels at the speed of sound in the fluid, which is virtually instantaneous compared to the speed of the flow, so at the point along the tube where the pressure falls below saturation, the water flashes and forms a volume of steam. If the pressure rises again and the steam condenses suddenly, a shock wave

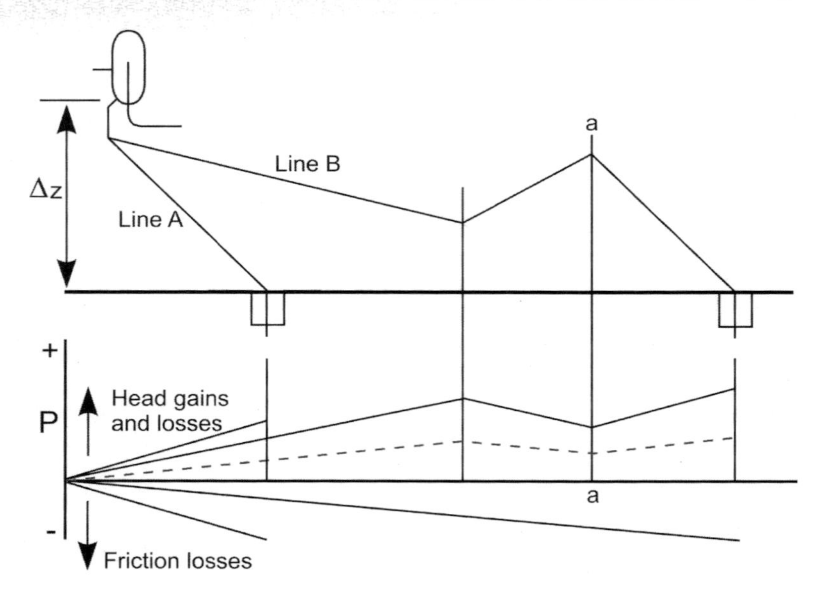

Fig. 12.8 Separator and injection line

may be produced which bursts the pipe or its fittings in the manner of water hammer. Dartnell [1985] quotes the transit time of water in his system as typically 25 s. In a 1.0 km injection line, it is typically 5 min, and measures are required to control the rate of fall of absolute pressure in the line. Gage [1951] had already proposed a design approach to avoid similar problems in power station boiler feed water pumping systems, and a solution for the geothermal steamfield problem was described by Watson et al. [1996].

The focus is on the separator pressure since it controls the absolute pressure distribution in the separated water pipeline containing the saturated water.

Figure 12.8 shows a simple separator drained by two injection lines. Consider first that labelled A, of length L which loses elevation Δz with a downward slope to an injection well. Under steady-state conditions, the pressure at the injection wellhead is

$$P_{wh} = P_{sep} + \rho g \Delta z - \Delta P_{fric} \tag{12.19}$$

where ΔP_f is the frictional pressure drop over the line. If the transit time of the water through the line is τ and if the separator pressure is changing at a rate $\left(\frac{dP}{dt}\right)_{sep}$, the pressure of an element of water reaching the injection wellhead will be

$$P = P_{sep,t=0} + \rho g \Delta z - \Delta P_{fric} - \tau \left(\frac{dP}{dt}\right)_{sep} \tag{12.20}$$

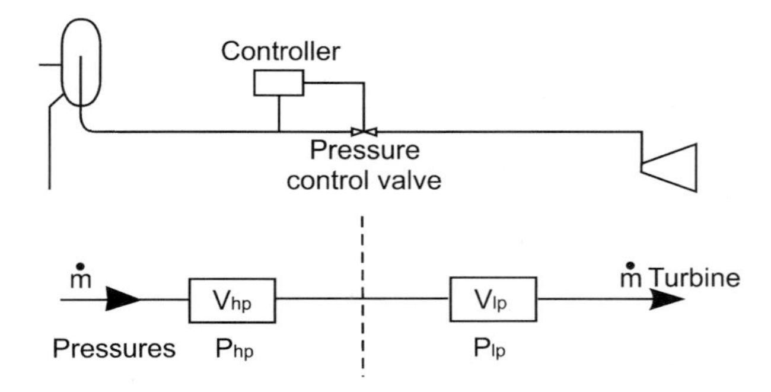

Fig. 12.9 Simplified single-separator and turbine steam line

If the line is insulated so there is no heat loss, then to avoid flashing, the water reaching the wellhead must be at $P = P_{sep,t=0}$ which requires that

$$\left(\frac{dP}{dt}\right)_{sep} = \frac{1}{\tau}\left(\rho g \Delta z - \Delta P_{fric}\right) \tag{12.21}$$

If the pipe friction factor is f, this equation can be rearranged as

$$\left(\frac{dP}{dt}\right)_{sep} = g\left(\frac{\Delta z}{L}\right)\left(\frac{\dot{m}}{A}\right) - \frac{2f}{\rho^2 d}\left(\frac{\dot{m}}{A}\right)^3 \tag{12.22}$$

The aim is to get the element of fluid through the pipe as fast as possible, since the separator pressure has the potential to decrease (increases are not a problem), but the axial pressure decreases due to friction increases with the speed of the flow. This equation balances the two effects and the pressure head graph in the lower part of Fig. 12.8 shows them exactly balancing each other. Cross-country injection lines have to follow the lie of the land and may rise and fall, as shown as line B of the figure; it reaches a highest elevation marked as "a". The pressure head graph shows a dotted line representing the net head, the difference between frictional and gravitational heads, which is at a minimum at a. Because of the extra length of line B, the allowable separator pressure decline rate will be smaller than for line A.

The separator is connected to three pipelines, steam, two-phase well discharge and separated water, and the following events could change its pressure:

- Steady change in turbine steam demand due to change in electrical load, e.g. a large increase in load
- Sudden loss of electrical load and steam flow
- Failure of other power station equipment that might alter the steam demand
- Failure of a bursting disc in a two-phase line

The separator pressure can be protected from changes in steam flow rate by introducing a control valve in the steam main. In Fig. 12.9 a single separator supplies a turbine, and an automatic control valve (a butterfly valve) situated in the steam line is able to open or close to maintain the separator pressure within the rate of change required to protect its injection line.

The line can be represented as two volumes, upstream and downstream of the control valve referred to as high pressure and low pressure (suffixes *hp* and *lp*, respectively), and a mass balance equation set up for each:

$$V_{hp}\frac{\partial \rho_{hp}}{\partial t} = \dot{m}_{hpin} - \dot{m}_{hpout} \tag{12.23}$$

$$V_{lp}\frac{\partial \rho_{lp}}{\partial t} = \dot{m}_{lpin} - \dot{m}_{lpout} \tag{12.24}$$

The volume of each part of the line is V_{hp} and V_{lp} and each has an inlet and outlet mass flow rate. It is assumed that the steam is saturated and an expression for its density is required; for the level of precision required of this calculation, it may be sufficient to assume P/ρ is constant and find the constant from steam tables over the range of pressure required. This allows the left-hand terms of the equations, rates of change of density, to be written as rates of change of pressure. The inlet and outlet mass flow rates into each section can now be defined. For the steam mass flow rate to the turbine, the Willans line, Fig. 11.13 and Eq. (11.1), relates the mass flow rate to the electrical load, and a linear relationship can be assumed between mass flow rate and turbine inlet pressure. These define $\dot{m}_{lpout}$. The mass flow rate into the low-pressure section is that flowing out of the high-pressure section and is governed by the control valve which has a pressure drop–mass flow rate relationship of the form

$$\dot{m} = \Psi\left(\theta, P_{hp}, P_{lp}\right) \tag{12.25}$$

where Ψ is a given function and θ is the valve angle; these being obtainable from the manufacturers of the valve being considered. The valve is driven in response to the upstream pressure, as indicated by the figure, which is representative of the separator pressure, and this is arranged for in the solution.

The two equations can be solved numerically starting from steady flow conditions up to time t = 0 after which one of the events occurs, and Watson et al. [1996] used a fourth-order Runge–Kutta solution procedure. There is no reason to limit the analysis by dividing the steam line into only two parts. A straight length of line can be divided into many segments and the mass flow rate between them related to the friction factor–Reynolds number relationship applicable, and the pressure drop being the difference between the two node pressures, similar to the finite difference solution of a conduction problem explained in Chap. 13. The solution then resembles a transient steamfield simulator and could in principle be used to design the steamfield control system.

References

ANSI (2001) American National Standard. Power piping; code for pressure piping, ASME B31.1-2001

Bangma P (1961) The development and performance of a steam-water separator for use on geothermal bores. In: Proceedings of UN Conference on new sources of energy, Vol 2 Geothermal Energy Agenda item II.A.2

Bergant A, Simpson AR, Tijsseling AS (2006) Water hammer with column separation; a historical view. J Fluids Struct 22:135

Brown K (2011) Thermodynamics and kinetics of silica scaling. In: Proceedings of International Workshop on mineral scaling, Manila, Philippines

Cerini DJ (1978) Geothermal rotary separator field tests. Geoth Res Trans 2:75–78

Ciurli M, Barelli A (2009) Simulation of the steam pipeline network in the Monteverdi geothermal field. Geoth Res Council Trans 33:1047–1052

Clayton JK, Cramer SH (1979) Development work on steam vent pipe silencing. Proc I Mech E 193(1):245–251

Dartnell L (1985) The thermal-hydraulic design of main feed water pump suction systems for large thermal power plant. Proc I Mech E 199(A4)

Engineering Sciences Data Unit (2002) Frictional pressure gradient in adiabatic flows of gas–liquid mixtures in horizontal pipes: prediction using empirical correlations and database ESDU 01014

Foong KC, Valavil J, Rock M (2010) Design and construction of the Kawerau steamfield. In: Proceedings of World Geothermal Congress

Fournier RO (1977) Chemical geothermometers and mixing models for geothermal systems. Geothermics 5(1–4):41–50

Fournier RO (1986) Geology and geochemistry of epithermal systems. In: Reviews in economic geology, vol 2, Society of Economic Geologists

Freeston DH (1981) Condensation pot design. Geoth Res Council Trans 5:421–424

Gage A (1951) Critére des conditions d'aspiration des pompes alimentaires IV Congres Int. du Chauffage Industriel, Paris

Garcia-Guttierez A, Martinez-Estrella JI, Ovando-Castellar R, Canchola-Felix I, Mora-Percy O, Guttierez-Espericueta SA (2012) Improved energy utilisation in the Cerro-Prieto geothermal field fluid transportation network. GRC Trans 36:1061–1066

Ghiaasiaan SM (2008) Two-phase, boiling, and condensation in conventional and miniature systems. Cambridge University Press, Cambridge

Harrison RF (1974) Methods for the analysis of geothermal two-phase flow. Master of Engineering Thesis, Department of Mechanical Engineering, University of Auckland

James R (1967) Pipeline transmission of steam-water mixtures for geothermal power. Geothermal Circ RJ-6 DSIR, New Zealand

Kotaka H, Mills TD, Gray T (2010) LP separator level control by variable speed and multi-stage brine reinjection pumps at Kawerau and Nga Awa Purua geothermal projects, New Zealand. In: Proceedings of World Geothermal Congress

Lazalde-Crabtree H (1984) Design approach of steam-water separators and steam dryers for geothermal applications. Geoth Res Council Bull 13:11

Mandhane JM, Gregory G, Aziz K (1974) A flow pattern map for gas–liquid flow in horizontal pipes. Int J Multiphase Flow 1:537

Marconcini R, Neri G (1978) Numerical simulation of a steam pipeline network. Geothermics 7(1):17–27

McAdams WH (1954) Heat transmission. McGraw Sill series in Chemical Engineering

Nishiwaki N, Hirata M, Iwamizu T, Ohnaka I, Obata T (1970) Studies on noise reduction problems in electric power plants utilizing geothermal fluids. Geothermics 2:1629–1631 (Special Issue)

Ovando-Castelar R, Martinez-Estrella JI, Garci-Gutierez A, Canahola-Felix I, Miranda-Herrera CA, Jacobo-Galvan VP (2012) Estimation of pipeline network heat losses at the Cerro-Prieto geothermal field based on pipeline thermal insulation conditions. Geotherm Res Council Trans, 36:1111–1118

Pointon A, Mills TD, Seil G, Zhang Q (2009) Computational fluid dynamic techniques for validating geothermal separator sizing. Geth Res Counc Trans 33:943–948

Purnanto MH, Zarrouk S, Cater JE (2012) CFD modeling of two-phase flow inside geothermal steam-water separators. NZ Geothermal Workshop

Schilling JR (1981) The diverging vortex separator: description and operations. Geotherm Res Counc Trans 5:445

Sharp BB, Sharp DB (1995) Water hammer: practical solutions. Elsevier, Amsterdam

Streeter VL, Wylie EB (1967) Hydraulic transients. McGraw-Hill, New York

Sulaiman S, Suwani A, Ruslan G, Suari S (1995) Scale in steam transmission lines at the Kamojang geothermal field. World Geothermal Congress

Takahashi Y, Hayashida T, Soezima S, Aramaki S, Soda M (1970) An experiment on pipeline transportation of steam-water mixtures at Otake geothermal field. UN Symposium on the Development and Utilization of geothermal resources, Pisa

Thain IA, Carey B (2009) 50 years of power generation at Wairakei. Geothermics 38:48

Watson A, Brodie AJ, Lory PJ (1996) The process design of steamfield pipeline systems for transient operation from liquid dominated reservoirs. In: Proceedings of 18th NZ Geoth Workshop

Wilkinson DH, Dartnell LM (1980) Water hammer phenomena in thermal power station feed water systems. Proc I Mech E 194(3):17–25

Zein A, Taylor PA, Indrinanto Y, Dwiyudha H (2010) Portable rock muffler tank for well testing purpose. In: Proceedings of World Geothermal Congress

The Resource Development Plan 13

This chapter sets the likely scene at the conception of a geothermal power project and then discusses how the power station capacity can be estimated using an elementary method. Numerical reservoir simulation is introduced as a planning and prediction tool, and its mathematical basis is outlined before returning to the identification of the main stages of a project and typical environmental impacts.

13.1 The First Stages of Planning

At the start of a development, all that will be known is that a geothermal resource exists beneath the land area available, that there is a demand for electricity and that transmitting it will be possible. The legislation governing geothermal resource use will have been identified. By the time the plan is complete enough to present to the authorities with the responsibility for making a decision, the power station site, transmission line corridors, land area on which wells are to be drilled and approximate pipeline routes will have been defined. The final prime movers are often still undefined at this stage, but all the major surface equipment needed will be known sufficiently for their environmental impact to be assessed. Enough wells will have been drilled, under an exploration permit, to convince the developer's bank to fund the enterprise, but the project will not be commercially risk free; the permitting authority will understand this, but so long as the development is in the national interest and the environmental impact is acceptable, then permits may be granted. This is the case in New Zealand (where permits are referred to as resource consents) and is probably a likely scenario anywhere.

The procedure would be the same if a fossil-fuelled plant was to be built, and the next step would be to design it, working from major parameters first to final manufacturing detail. The way the equipment functions is known in detail—every component is designed for a specific purpose, and its performance is known from engineering calculations and laboratory tests. In contrast, a natural geothermal resource is unique in every detail—it may be the result of recognisable physical processes, but how the fluids circulate and the response to drilling, production and

A. Watson, *Geothermal Engineering: Fundamentals and Applications*,
DOI 10.1007/978-1-4614-8569-8_13,

injection have to be explored and measured. Some risk that the resource will not supply heat and fluid at the estimated rate must be accepted. Extreme caution is unhelpful as the economic advantage of a particular geothermal plant over a fossil-fuelled alternative might be slim, but not enough caution will result in idle plant which has to be sold at a loss.

The chronological steps in formulating a plan are typically as follows:

(a) Explore the surface geology and conduct geophysical studies to help estimate geological stratigraphy and structure and determine possible boundaries of the system.
(b) Conduct geochemical investigation of surface discharges to help determine its hydrology and temperature distribution; this will include planning the sites and numbers of wells needed to give sufficient information and will involve scientists and engineers. Well sites positioned for easy use later will be preferred.
(c) Knowing the surrounding area and its uses, plan a strategy for extracting the hot resource fluid and injecting it at lower temperature, including possible schemes for how to respond to future reductions in energy output from the wells. This is the task for which numerical reservoir simulation was primarily developed, but it is also useful for step (b).
(d) Decide on the power station capacity and type. Reservoir simulators now play a part in this step, but a simpler methodology was developed earlier, called a stored heat estimate.

13.2 A Stored Heat Estimate

Resource assessment in general attempts to answer the question "what installed generating capacity can the resource supply for at least the term of the loan taken out to fund the project?". The question may be asked before many wells have been drilled and tested. A stored heat estimate was the usual way of providing an answer until the 1980s and remains a worthwhile exercise.

The UK consulting company Merz and McLellan [1956], under contract to the NZ government in relation to the development of Wairakei, reported at one stage on the generating capacity that might be installed over and above the 69 MWe that had already been committed to. It was noted that drilling success had fluctuated since 1953 and that initially deep drilling had been successful but then had fallen behind shallow drilling in terms of output per well, before finally returning to being more successful. Enough steam to support 150 MWe was available at wellhead at the time. A high drilling success rate, it was said, must not be expected to continue indefinitely. A capacity of 250–500 MWe had previously been suggested by the NZ Department of Scientific and Industrial Research, and Merz and McLellan advised the lower of this range as the maximum and that 20 % of capacity should be kept available as steam at wellhead, i.e. stand-by wells ready to produce. Their report is cautiously worded but without any methodology at all and contains no discussion on possible resource decline—the resource is still producing 150 MWe with 95 %

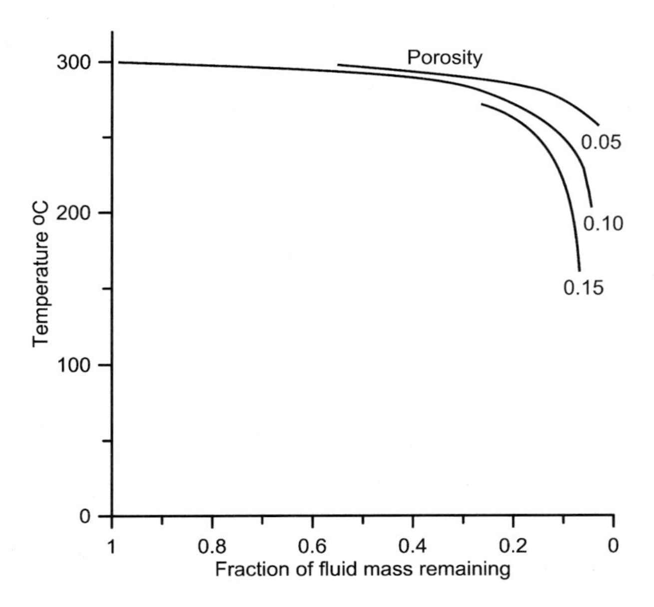

Fig. 13.1 Showing the decline in steam conditions as mass is withdrawn from a resource considered as a single tank of saturated water (*Watson and Maunder* [1982])

availability 60 years later, so the estimate might be said to have been valid, but the fact remains, they produced a resource capacity recommendation without any methodology.

The central problem of resource assessment is that hot water, steam or both are to be extracted from a body of fractured hot rock at a rate very much higher than the rate of replenishment. When viewed as a whole, the mass of fluid contained declines by 2/3 before any significant loss of steam temperature is evident, suggesting that well outputs can decline with little warning after a long period of slow steady decline. This was demonstrated by Whiting and Ramey [1969] using the heat and mass balance equations for the whole resource, an exercise repeated by Watson and Maunder [1982] and shown here as Fig. 13.1.

The stored heat estimate was described by Muffler and Cataldi [1978]. The area of the resource was defined in terms of its resistivity boundary, and its thickness was taken as the distance between the depth at which the rock temperature was 180 °C and 500 m beneath the drilled depth; 180 °C was taken as the minimum useful temperature. A certain proportion of the heat stored in this body of rock was assumed to be recoverable, and a percentage conversion of recovered heat to electricity was stipulated. The stored heat method was used by several organisations, as reviewed by Watson and Maunder [1982], who noted that the minimum temperature adopted varied between 180 and 200 °C, that recovery factors varied from 25 to 100 % and that conversion percentages ranged from 7.5 to 10 %.

A typical calculation is as follows. The fully heated temperature distribution of a well is assumed to characterise the formation temperature in the vicinity. Figure 13.2 is a simple example in which the temperature distribution is

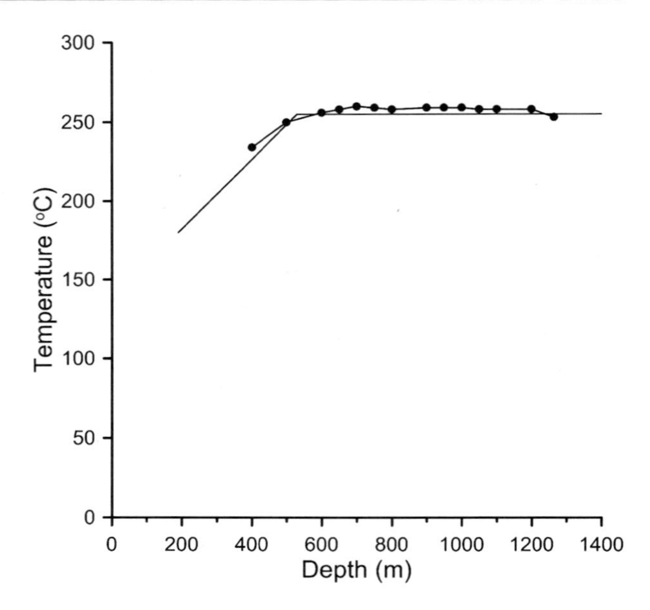

Fig. 13.2 A measured well temperature distribution approximated as two linear portions for a stored heat calculation

approximated as constant at 255 °C from 540 to 1,800 m (500 m below the well depth of 1,300 m) and with a linear variation from 180 °C at 200 m to 255 °C at 540 m, to which an equation can be fitted:

$$T = 0.2206z + 135.9 \tag{13.1}$$

where z is the depth below the surface (m).

Although the resource is usable from a depth of 200 m if temperature is the criterion, assume that the minimum practical depth from which to produce is 400 m. The resource can be considered as two slabs; the lower one from 540 m downwards with a uniform temperature of 255 °C, and the other from 400 to 540 m which, based on Eq. (13.1), has an average temperature $\overline{T}$ of 239.6 °C.

Each cubic metre of resource rock of porosity ϕ has stored heat components due to the rock and to the water in the pores, and for rock of average temperature $\overline{T}$, the total stored heat above 180 °C is Q kJ/m^3 where

$$Q = ((1-\phi)\rho_R Cp_R + \phi\rho_f Cp_f)\left(\overline{T} - 180\right) \tag{13.2}$$

$$= \left(1 + \frac{\rho_f Cp_f \phi}{\rho_R Cp_R (1-\phi)}\right)\rho_R Cp_R (1-\phi)\left(\overline{T} - 180\right) \tag{13.3}$$

$$= \Lambda \rho_R Cp_R \left(\overline{T} - 180\right)\left(\mathrm{kJ/m^3}\right) \tag{13.4}$$

The (dimensionless) group Λ has been gathered because if it is assumed to be unity, then the stored heat can be calculated using the rock properties only.

Table 13.1 Showing the variation of factor Λ with pressure and porosity

Pressure (bar abs)	ϕ	0.05	0.10	0.15
100		1.037	1.074	1.111
150		1.035	1.070	1.105
200		1.034	1.067	1.100

The variation of Λ over a range of porosity and water pressure is shown in Table 13.1, using typical rock properties.

The value of Λ lies between 1.0 and 1.1, and assuming it to be 1.0 represents a small uncertainty compared to that of the recovery factor, which is a pure guess, and the utilisation factor. In other words, the stored heat in the resource can be assumed to be the heat in the rock.

Returning to the example of Fig. 13.2 and assuming $\phi = 0.05$, $\rho_R = 2{,}600\ \text{kg/m}^3$, $Cp_R = 0.9$ kJ/kgK and $\Lambda = 1.0$, then using Eq. (13.4) the heat stored in a column of the resource with a surface area of 1 km^2 is

$$2,600 \times 0.9 \times 1.0\text{E}6[(239.6 - 180) \times (540 - 400) + (255 - 180) \times (1,800 - 540)]$$
$$\text{or} \quad 240.6\text{E}12\ \text{J/km}^2$$

If it is assumed that 50 % of this heat is recovered and 10 % is converted to electricity, then the total energy in the form of electricity is 12.03E12 J/km^2. Assuming the energy is extracted steadily over 25 years, then for every square kilometre of the resource, which has the temperature distribution shown in Fig. 13.2, the electrical generation rate will be

$$12.03 \times 10^{12}/(25 \times 8,760 \times 3,600)\ \text{We/km}^2 = 15.3\ \text{MWe/km}^2$$

Circles of area 1 km^2 can be drawn around each well drilled and a capacity figure found for each one based on the heat-up temperature distributions. Wells with a downflow complicate the calculation by obscuring the formation temperature.

The stored heat calculation is very simplistic, but it is so easy to carry out that there can be no reason not to do it, even today. The conclusions presented by Watson and Maunder [1982] are still relevant 30 years after they were written, with the exception that lumped parameter models have been superseded by numerical simulation. They were written while the authors were grappling with the task in reality and they wrote

"*Arriving at a decision*
When the decision point is reached the developer will find himself in possession of several weakly connected sets of information. A stored heat calculation will have been done because it is relatively cost free. Simulation may have been carried out, and may have produced a prediction of field output decline which can be combined with known plant performance to give electrical output over the plant life. Both of these assessments will have incorporated unverifiable assumptions. Unrelated to these are well power output measurements, which give no direct indication of gross power output of the field but

serve to show how many wells will be needed to power a given plant. (Installation of a small turbine, say 1.5 to 3 MWe may give enough information over say 2 years to allow a lumped parameter model to be constructed). The only way to connect these sets of data is via *an economic analysis of the project. At its basic level this should be a sensitivity study based on various plausible patterns of field behaviour—how frequently new wells will be required, workovers, turbine outage, reinjection problems, etc. At the same time as bringing all the information together, an analysis of this type decreases the importance of assumptions made in modeling or stored heat assessment. It is not sufficient to attempt to apply conservative scientific and engineering assumptions to the reservoir assessment. The uncertainty in reservoir capacity must be translated and quantified as a financial risk*".

Recently, Sanyal et al. [2011] presented what amounted to a risk assessment of geothermal resource development in Indonesia, where enough wells have been drilled to provide a reasonable statistical sample. Their paper related to a plan by the Indonesian government to expand geothermal electricity generation in the country and examined 215 wells which had been drilled at 100 resource sites. It concluded that there was sufficient experience available in assessing and drilling Indonesian resources that the overall resource risk "should be" lower than in other countries—the qualification in parenthesis is a necessary part of reporting on resource and risk assessment.

13.3 Numerical Reservoir Simulation

To simulate a geothermal resource is to describe it by a set of equations written with space and time as independent variables and pressure and temperature as dependent ones and solve them for appropriate boundary conditions. The general process is used in many fields of engineering and consists of writing the governing equations in finite difference form and solving them numerically. Computers able to carry out very fast numerical procedures became available to researchers around 1960; numerical solution procedures for sets of partial differential equations had been developed earlier and had to be carried out by hand, although at least one mechanical analogue computer was developed in the 1930s for aerodynamic design by Hartree. Numerical simulators were developed for petroleum reservoirs towards the end of the 1970s. For petroleum, groundwater and geothermal reservoir simulation, the equations are those describing the flow of fluids through permeable (porous) media, incorporating Darcy's law. As with numerical simulators in other engineering fields, the whole programme is so large and complex that the set of governing equations is invisible to the user, who is presented with an interface allowing the data prescribing the problem to be written as an input file to the main computer programme. An initial pressure and temperature distribution are provided, and changes to it in response to well discharge and injection are calculated, advancing in time. The spatial distribution of variables is available for inspection at times after the start of the calculation requested by the user. The calculation proceeds automatically in time steps that are adjusted to ensure accuracy in the solutions. Although the use of a reservoir simulator as a planning tool is the focus here because it is the only available method of predicting how a resource will respond to the extraction and injection of fluid, they can also be used to examine detailed processes, for example, as part of the transient pressure well testing described in Chap. 9—see, for example, O'Sullivan [1987], who modelled the two-phase flow in the formation and determined formation properties by interpreting changes in flowing enthalpy.

In 1979, the US Department of Energy arranged for the principal research groups internationally to define six problems involving single- and two-phase flow in geothermal resources and then invited the groups to take part in a "competition" to see which simulator was best. Molloy and Sorey [1981] describe the process and the results, which showed that several different reliable simulators were available at that time, in other words, that the general validity of the invisible processing was established for some programmes. An historical perspective of simulators and their predecessors, lumped parameter models, is provided by O'Sullivan et al. [2009] in describing the numerical modelling of Wairakei.

To fulfil the aims of this book, it is necessary to examine the way the simulator functions. There are several textbooks on petroleum reservoir simulation, e.g. Critchlow [1977], which give details of the numerical procedures, but having been developed and proven over 30 years, most of the literature on the subject deals with applications. Given in the next section is a basic introduction. Gelegenis et al. [1989] provide a detailed insight into the calculation procedure used in their own geothermal reservoir simulator, and their paper might form the next level of study. The TOUGH2 simulator is probably the most often used today, and the various manuals related, e.g. Pruess [1987], give user information that includes some calculation details.

13.3.1 The Mathematical Basis of a Geothermal Reservoir Simulator

The governing equations are a partial differential set which must be changed into a numerical set, or "discretised", and then solved simultaneously. The process can be illustrated by examining the solution of Eq. (13.5), which describes the thermal conduction taking place in a long metal bar, thermally insulated and initially at uniform temperature, when a heat source is applied to one end at time $t = 0$. The temperature distribution along the bar at any time is to be found. A more general form of this equation was given as Eq. (4.65), which without the heat source term and reduced to one dimension becomes

$$\frac{\partial T}{\partial t} = \kappa \frac{\partial^2 T}{\partial x^2} \tag{13.5}$$

The formal method of converting this equation into numerical form is to write a Taylor series expansion of T and its gradients with x:

$$T' = \frac{\partial T}{\partial x}, T'' = \frac{\partial^2 T}{\partial x^2}, T''' = \frac{\partial^3 T}{\partial x^3}, etc$$

as

$$T(x+\Delta x) = T(x) + \Delta x.T'(x) + \frac{1}{2}(\Delta x^2)T''(x) + \frac{1}{6}(\Delta x^3)T'''(x)\cdots \tag{13.6}$$

$$T(x-\Delta x) = T(x) - \Delta x.T'(x) + \frac{1}{2}(\Delta x^2)T''(x) - \frac{1}{6}(\Delta x^3)T'''(x)\cdots \tag{13.7}$$

Neglecting all terms containing Δx^3 and higher powers, the remaining terms of these two equations can be eliminated or rearranged to obtain three expressions for the temperature gradient T′ and one for the second derivative T″, and these are named as shown:

$$T' = \frac{(T(x+\Delta x) - T(x))}{\Delta x} \quad \text{forward difference} \tag{13.8}$$

$$T' = \frac{(T(x) - T(x-\Delta x))}{\Delta x} \quad \text{backward difference} \tag{13.9}$$

$$T' = \frac{(T(x+\Delta x) - T(x-\Delta x))}{\Delta x} \quad \text{central difference} \tag{13.10}$$

$$T'' = \frac{(T(x+\Delta x) + T(x-\Delta x) - 2T(x))}{\Delta x^2} \tag{13.11}$$

The same approach can be adopted for gradients of T with time. Equation (13.11) can now be substituted into the right-hand side of Eq. (13.5), but there is a choice of expressions for the first derivative of temperature with time, analogous to Eqs. (13.8) and (13.9). How the temperature develops with time is to be found, so a forward difference form is the intuitive choice, for which the discretised form of Eq. (13.5) is

$$\frac{T_i^{n+1} - T_i^n}{\Delta t} = \kappa \frac{(T_{i+1}^n + T_{i-1}^n - 2T_i^n)}{\Delta x^2} \tag{13.12}$$

where T_i^n is the temperature at location i at time n. The locations are referred to as nodes, the distance between nodes is Δx and the time step from n to n + 1 is Δt. Figure 13.3 shows how the solution advances in time.

The rationale for choosing the values of Δx and Δt is based on a reorganisation of Eq. (13.12) into the form

$$\begin{aligned} T_i^{n+1} &= \left(\frac{\kappa \Delta t}{\Delta x^2}\right)(T_{i+1}^n + T_{i-1}^n - 2T_i^n) + T_i^n \\ &= F\left(T_{i+1}^n + T_{i-1}^n + \left(\frac{1}{F} - 2\right)T_i^n\right) \end{aligned} \tag{13.13}$$

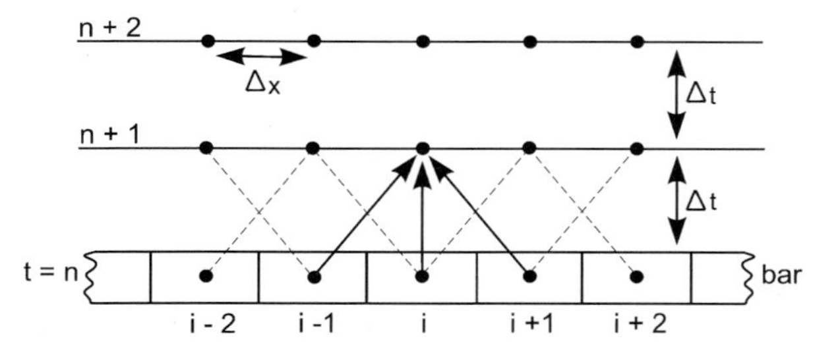

Fig. 13.3 Showing the influence of nodal values in a finite difference solution for one-dimensional transient thermal conduction

The equation has been rearranged in this way because F is the Fourier number of Eq. (4.67), written here as

$$F = \frac{\kappa \Delta t}{\Delta x^2} \tag{13.14}$$

Equation (13.13) shows that if F = 1/2, the influence of the temperature at node i is lost, which is clearly unreasonable—see Fig. 13.3—so the solution only works with F less than this. A choice of Δx then leads to Δt or vice versa. Boundary conditions can be applied, either a fixed temperature at a node or a fixed gradient by means of an "artificial" node at the end of the bar. This method can be set up using a spreadsheet (an instructive learning exercise) because it is explicit, that is, the values at the next nodes in time depend only on known information and not on the neighbouring values at the new time. The values at the new times are calculated individually.

It is physically unreasonable to expect that the value T_i^{n+1} will not be affected by T_{i-1}^{n+1} and T_{i+1}^{n+1}, and correcting this makes the solution implicit, that is, the values for every node at the new time step must be solved simultaneously because they depend on each other. Equation (13.5) can be written in terms of the unknown node values at the new time step, with the influence of the previous time step still present through the nearest neighbour, T_i^n:

$$\frac{T_i^{n+1} - T_i^n}{\Delta t} = \kappa \frac{\left(T_{i+1}^{n+1} + T_{i-1}^{n+1} - 2T_i^{n+1}\right)}{\Delta x^2} \tag{13.15}$$

and this equation can be generalised as

$$a_{i-1}T_{i-1}^{n+1} + b_i T_i^{n+1} + c_{i+1}T_{i+1}^{n+1} = d_i \tag{13.16}$$

in which a, b, c and d are known coefficients for every node, incorporating the previous values of T and the Fourier number. A set of equations of this type, centred around each node, can be written as

$$\begin{array}{llllll} a_{i-1}T_{i-2}^{n+1} & + b_{i-1}T_{i-1}^{n+1} & + c_{i-1}T_{i}^{n+1} & & & = d_{i-1} \\ & a_{i}T_{i-1}^{n+1} & + b_{i}T_{i}^{n+1} & + c_{i}T_{i+1}^{n+1} & & = d_{i} \\ & & a_{i+1}T_{i}^{n+1} & + b_{i+1}T_{i+1}^{n+1} & + c_{i+1}T_{i+2}^{n+1} & = d_{i+1} \\ & & & & \cdots & = d_{i+2} \end{array}$$

etc.

(13.17)

Computational techniques have been developed for solving sets of equations like this, which form tri-diagonal matrices, and also for sets which are not quite so regular, like the equations describing flow in the resource.

This introduction has dealt with the solution of only one equation, the energy equation, because it is for a solid material, but for flow in permeable materials, two equations are to be solved, one representing continuity of mass and momentum combined by the incorporation of Darcy's law, like Eq. (4.78), and the other the energy equation, and they need to be written for a permeable material and two-phase fluid. To illustrate the equations and express them in terms found in the literature on simulation, Eq. (4.77) is a suitable starting point, an equation expressing mass conservation:

$$\phi\frac{\partial\rho}{\partial t}+\left(\frac{\partial(\rho u)}{\partial x}+\frac{\partial(\rho v)}{\partial y}+\frac{\partial(\rho w)}{\partial z}\right)=0 \tag{13.18}$$

The first term represents the mass of fluid per unit volume contained in the control volume and the second the net flux (mass flow rate) in each of the three directions. The control volume can be considered as a block. To include the possibility of production or injection, a sink or source term must be included, and the equation is usually written:

$$\frac{\partial A_m}{\partial t}+\nabla F_m+\dot{q}_m=0 \tag{13.19}$$

where

$A_m = \phi(S_v\rho_v + S_l\rho_l)$ is the mass of fluid per unit volume in the block;
S_v and S_l are the saturations (volume fractions) of vapour and liquid, respectively;
∇ is the operator $\frac{\partial}{\partial x}+\frac{\partial}{\partial y}+\frac{\partial}{\partial z}$
F_m is the mass flux (flow rate per unit area) crossing the boundaries of the block then $\dot{q}_m$ is the rate of production of the mass source.

Darcy's law is next introduced to couple momentum and continuity of mass equations. The gravitational term in the momentum equation has been neglected as the main application so far has been to horizontal formations of moderate thickness, but when considering the whole resource, it must be included, and it might as well be included in the equation set for every direction, for regularity, since numerical solutions and not analytical ones are being applied and g set to zero in two of the three directions. Hence the mass fluxes can be written for vapour and liquid phases:

$$F_{mg} = -\rho_g k \frac{k_{rg}}{\mu_g}\left(\nabla P - \rho_g g\right) \tag{13.20}$$

$$F_{mf} = -\rho_f k \frac{k_{rf}}{\mu_f}\left(\nabla P - \rho_f g\right) \tag{13.21}$$

The energy equation can be similarly treated, using the specific internal energy version similar to Eq. (4.23), but adopting the format of Eq. (13.19). However the energy in the rock must be included:

$$\frac{\partial A_e}{\partial t} + \nabla F_e + \dot{q}_e = 0 \tag{13.22}$$

where

$$A_e = (1-\phi)\rho_r U_r + \phi\left(S_g \rho_g U_g + S_f \rho_f U_f\right) \tag{13.23}$$

and

$$F_e = -\rho_g h_g \frac{k k_{rg}}{\mu_g}\left(\nabla P - \rho_g g\right) - \rho_f h_f \frac{k k_{rf}}{\mu_f}\left(\nabla P - \rho_f g\right) - \kappa \nabla T \tag{13.24}$$

Equations (13.19) and (13.22) must now be discretised, a pair for every block (node) of the resource model, and because they are implicit in the variables to be determined, and non-linear, with coefficients that are functions of P and T, they must be solved iteratively. For this reason, they are written with residuals for every node and time step. Thus for node i at time step $n + 1$,

$$\left[\frac{\partial A_m}{\partial t} + \nabla F_m + \dot{q}_m\right]_i^{n+1} = [R_m]_i^{n+1} \tag{13.25}$$

$$\left[\frac{\partial A_e}{\partial t} + \nabla F_e + \dot{q}_e\right]_i^{n+1} = [R]_i^{n+1} \tag{13.26}$$

By adjusting the time step, the residuals can be minimised until they reach acceptably small values.

The inclusion of a gas dissolved in the reservoir fluid calls for a third equation, and CO_2 was added by Zyvoloski and O'Sullivan [1980], increasing the numerical complexity significantly.

The explanation so far has assumed a regular pattern of nodes throughout the region of interest. A large-volume resource divided at the close spacing required for those parts of the resource where the flow is most complex would result in an excessive number of equations to be solved. A numerical form capable of dealing with whatever irregular node distribution was chosen was introduced by Narasimhan and Witherspoon [1976], called the integrated finite difference method. It was particularly convenient for a system of embedded blocks matching the concept of a dual porosity system as described in Sect. 9.4.3 (see Pruess [1990]). Pruess [1991] explains that the TOUGH2 simulator, which uses the method, has blocks defined in terms of their volume, area of interface with neighbouring blocks and distance from node to interface. In TOUGH2, time is discretised implicitly using backward differences, CO_2 is incorporated and the set of non-linear equations has coefficients which are often strong functions of the independent variables. An iterative simultaneous solution is necessary, and the Newton–Raphson method is used. The programme has automatic step length control; Δt is initially set by the user but is varied automatically to retain accuracy or to avoid wasting computing time.

13.3.2 Reservoir Simulation in Practice

Figure 13.4 shows the general arrangement of a recent model of the Wairakei resource.

The map on the left shows a plan view of the division of the resource into blocks—note that they vary in shape, as permitted by the integrated finite difference method. There are many more small blocks in areas of particular interest. The column on the right of the figure shows the vertical distribution of layers, varying in thickness according to their influence on the results and the degree of interaction with the fluid flow. The model extends to a depth of almost 3,500 m, but the average well depth at Wairakei is of the order of 1,100 m, so most of the production and injection flows take place in formations above this depth and the drilling and well measurements information allows greater definition of the model in this range.

The model is initially constructed using information about the geological formations based on drilling results, coupled with an indication of the area of the resource in plan from the resistivity boundary. Major faults intersecting the resource may be represented in the block structure of the model, although this is not illustrated in Fig. 13.4. A preliminary model of the natural evolution of the resource from emplacement of the heat source up to the relatively steady state of an undisturbed resource is made first. This provides the initial distribution of pressure and temperature and hence fluid state. The results of well measurements to determine the undisturbed reservoir temperature and pressure distributions are next compared with the model results. In the case of a resource which has had many years of production, the most recent well measurements can be compared; careful measurement interpretation is required at this stage, to make sure that internal flows

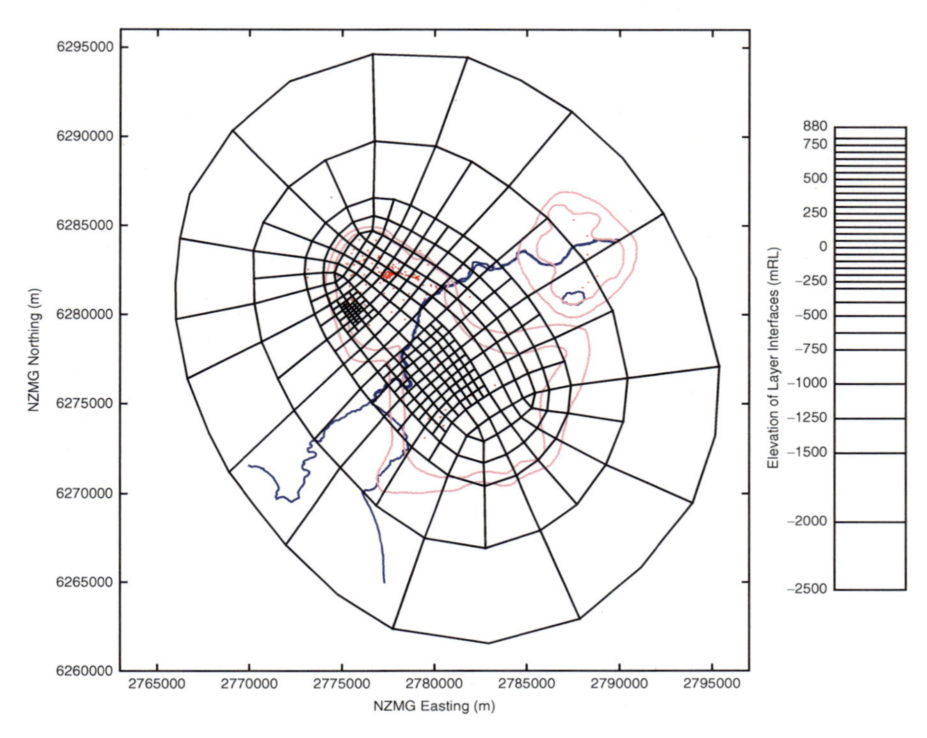

Fig. 13.4 A numerical reservoir simulation model of the Wairakei geothermal resource, using TOUGH2 (*reproduced from O'Sullivan et al.* [2009] *by permission of Elsevier*)

in the well are not misleading. Figure 13.5 shows an example of the comparison between measurements and predictions for a particular Wairakei well. This represents good agreement. Modelling near-surface formations and surface discharges is particularly difficult, perhaps because of the greater diversity of flow paths and material properties close to the surface where lithostatic pressures are relatively small compared to greater depths.

The Wairakei model shown above is the outcome of continuous development since the 1980s, in both simulators and field measurements used as input data, and it is at the extreme end of the spectrum of resource models as regards detail. Numerous examples of models for various fields can be found in the literature for all stages of resource use (see IGA [2012]). The starting point for modelling is an idea of the flow paths available in the resource, both permeable formations and faults, the boundary conditions that influence the flow, perhaps providing an impermeable barrier or very good permeability and hence a constant pressure. This idea of how the resource functions is referred to as a conceptual model—it is the equivalent of a diagrammatic sketch showing the flow paths and heating surfaces in a fossil-fuelled boiler or a piece of machinery. The chemical interaction between the resource rocks and the geothermal fluid is an important feature of any

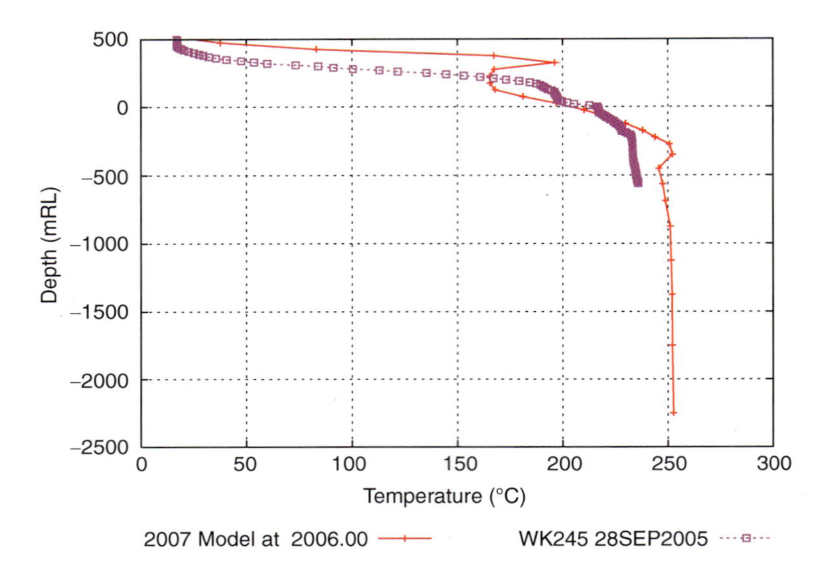

Fig. 13.5 Demonstrating the comparison between vertical temperature measurements in an individual well and model predictions for the block in which the well is located, for Wairakei (*reproduced from O'Sullivan et al.* [2009] *by permission of Elsevier*)

geothermal resource, and sometimes the outcome of this interaction is a modification of the fluid mechanics of the resource, for example, the deposition of solids blocking pores and sealing faults. Chemical interactions and measurements provide clues as to the physical processes taking place and contribute directly to the understanding of the full workings of the resource but currently only indirectly to the conceptual model; modelling including chemical reactions is at the development stage. A change in the concentration of carbon dioxide in solution, however, is a physical change that is routinely modelled.

O'Sullivan et al. [2001] reviewed the state of the art and reported that models with up to 6,000 blocks were being used, with minimum block thicknesses of 100 m and minimum horizontal dimensions of 200 m. They refer to an established method of constructing a model, or more precisely, of adjusting the parameters of the concept to make the simulation reproduce the observed behaviour; this involves running the model from a starting point in the distant past and ensuring that the trend of changes ends up with the distribution of measured variables (pressure, temperature and fluid state) used to arrive at the conceptual model.

It is no fault of modelling that the forward predictions are most reliable halfway through the working life of a resource or beyond and least reliable at the beginning, when the total production and injection amounts are small and the operating history short or even non-existent. Despite this, models are used in the initial planning of a project. Adjustment of formation properties to match the flowing enthalpy of a discharging well has already been mentioned, and O'Sullivan et al. [2001] mention the same approach used with measurements of decline in well output—if there has been insufficient production and injection to result in resource wide changes, the behaviour of individual wells provides the only alternative. The paper notes the difficulty of correctly modelling changes to surface activity.

After about 35 years of development, reservoir simulation is now an essential part of managing the large-scale use of a geothermal resource. The numerical details of the programmes can justifiably remain invisible to the user, and as a by-product, the literature on the topic is more generally readable, focusing on the comparison of measurements and predictions because the physics are correctly represented in the software and the mathematics is reliable. There is always a risk with software packages that because the input data are well defined, it is assumed that once they are provided, no further skill is required. To use a simulator and obtain reliable results requires an intuitive feel for the physics of flow in the resource, together with experience and aptitude.

13.4 The Overall Project Plan

Some information must be available before any large-scale geothermal resource use can be planned for, so funds must be committed without any guarantee that the expenditure will be recovered. These might be referred to as establishment costs. The developing organisation is most likely to have had previous experience in geothermal resource development or be a government organisation set up to initiate resource use. It is likely to have its own qualified and experienced scientific and engineering staff but unlikely to have immediate access to cash at the flow rates needed for a project of even a few tens of MWe, so other parties must be involved.

Staged development is a common way to minimise financial risks. A 25 MWe development may provide enough information in 5–10 years for a second stage to proceed with less risk than the first. Field measurements will have allowed a reservoir model to be developed, and experience will have been gained in dealing with the chemistry of the well discharge. On the other hand, the security of a moderate-sized first stage may cost a premium; the optimum development may be larger. Modular ORC plant will be included in the list of contenders for prime movers.

13.4.1 The Parties Involved

Most often, the development is the work of a single company. Sometimes the arrangement is as shown in Fig. 10.5, where a Resource Company is set up to sell steam to a Power Station Company. In either case, discussions with the Electricity Supply Company would take place early in the planning, to make sure that there is a potential market. The marketability of the electricity generated will depend on the cost of generation, but it is too early to make a proper estimate of this.

Figure 10.5 shows many third party opinions being sought by the developers and the banks. This is because the resource is as yet unexplored—a few wells may have

Table 13.2 Relative cost components for a 50 MWe development in 2007 US$ (*abstracted with permission from a New Zealand Geothermal Association report 2009*)

Cost item	Details	US$ million
Establishment	Geoscientific exploration, resource consents, site preliminaries	2.5
Drilling	2,500 m production well	3.6 per well
	2,000 m injection well	2.9 per well
Steamfield	Dual flash	27
Power station	Single unit, pass-in condensing steam	74
Annual operations and maintenance		2.7 pa

been drilled but not many, because of the need to minimise expenditure of funds if there is no ready way of recovering them. Table 13.2 illustrates the cost components of a 50 MWe geothermal power station development in US$ million, based on a report commissioned by the New Zealand Geothermal Association [2009].

The output of a production well can vary between 3 and 25 MWe, although the upper figure is exceptional, so the number of wells cannot be estimated very precisely.

A first estimate of the type of prime mover most suitable will be made, steam or ORC, and whether there are any major obstacles in the form of difficulty in landing heavy machinery at the available ports, transporting it overland, or problems with the actual siting of the station, bearing in mind local geotechnical conditions and environmental impact. The electricity transmission route will be determined. Such studies will result in short, broadly scoped reports at first, with more detail added as discussions progress.

A financially orientated project analysis will be started, using the methods described in Chap. 10. The banks will lend more easily to a company which has previous experience in geothermal engineering, or at least in the energy supply industry, and also has substantial financial resources and assets so that the loans can be recovered, if necessary by the sale of assets. A long-term electricity sales contract which assures the developer's income is likely to be required. Lenders are generally cautious about new technology, so the plant and equipment, the construction and the operating procedures should all be well proven.

13.4.2 Stages of the Project

The project can be broken into stages. The order is important, but it may not be possible to define it at the outset, so points of no return need to be identified and critical paths. A project team must be set up under the direction of a project manager. On completion of the project, an operating organisation will be required, and the project will be staffed as required during construction, with a view to some construction staff being retained as operating staff. Because every activity has costs associated with it, whether carried out in the field or at a desk, the scheduling of the steps has to be under continuous scrutiny; there will be a final commitment date for every step, after which changes of mind may be costly. A typical list of the steps is shown:

- Surface exploration
- First estimate of resource capacity
- First drilling
- Further drilling after reviews of first results
- Preliminary resource assessment
- Arrange land access or purchase
- Preliminary engineering design
- Seek permissions to develop the resource
- Plan the project, cash flows, activities and decision points and produce a schedule in the form of a Gantt chart
- Make commercial arrangements and seek finance
- Carry out further design, call tenders or negotiate directly with suppliers
- Tender assessment if required and manpower planning
- Carry out baseline environmental surveying
- Start construction site work

13.5 Environmental Impact Assessment

Permits or resource consents are usually required by law before the large-scale use of a geothermal resource is allowed in any country. Environmental impact is high on the list of items to be examined and necessarily involves scientists and engineers but in an unfamiliar role. Chapter 14 provides examples of the interaction between the authorities and the developers in New Zealand resulting from the legislation there. The major environmental impacts of geothermal resource use are by now well known, but some of them are country and resource specific. The remainder of this section introduces the most important ones.

13.5.1 The Effect on Geothermal Surface Features

Surface discharges of hot water and steam are quite rare features worldwide, so they often provide interest for locals and tourists; the natural features and spas of Japan are a good example. Extracting fluid from a resource decreases the pressure within the resource, often resulting in a lowering of the water level. Heat may still break the surface as steam because the rocks above the new water level are still hot, but the chemical composition and appearance of springs may change totally; springs discharging neutral pH chloride water may change to discharging acid pH sulphate water as a result of the upper formations becoming filled with steam and geothermal gases, H_2S being particularly significant. Fumaroles may replace springs. Surface features are valued by scientists as there is a good deal still to learn about the natural evolution of the land in response to the arrival of a pluton and by sections of the

Table 13.3 Comparison of the CO_2 emission rates for various fuel types as quoted by various authors

Source	Coal	Oil	Natural gas	Geothermal
Bloomfield and Moore [1999]	967	708	468	82
Thain and Dunstall [2001]	915	760	345	69–100
Contact Energy [2003]	930		340	10

Units are tonnes of CO_2/GWh

population with an ancestral record of using them for fundamental purposes such as heating, cooking, bathing and a range of cultural activities which rely on the existence of the features in their present general form rather than a particular level of discharge. In some instances, they have already been changed by previous generations. Since surface discharge is a feature common to most geothermal resources, a development proposal must address the effect on them, a difficult task since the fluid mechanics of surface features is difficult to describe in reservoir simulation models. Injection of separated water is necessary to avoid the contamination of waterways and maintain resource pressure, although it is not necessarily sufficient to achieve the latter.

13.5.2 Effect on Global and Local Air Quality

H_2S and CO_2 are produced at the surface where there is geothermal activity, and they are produced at a much greater flow rate by discharging wells. These are non-condensable gases which collect in the power station condensers and have to be continually removed to maintain vacuum. CO_2 is a global warming concern, but both gases can kill in sufficient concentrations and H_2S is noticeable and unpleasant even at very small concentrations.

Various authors have cited the CO_2 output from power stations using different heat sources, and Table 13.3 compares their data.

There are variations due to the efficiency of particular plant and the quality of the fuel. Contact Energy's [2003] figures are for individual New Zealand power stations, the geothermal one being Wairakei, whereas Bloomfield and Moore [1999] show the results of a nationwide US survey. Thain and Dunstall [2001] give the average output from 580 MWe of Indonesian geothermal power plant as 69.2 tonnes CO_2/GWh and 1,124 MWe of Philippines geothermal plant as 94.1 tonnes CO_2/GWh. Electricity generated from geothermal resources produces on average 11 % of the CO_2 produced by burning coal and about 25 % of that produced by burning gas. It has been claimed that geothermal binary cycle plants emit no CO_2, but those that condense geothermal steam in their heat exchangers release the non-condensable gases, often injecting them into the rising plume above the air-cooled condensers.

The gas concentrations in air in the locality of the resource may be of concern. At atmospheric pressure and temperature, CO_2 is denser than air and can be dangerous in confined spaces with only high-level ventilation and in depressions in the ground, but H_2S is much more of a threat in similar circumstances. Wellhead cellars need low-level ventilation, and diggings in the resource area may need to be monitored. At low concentrations, the odour of H_2S can be a nuisance. Fisher [1999] reviewed natural levels in New Zealand and permitting considerations. A concentration of 7 $\mu g/m^3$ is a typical "not to be exceeded" guideline in areas without any natural sources of the gas and 70 $\mu g/m^3$ for those which have, such as geothermal surface discharge areas. He reported that the concentrations in Rotorua, a New Zealand town built over ground with extensive hot spring discharges, are typically in the range 50–400 $\mu g/m^3$.

The evidence presented in New Zealand resource consent hearings usually includes atmospheric modelling incorporating wind measurements at the site. The compressed gases from the condenser are usually discharged from one or more vertical pipes immediately above the cooling towers, within the rising plumes of steam or hot air. The exact location and the format of the cooling towers must be specified for the model, which can be used to choose sites. Background (baseline) readings before development are required. At Ohaaki, the site of the only concrete natural draft geothermal cooling tower so far as is known, discharging the non-condensable gases into the tower has resulted in some acid damage to the reinforced concrete.

13.5.3 Ground Subsidence

Ground subsidence is the lowering of elevation as a result of the compaction of geological formations. Where it occurs, it is a difficult issue to assess because the physical processes taking place are poorly understood. It is not confined to geothermal resource development but has occurred in response to taking groundwater and oil—Bloomer and Currie [2001] reviewed international experience in all three activities. It has been a problem in New Zealand at geothermal development sites. At Kawerau, the geothermal resource was developed in the 1950s as an energy supply for the pulp and paper industry located there. Paper is manufactured in very long lengths by passing through a sequence of rollers many tens of metres in length at high speed. The alignment of the rollers must be precise and stable, but the ground was found to be subsiding in response to fluid production from the wells. The Kawerau subsidence was relatively small and uniform throughout, whereas at Wairakei, it was small and uniform over large areas but extreme in a few localities, as much as 20 m (it is discussed in more detail in the next chapter).

Reddish et al. [1994] addressed subsidence over oil and gas reservoirs and developed a numerical calculation procedure, noting that subsidence requires at least one of the following circumstances:

(a) A significant reduction in reservoir pressure
(b) A very thick reservoir
(c) Weak and poorly consolidated reservoir rocks
(d) A reservoir with a considerable areal extent compared with its thickness

The modus operandi of geothermal developments, up to the fairly recent past, has allowed (a) to occur. Many being composed of stratified eruption debris, circumstance (c) is often met but (b) and (d) less so. Compaction can follow a reduction in pore pressure or a reduction in pore fluid temperature. Geertsma [1973] developed a theory and provided an analytical solution to predict the subsidence resulting from the compaction of a uniform thickness circular disc of compactable material buried at a specified depth. The compaction of the disc allows the overburden to sink, forming a smooth-sided bowl. The edge of the bowl has a bigger radius than the disc, and the size ratio is a function of the depth of burial and compressibility. The compaction of the disc is much greater than the final subsidence. The characteristics of this solution are responsible for Reddish et al.'s circumstance (d), and Geertsma concluded that for subsidence to equal compaction, a reservoir at a depth of 1,000 m would need a surface area of not less than 50 km^2. In summary, wide, shallow subsidence could be formed by localised compaction much greater than the subsidence, in homogenous ground. Geertsma's theory has been applied to geothermal resources, despite their inhomogeneity.

13.5.4 Induced Seismicity

Any faulted and fractured geothermal resource is prone to move (as opposed to compact) in response to thermal contraction and hydrostatic pressure reduction, producing micro-earthquakes, but associating particular events with any degree of certainty to the activities of a geothermal development is difficult. Most geothermal developments are in tectonically active locations naturally prone to earthquakes. Many cases have been cited of micro-seismicity in response to geothermal resource development, for example, Bromley et al. [1983] for Puhagan (Philippines). At the latter, it was reported that during the first few years of production, induced events occurred at rates of about 100 per day, but they then reduced to pre-development levels of typically 1 per day, at magnitudes up to 2.4, which can be felt locally but would cause no damage. Micro-seismicity can be easily monitored. It is not usually regarded as a serious environmental impact.

13.5.5 Effects on Local Groundwater Resources

Several effects are identifiable. Where ground has been cleared of vegetation, for construction, storm water can cause erosion and silting of natural waterways with damage to flora and fauna. Local knowledge and attention to storm water control can avoid problems, and constraining conditions can be defined and attached to permissions to develop a resource.

Large flow rates of contaminated water must be disposed of occasionally, for example, cooling tower ponds may need to be emptied and refilled with clean water, and drilling mud and preliminary well discharge may be held in ponds that must be emptied. In the event of a long power plant outage, injection pipelines must be emptied of separated water to avoid deposition of silica when the water cools down, and the holding ponds must be emptied in turn, to provide for future outages. Ground soakage is often possible in geothermal areas because the groundwater has already been in contact with geothermal contaminants; alternatively, it may be discharged from a lined holding pond into a waterway at a controlled rate at which the concentration of contaminants does little damage. Disposal of the entire separated water production into a river was allowed at Wairakei from the start of production in the 1950s until the present time, although it is about to end under recently renewed resource consents. At Tiwi, Philippines, separated water and condensate were discharged into the sea from start-up in 1979 to the mid-1990s. At Cerro Prieto, Mexico, separated water is reduced by evaporation before being injected. In the present era, many organisations are adopting a zero discharge policy, with deep injection of all liquid waste.

Continuous injection of waste water is not free of problems. Injection is an integral part of extracting as much heat as possible from the resource—it is a way of maintaining pressure and ensuring that heat is transferred from the rock to water, which is more effective because the heat transfer coefficient is higher and the heat capacity of water is greater than that of steam. The resource will have been defined in area from resistivity surveys, and using drilling results, a plan will have been made of which areas to designate as production areas and which as injection areas. Injected water returning to production wells is to be avoided, because it may be cooler and have higher concentrations of dissolved solids. Freshwater may reside over a geothermal resource, perhaps separated by impermeable formations, although total impermeability cannot be assured because geothermal areas are typically faulted. Injection should not result in large increases in pressure because of the potential for contamination of groundwater resources as well as the risk of induced seismicity.

13.5.6 Ecological Effects

Certain species of vegetation have adapted to growing in areas with an above-average surface heat flux and are referred to as thermotolerant species. Since the areas of geothermal surface activity are rare in themselves, thermotolerant species growing there are also rare and considered worthy of protection. Hot water discharge from springs enters streams which may also be sites of thermotolerant species along their length. Reductions in the discharge of water and increase of surface temperature by increase in steam flow combine to reduce the area of habitat for the thermotolerant species, the ground either becoming cool and normal, allowing invasion of less specialised species, or so hot that nothing will grow.

The secondary effects of air and water pollution by geothermal contaminants must be considered in respect of ecology.

13.5.7 The Potential for Significant Effects Due to Noise, Social Disturbance, Traffic and Landscape Issues

The sources of noise vary between the construction stage and the operational stage of the development. Drilling is a significant source of noise which cannot be halted during the night and may last for 4 weeks or more per well. Discharge testing is likely to be noisier for short discharges than for long-term discharge testing, when there is sufficient justification for the installation of more sophisticated silencing equipment. Earthmoving equipment and the noise typical of heavy engineering facilities construction can be limited to daylight hours if necessary. The background noise level of a proposed site would be monitored to form a datum for comparison. During power station operation, the noise from forced draft cooling towers is likely to dominate other sources. It may be possible to reduce noise in neighbouring areas by landscape modifications.

Social disturbance is very site specific. In some developing countries, areas with a high enough population density to make electricity supply beneficial are nevertheless remote. The resource area may be covered in dense forest, and roads cut to allow access by large trucks carrying a drilling rig also allow deeper access to the area by the local populace. Damage to local ecology may follow. Local people used to isolation may be exposed to an influx of workers. Social change would follow electricity supply in any event, but the rate of change may be more abrupt than is desirable.

Traffic problems occur in already developed countries, and the solutions may amount only to widening road junctions to allow safer passage of large vehicles or similar.

The effect of a development on the landscape is essentially the change in view caused by a power station, pipelines, wells and associated steam columns. Land must be cleared to make drilling sites and pipeline routes, but landscaping for appearance is possible.

References

Bloomer A, Currie S (2001) Effects of geothermal induced subsidence. In: Proceedings of the 23rd New Zealand geothermal workshop, University of Auckland, Auckland

Bloomfield KK, Moore JN (1999) Production of greenhouse gases from geothermal power plants. Geoth Res Counc Trans 23:221–223

Bromley CJ, Rigor DM (1983) Microseismic studies in Tongonan and Southern Negros. In: Proceedings of the 5th NZ Geoth Wkshop. Univ of Auckland, pp 91–96

Contact Energy (2003) Geothermal energy: developing a sustainable management framework for a unique resource. Company Brochure, Aug 2003

Critchlow HB (1977) Modern reservoir engineering: a simulation approach, Prentice-Hall

Fisher GW (1999) Natural levels of hydrogen sulphide in New Zealand. Atmos Environ 33(18):3078–3079

Geertsma J (1973) Land subsidence above compacting oil and gas reservoirs. Jnl Pet. Tech, June:734–744

Gelegenis JJ, Lygerou VA, Koumoutsos NG (1989) A numerical method for the solution of geothermal reservoir model equations. Geothermics 18(3):377–391
IGA (International Geothermal Association) (2012) http://www.geothermal-energy.org
McDonald WJP, Muffler LJP (1972) Recent geophysical exploration of the Kawerau geothermal field, North Island, New Zealand. NZ Jnl Geol and Geophys 15(3):303
Merz and McLellan (London) (1957) Wairakei geothermal project: report on power plant extensions. Govt. Printers, Wellington, New Zealand
Molloy MW, Sorey MJ (1981) Code comparison project - a contribution to confidence in geothermal reservoir simulators. Geoth Res Counc Trans 5:189–192
Muffler P, Cataldi R (1978) Methods of regional assessment of geothermal resources. Geothermics 7(2–4): 53–89
Narasimhan TN, Witherspoon PA (1976) An Integrated Finite Difference Method for Analyzing Fluid Flow in Porous Media. Water Resources Research 12(1):57–64.
O'Sullivan MJ (1987) Modeling of enthalpy transients for geothermal wells. In: Proceedings of the 9th New Zealand geothermal workshop
O'Sullivan MJ, Pruess K, Lippman MJ (2001) State of the art of geothermal reservoir simulation. Geothermics 30(4):395–429
O'Sullivan MJ, Yeh A, Mannington W (2009) A history of numerical modeling of the Wairakei geothermal field. Geothermics 38(1):155–168
Pruess K (1987) Tough user's guide. LBL-20700. Lawrence Berkeley Laboratory, University of California, Berkeley, CA
Pruess K (1990) Modeling of geothermal reservoirs: fundamental processes. Computer simulations and field applications geothermics 19(1):3–15
Pruess K (1991) TOUGH2- A Gerenal-Purpose numerical simulator for multiphase fluid and heat flow. Lawrence Berkeley Laboratory, University of California, LBL-229400, UC-251
Reddish DI, Yao XL, Waller MD (1994) Computerised prediction of subsidence over oil and gas fields. SPE 28105, Rock mechanics in petroleum engineering conference, Delft
Sanyal SK, Morrow JW, Jayawardena MS, Berrah N, Fei Li S, Suryadarma (2011) Geothermal resource risk in Indonesia – a statistical enquiry. In: Proceedings of the 36th workshop on geothermal reservoir engineering, Stanford University, Stanford, CA
Thain I, Dunstall M (2001) Potential clean development mechanism incentives for geothermal power projects in developing countries. In: Proceedings of the 5th INAGA Annual Science Conference and Exhibition, Yogyakarta
Watson A, Maunder BR (1982) Geothermal resource assessment for power station planning. In: Proceedings of pacific geothermal conference 1982, incorporating the 4th New Zealand geothermal workshop, University of Auckland, Auckland
Whiting RL and Ramey HJ (1969) Application of material and energy balances to geothermal steam production. Jnl Pet TEch. July 893
NZGA (New Zealand Geothermal Association) (2009) Report prepared by SKM "Assessment of current costs of geothermal power generation in New Zealand (2007 basis)". http://www.nzgeothermal.org.nz
Zyvoloski GA, O'Sullivan MJ (1980) Simulation of a gas dominated two-phase geothermal reservoir. Soc Pet Eng J 20:52–58

14 Struggles Between Commercial Use and Conservation: Examples from New Zealand

The large-scale use of a geothermal resource anywhere is likely to be a matter involving the government of the country. In a legal history of geothermal resource issues in New Zealand, Boast [1995] refers to them as "legal battles" fought out before courts and tribunals. The weapons are scientific and engineering ideas, presented by expert witnesses. For many years now, obtaining consents (permits and licences are generally called resource consents in New Zealand) has been a public process which is adversarial, that is, those for and against the proposal present evidence to a committee or court in support of their views. The developer normally presents all the scientific background to show that the resource is understood sufficiently to allow the effects to be predicted. There are often three separate groups: those in favour, those against on commercial grounds and those against on environmental conservation grounds. The expert witness, although paid by one particular group, has a professional duty to the court—they are not advocates, as are the lawyers, but experts whose opinion the court expects to be unbiased towards their client.

Historically in New Zealand, geothermal conservation has been poorly represented in the resource consent process. This might be due solely to the long time that geothermal resources have been used there, starting before concerns about the environment came into international prominence. In any event, it led to some public hearings with interesting issues in the context of this book. The aim of this chapter is to review New Zealand geothermal legislation and to give examples of particular scientific and engineering issues from developments at Rotorua, Wairakei and Ngawha.

14.1 Background to New Zealand Legislation Governing Geothermal Resource Use

The land area of New Zealand is similar to that of Japan or the UK, but it has a population of only 4.5 million. There is no evidence of any habitation before about 1200 AD; the nearest large land mass is Australia, 2,000 km away, and it is a similar distance from Polynesian islands. It was eventually settled by

A. Watson, *Geothermal Engineering: Fundamentals and Applications*,
DOI 10.1007/978-1-4614-8569-8_14,

a Polynesian people, the Maori. King [2003] explains that as the land mass of the Pacific Ocean consists of very small scattered islands, Polynesians of necessity developed the ability to make long voyages. New Zealand was settled last in the Pacific, being a long way south, and he ventures the opinion that on arriving there, Maori found resources in plenty for the size of their population and abandoned long-distance voyaging. Having thus had no contact with the more populated parts of the world or competition for resources, Maori had an un-unified tribal society without written language or metal technology by the time of arrival of the British explorer Cook in 1769. European immigration proceeded only slowly in the nineteenth century, and New Zealand was governed on British principles as a Dominion, the same status as Australia, but from the outset, according to King [2003], with an approach more sympathetic to the welfare of the indigenous people than in Australia. The Treaty of Waitangi was drawn up by the first British-appointed Governor and accepted in 1843 by Maori tribal chiefs and the British. It has a bearing on geothermal resource use. New Zealand has a Governor General who is the representative of the British monarch. In 1975, the Waitangi Tribunal was set up to address Maori grievances that the principles of the Treaty had been ignored in some dealings.

Maori had traditionally lived near springs and made use of them for cooking, bathing and medicinal purposes and for the preparation of flax for weaving. They dug bathing pools, e.g. at Ngawha, and seem to have channelled spring water over short distances, e.g. at Rotorua and the more populated centres of the Taupo Volcanic Zone. Significantly for modern considerations, geothermal surface activity assumed importance in terms of their spiritual beliefs. Otherwise, the use that early European immigrants made of geothermal springs was very similar to that of Maori. The geothermal surface activity and features such as the Pink and White Terraces, similar to those existing at Pamukkale, Turkey, attracted European tourists. The association of geothermal heat with power did not come until very much later.

New Zealand was never a developer of new power generation methods but was quick to adopt available technologies, including hydroelectric generation for which it has a high potential. A hydroelectric station at Bullendale was built as early as 1886 (Martin [1998]) because of gold mining activity, and large hydroelectric dams and schemes were under construction by the 1920s. Although the first electricity-generating plants were installed and owned by private companies, government played the major role until privatisation in 1987. As the population and level of commercial activity grew so did the demand for electricity and hence the need for a structured approach to supply, and the New Zealand Electricity Department was established in 1959.

Electricity supply was restricted by WW2, as plant was unavailable for purchase overseas. Because of the long planning and construction period for new hydro-stations in particular, electricity continued to be in short supply into the early 1950s. The generation of geothermal electricity in Italy mentioned in Chap. 1 had not gone unnoticed and the idea of using the central North Island geothermal resources for the same purpose was proposed in 1924; however, good hydro-generation opportunities still existed at that time. Scientific interest led to the drilling of wells in the Taupo Volcanic Zone at Rotorua, Taupo and Tokaanu in the 1930s, and the discharges were used for heating and bathing. The shortage of electricity caused the government to set up a Geothermal Advisory Committee in 1949, and further issues led to the immediate choice of Wairakei for development. A survey of the entire Taupo Volcanic Zone resources was not made until the 1970s.

14.2 Acts of Parliament Relating to Geothermal Energy

Boast [1995] presented a history of geothermal legislation from a legal point of view, as mentioned above, but this section is directed to aspects relevant to geothermal engineering.

14.2.1 The Geothermal Steam Act 1952

The purpose of the Act was to enable the generation of electricity from geothermal energy to be controlled by the government; it provided for the private generation of electricity by means of a licence in the form of a contract between the government and the licensee. That it came at the very beginning of geothermal electricity development is evident from its wording—it displays an attempt to cover every eventuality without knowing what was possible. Thus, steam is defined as "steam, water, water vapour, and every kind of gas, and every mixture of all or any of them, that has been heated by the natural heat of the earth", and "bore" is defined as "any well, hole or pipe which is bored, drilled, or sunk in the ground for the purpose of investigating, prospecting, obtaining, or producing geothermal steam, or which taps or is likely to tap geothermal steam: and includes any hole in the ground which taps geothermal steam". Geothermal steam (using the definition of the Act) was already being used for non-electric commercial purposes and had been since 1881, when the government deliberately encouraged it with the Thermal Springs Districts Act of that year, aiming to promote tourism. The 1952 Act dealt with this existing use in a non-technical manner by defining a "Geothermal Steam Area" without mention of any physical attributes. Any area of land that was, or was believed to be, a source of geothermal steam under the definition of the Act could be declared a Geothermal Steam Area by Proclamation of the Governor General, after which the use of geothermal steam and bores there required a licence from the Minister, revocable at his sole discretion and with conditions of use specified by him. Existing users were to be allowed to continue without consent but only if the Minister agreed, "having regard to the public interest".

14.2.2 The Geothermal Energy Act 1953

The Geothermal Steam Act was repealed after 1 year and replaced by the Geothermal Energy Act 1953, a 16-page document in place of the previous 6, representing more extensive and well-defined government control. This Act remained in force until the introduction of the Resource Management Act 1991. The primary purpose of the new Act was the same—to retain for government the sole rights to geothermal energy and control of its use. The clumsy definition of geothermal steam in the 1952 Steam Act was replaced by a definition of geothermal energy, which, perhaps only by chance, seems scientifically accurate enough to capture for the government all matter that comes out of the bore, including matter emerging from natural vents. It thus appears to cover precious metals and thermophilic bacteria which were not considered at the time (although this is the opinion of an engineer and not a lawyer). Drilling for geothermal energy and using it requires a licence from the Minister, the Minister of Works in this Act, with a few exceptions. In 1953 geothermal drilling practices were still being developed, and a Code of Practice for Geothermal Drilling did not appear as a NZ Standard until 1991. The Minister had the power to require an unsafe well to be sealed—the present Code of Practice defines a sealed well as having been "abandoned" and sets out the engineering requirements.

The wording of the Act provided for the introduction of fees for the geothermal energy used, defined in a set of regulations which could be changed from time to time. The Act gives the Minister the power to revoke a consent in response to non-compliance with any conditions attached to it, or if operations constitute a threat of danger to the public. Geothermal energy use was in its infancy in 1953, and it was to be several decades before Codes of Practice such as the NZ drilling code already mentioned and codes by the American Standards for Testing and Materials became available for use by NZ authorities.

The Act has been amended several times, the first in 1957 apparently because the 1953 Act provided continuation for those already using geothermal energy at the time it came into force, but excluded those who had a well, had been using it earlier but were not actually using it at that time. Later amendments came in response to events at Rotorua, which are to be explained.

14.2.3 Environmental Protection Versus Development

The government organised a survey of the nation's geological assets, including geothermal energy, which resulted in a report by the New Zealand Geological Survey—NZGS [1974]. It includes a section entitled "Assessment of geothermal energy resources". Guidance on environmental values appears to have been generated only informally, from within the geoscientific community. The protection of geothermal surface features from the effects of development became of concern in the late 1970s as a result of Rotorua and Wairakei experiences. Houghton et al. [1980] eventually produced a report which placed the geothermal resources of the TVZ in a ranking according to their state of development and priority for preservation, and the Resource Management Act was eventually to require regional planning incorporating environmental values.

In 1978 the Ministry of Works and Development applied under the Water and Soil Conservation Act, which at that time controlled the use of "natural water", to drill a well at Rerewhakaaitu in the TVZ. A suitable site was available at a privately owned timber processing yard, and the application included taking 2,500 tonnes per day for up to 5 years. The area lies on the edge of the Kaingaroa forest, a major source of timber for domestic and export markets, and it had been decided that sawmilling and timber treatment should be based around the perimeter of the forest to reduce transportation costs and effects. If the well would produce it was intended that the Minister of Energy would supply the discharge under the Geothermal Energy Act 1953 to the owner of the site, to tanalise logs by delivering the well discharge directly to a pressure vessel containing the logs and the necessary chemicals. Practically, the project went ahead, the well was drilled and turned out to be a producer, and timber treatment began. The site was about 3 km away from the Waimangu Valley, a geothermal tourist attraction untouched by development and with many spectacular surface features. In the meantime R. Keam, a co-author of the Houghton et al.'s paper [1980] started a complex appeals process—it was complex by being taken to increasingly higher courts. As a result of the first appeal, the consent to drill the well and take the fluid was cancelled on the grounds that the project was a random exploration at one point in a large field and that the benefit which might follow was not sufficient to justify the possible detriment to the Waimangu surface discharge. The Geothermal Energy Act defined geothermal energy as including water heated to over 70 °C, while the Water and Soil Conservation Act simply addressed natural water, and there was uncertainty about which was appropriate for the issuing of geothermal resource consents. The matter was finalised by the Court of Appeal [1982]. A licence under the Geothermal Energy Act was required by a developer as well as a consent to take water under the Water and Soil Conservation Act. The well at Rerewhakaaitu was abandoned, and protection of the resource was given priority; however, later events at Rotorua would again illustrate lack of coordination between various government departments and local government on the question of environmental protection versus development.

14.2.4 The Resource Management Act 1991

The Resource Management Act 1991 (RMA) regards geothermal resources as water resources; it relates to the use of land, air and water and specifies the principles under which they are to be managed and their use allocated. The need for sustainable use is emphasised. Responsibility is delegated to regional government, which issues and administers all consents to use geothermal resources. Under the Act each regional government is required to prepare a regional plan which identifies the particular resources in its area and how they are to be managed; it is a definition of regional policy. There have been various amendments to the Act, a significant one being an acknowledgement that although administered regionally, the allocation of resources can be made in consideration of the national interest. This means that an application for a geothermal power station project in the TVZ can be assessed considering the benefits to the nation as a whole. For those whose role in the process of seeking consents is to provide expert witness, Schedule 4 lists matters that should be considered in an assessment of effects on the environment.

14.3 Rotorua

The conflict between the use of the resource by means of wells and the conservation of its natural features is the single most important issue in presenting any history of geothermal use in Rotorua. The conflict was raised as early as 1938 but did not become a matter of public debate until the 1970s, reaching a climax in the 1980s. The same conflict eventually arose in respect to other NZ geothermal resources, but nowhere else were there such a large number of established small users as at Rotorua. There were in excess of 350 bores (wells) taking a total of more than 25,000 tpd of discharge to supply individual households, hotels, motels, two hospitals, a Maori Institute and a Government Research Institute. Use of the heat was not very efficient and the discharged water was returned as ground soakage rather than injection.

The early history of geothermal use at Rotorua is set out by the former NZGS Rotorua District geologist J.F. Healey [1980]. He records that by the 1880s famous European spas with considerably fewer natural resources had become very popular tourist attractions. The Thermal Springs Districts Act (1881) was introduced and the Minister of Lands announced that land in Rotorua would be the first offered for public selection "in order to lose no time in rendering available the wonderful curative properties of the mineral springs in the vicinity of Lake Rotorua".

The Act preamble states that it would be advantageous to the colony and beneficial to the Maori owners of land with geothermal springs for the land to be opened up to colonisation and made available for settlement, and it sets out the powers given to the Governor to achieve this end, which included the purchase of land from Maori by the government, or assisting Maori to sell or lease land to settlers. Maori rights were sometimes overridden, the reason for the establishment of the Waitangi Tribunal in 1975 to correct injustices. The three-page Act contains nothing to restrict the way that the springs might be used and is devoid of any scientific wording. Encouraged by the Italian efforts, the NZ Department of Agriculture suggested drilling for hot water and steam in 1933, and the DSIR Geological Survey opinion was that this could be successful throughout the Taupo Volcanic Zone. Two wells 23 m and 59 m deep were drilled in Rotorua, which prompted the Ministries of Health, Industries and Commerce and Tourism and Publicity into debate about whether production from wells would reduce the output of the springs that supplied the sanatorium (hospital). Thus, the concern that was to become nationwide by the 1980s had been placed on record by 1938. Healey records that the DSIR had provided a quantitative response that since about 2,700 tonnes per day was being pumped from the springs without any signs of "overdrawal", considerably more could be extracted without depleting the resources, an assertion that is unlikely to have been supported by any evidence that would be acceptable today.

Electricity shortages gave rise in 1948 to the formation of a Rotorua Borough Council committee charged with the responsibility of developing the use of geothermal heating in the town. The committee produced a draft scheme for hot water reticulation, and the Council drafted legislation under which it could drill wells and construct reticulation schemes and also raise money for these activities through rates. Healey states that a delay in presenting the draft legislation to the government was incurred by the Council's Tourist Department, on the grounds that it should, but did not, contain provisions for protection of the springs. He records that in 1945 the Tourist and Health Resorts Control Act had been modified by the addition of clauses giving the Governor General the ability to declare any geothermal spring area a "thermal water area" in which no geothermal wells could be drilled without the written permission of the Minister, granted after consultation with the Minister in charge of the DSIR. Apparently the modification to the Act was never used. Developments at Wairakei precipitated the Geothermal Energy Act, 1953, through which control over geothermal drilling and production throughout New Zealand was exercised. This Act enabled the Minister to delegate some powers, and for Rotorua they were eventually delegated to the District Council under legislation called "The Rotorua City Empowering Act, 1967". The resource was not well managed; the Empowering Act gave the Council the power to issue licences to drill wells but none were issued despite many bores being drilled in the period up to the late 1970s. When the matter of protection of the important surface features finally came before the courts, there was criticism that the Council had missed an opportunity to collect revenues from licences which could have been spent on resource investigations. The community was divided into conservationists and "bore" users. With the benefit of hindsight, if the groups ever existed, neither should be blamed. Successive governments had passed legislation and funded research programmes aimed at encouraging the extraction and use of geothermal energy, but without providing guidance on conservation. Deeper and larger output wells than the usual for Rotorua had been drilled to supply the Forest Research Institute, a government establishment for research into timber production, which was a large export industry. The primary tourist feature was a geyser, Pohutu, which by the late 1970s was becoming less regular in its performance; it was the main one of a small group. The extraction from bores was blamed. The geyser had ceased discharging earlier in the century, before wells had been drilled and before scientific instrumentation and understanding was available to interpret events, but it had restarted again without any intervention. Local government scientists appear not to have had the ear of the authorities; when attention was at last gained in the late 1970s, action was delayed because the historical measurements of surface activity and resource use were considered inadequate. Public concern for the geysers produced a government reaction, and a lengthy new scientific programme was undertaken after the inevitable political debate about who should pay. By 1986 evidence of annual variations in the daily take had been correlated with changes in resource pressure in the producing formations, these being measured as water level changes in several monitor wells, one of them fairly close to the Pohutu geyser. It was decided to impose restrictions on well discharge within a radius of 1.5 km of Pohutu, originally only for summer use, with exceptions permitted if there was no alternative source of energy, but eventually total closure was required of all wells within the 1.5 km zone. Punitive energy licence charges were imposed which made geothermal heating uneconomic.

The Rotorua Geothermal Users Association, a group of domestic well owners, challenged the Minister's authority to introduce these changes by a High Court action. In preparation for these proceedings, the association engaged the consulting firm of KRTA to examine the evidence on which the Minister was acting and asked whether bores needed to be closed. The report concluded that the Rotorua draw-off was certainly too large and that some wells must be closed; however, several weaknesses in the evidence presented by government scientists were identified. The first was that the pressure in the producing formations was measured as a change in water level of certain wells which would stand open without discharging—the monitor wells already mentioned. Using a well as a manometer relies on the temperature and hence fluid density distribution being constant; no checks on this had been carried out. The second was that the casing of an old well within the 1.5 km radius had recently failed, resulting in the formation of a crater and a continuous blow-out discharge. This was catalogued by government scientists as a spring not a well, and its influence on formation pressure was ignored—in fact, the influence of wells on formation pressure was never discussed in any of the published government investigations. An examination of the monitor well water level variations at and shortly after the blowout, and application of the Theis solution suggested that the blowout had contributed to the reduction in the well's water level. Thirdly, no scientific reason for adopting a 1.5 km radius of closure was given. The High Court challenge by the association was on the grounds of arbitrariness of the 1.5 km radius closure zone and that the Minister had exceeded his authority. The law gave the Minister authority to close wells (bores) that were "in the opinion of the Minister, affecting detrimentally other specified bores...or a specified tourist attraction". Courts are skilled at identifying critical issues, but need the guidance of expert witnesses; no expert witness was called by the Users Association. The written judgement (RGUA [1987]) indicates the level of understanding that the court was left with when it states:

> *I can imagine that something can be made of the efficiency of certain bores within the 1.5 km area which gives rise to different considerations as to likely impact.*

and in discussing the complaint that the 1.5 km radius was arbitrary:

> *...Proximity* (to Pohutu) *is the most relevant factor, and once that is established in my view questions of arbitrariness disappear.*

How the court was persuaded that proximity to Pohutu had been established as the most relevant factor in assessing the effect of well discharge is unknown, as the court proceedings are retained as confidential. The effect of the production from a number of wells on formation pressure at a particular location is a standard well testing issue, resolved by superposition of the Theis solution for each well at its discharge rate. This had been addressed in the KRTA report, which had concluded that certain large output wells 2 km away from the geysers, some owned by government organisations, could be producing a bigger pressure reduction at Pohutu than many of the private wells within 1.5 km. It was this aspect that provoked an appeal by the association against the proposal to apply the 1.5 km radius closure zone; many of its members were concerned about environmental impact. From a purely geothermal engineering point of view, a more equitable solution could have been designed to fulfil the need to protect the resource and was proposed by KRTA; however, the appeal was declined and all wells within a 1.5 km radius were closed.

The reduction in the total production from the resource produced a recovery of water level (resource pressure) over 2–3 years. The discharge frequency of Pohutu and the adjacent geysers, which had never been absolutely regular, increased to the point of discharging almost continuously. Areas of hot ground in Rotorua which had cooled off as a result of excessive production and been built on, now became hot again, causing hydrothermal eruptions, ground collapses and some property damage. Surface springs in many areas increased their discharge rate and new ones appeared. The well closures were hailed as a success, a victory for environmental protection over commercialism. The situation had been rescued, but clumsily, and a financial loss, and hence a loss of opportunity, had been unnecessarily incurred by individuals and taxpayers. The sad history of resource use at Rotorua from the 1960s to 1986 might justifiably be seen as the result of a failure by scientists and geothermal engineers to recognise impending problems, design solutions and exercise their influence on local government.

14.4 Wairakei

14.4.1 The Original Development 1956–2001

The origins of the Wairakei resource development are explained by Bolton [2009], who played a significant engineering part in it. It began as a combined project between the British and New Zealand Governments, the former wanting a cheap source of heat and power for the production of heavy water for its nuclear programme and the latter simply wanting more electricity. Not long after the turbines for the combined generation and processing plant had been ordered, the British need for heavy water decreased and the New Zealand Government continued with the development alone. This partly accounts for the many small turbo-generators which make up the power station, which operates on a double flash system but has three different turbine inlet pressures. Of the six 11.2 MWe capacity turbines, two are operated as back-pressure turbo-alternator sets and the others as condensing sets. In addition there are three pass-in condensing sets of 30 MWe capacity each, one 4 MWe back-pressure set and an Ormat binary plant. The steam-powered units were commissioned between 1958 and 1963 and the 14 MWe Ormat ORC plant in 2005 (Thain and Carey [2009]).

By the 1990s 140,000 tonnes per day of geothermal fluid was being produced from the wells. Up to that period, none of the produced fluid had been returned to the resource with the result that the nature of the surface activity was irreversibly changed. The main area of boiling water discharge had been Geyser Valley, a tourist attraction said to have tens of geysers and many more named springs;

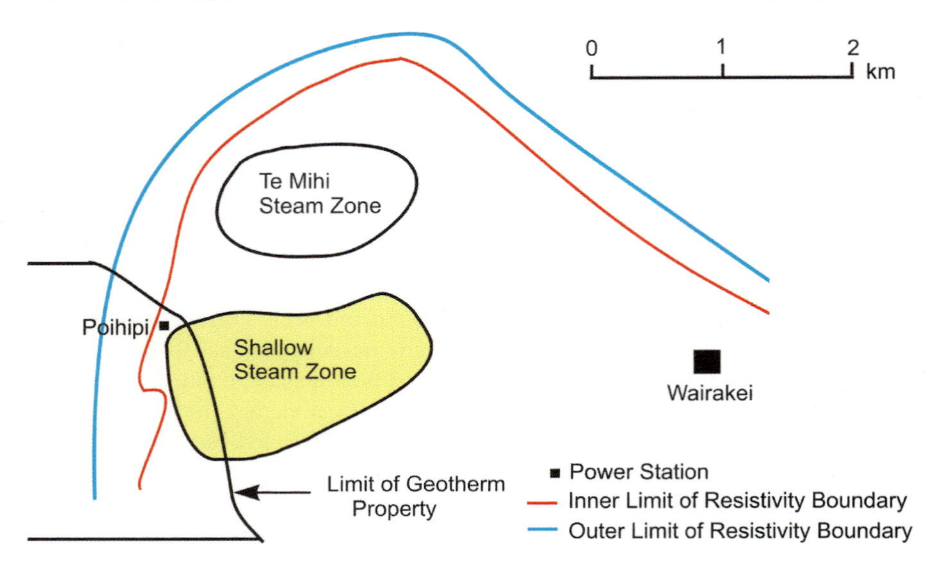

Fig. 14.1 Sketch map showing the Wairakei resource, two steam zones and the Wairakei and Poihipi power stations. The uncertainty in the resistivity boundary is indicated, as is the extent of the Geotherm property

Glover [1998] gives a detailed account, stating that neutral chloride water from the deep reservoir ceased to flow as a result of large-scale production. The surface activity changed from boiling springs and steam-heated features to acidic springs, and steaming ground increased in area and heat output. Various modifications were made to the scheme for flashing separated water, to improve efficiency. The high-pressure steam supply began to run down and in 1982 the high-pressure sets were removed for use elsewhere, reducing the installed capacity to 157.2 MWe. The resource and its operations represented a continuous research and development project for earth scientists in particular, and reservoir simulation at the University of Auckland was being developed, while it was being used to make predictions.

14.4.2 The Poihipi Development 1988–1997

In 1987 the electricity industry in NZ underwent a major change. The government-owned electricity department became a state-owned enterprise called the Electricity Corporation of New Zealand (ECNZ), and electricity generation by commercial companies was permitted. Many of the nationally owned regional electricity distribution authorities became private companies, and some had funds to invest. Mr. Alistair McLachlan, a local businessman and entrepreneur, had a small farm on land situated on the edge of the Wairakei resource. The resistivity boundary bisected his land and it was estimated that about 1 km^2 lay over the shallow steam zone (the largest of the two steam zones shown in Fig. 14.1) which had been extended by the extraction of water at Wairakei; a shallow well used for greenhouse heating produced steam consistent with it accessing the shallow steam zone, and Mr. McLachlan had consent to take 1,800 tpd of steam. In 1988, he applied for resource consents for his company Geotherm Energy Ltd to extract up to 44,000 tonnes per day (tpd) and inject up to 40,000 tpd, with smaller amounts of freshwater to be taken for cooling and disposed of by injection along with steam condensate, outside the resistivity boundary. There were four objections, from the Government Tourist and Conservation authorities on the grounds that surface activity had tourist and conservation values which would be diminished, from the local Maori Tribal Trust on cultural and loss-of-opportunity grounds and from ECNZ on commercial grounds that it would affect its production.

Consent was given by the regional authorities for the production of 10,000 tpd, of which not more than 3,000 tpd was to come from the shallow steam zone because this was an existing source of production for ECNZ and was the source of steaming ground activity of tourist and conservation value.

This consent was appealed by ECNZ, which meant that the application was referred to the Planning Tribunal (which has since become the Environment Court). Decisions of the Environment Court can only be appealed on grounds of law, and not on scientific or engineering matters. ECNZ had concern about the injection of the separated water, which had to be injected into the resource and not into freshwater formations surrounding it. The locality is heavily faulted and there was considerable debate about where the cool injected water would flow to; it would reduce Wairakei output if it flowed back to Wairakei production areas. The Tribunal's decision was that Geotherm's injection area must lie adjacent to the resistivity boundary and just within it, and its production wells must lie between that and ECNZ's part of the resource area and be cased to below the bottom of the steam zone. If injected fluid was to move in the direction of ECNZ's production, then it would affect Geotherm's production first. The production was allowed to remain at 10,000 tpd, with 3,000 tpd to come from the steam zone. No well was permitted to cross the vertically projected boundaries of Geotherm's property, a standard clause in New Zealand consents.

Geotherm proceeded with drilling wells into the steam zone, which turned out to be good producers. It had the option of purchasing from the Geysers field (California) a reconditioned steam turbine power station with a full-load demand which could be met by the 4,800 tpd of steam permitted from the shallow steam zone (the original 1,800 tpd plus the 3,000 tpd permitted by the Planning Tribunal). The payback period for the investment was less than 10 years, an attractive proposition, but subject to the risk that the steam zone pressure would decline beneath the Geotherm property before the investment was paid back. The only significant reservoir simulations carried out were for ECNZ, and they suggested that the steam zone pressure would decrease drastically in only a few years. Acting for Geotherm, the consulting firm of KRTA was limited to estimating the flow to Geotherm's production wells based on the thickness (about 125 m) and permeability of the steam zone, assuming that the source of steam was within ECNZ's part of the field. The indications were positive for Geotherm but there was uncertainty about the calculations as too little was known about the source. In the meantime, Mercury Energy, the Auckland area electricity distributor, formed a partnership with Geotherm and the 55 MWe (gross) Poihipi power station was built, a single turbine with a separate (tubular) condenser. The decision to install this plant cannot have been based on the consents that were available to Geotherm, as only the steam zone production was proven, and no doubt wider commercial considerations were involved. Its base load steam demand at full output is 10,700 tpd or just over twice the 4,800 tpd that was permitted, and its output when operated by Mercury Geotherm was varied during the day. ECNZ, which owned all major NZ geothermal and hydroelectric stations, was subsequently broken into smaller units and sold, and Wairakei and Poihipi are currently owned by Contact Energy.

Very little information on which to base a resource simulation was available for the Geotherm part of the resource where no well had been drilled into the steam zone previously. There was uncertainty about where the edge of the steam zone was, and eventually there was direct evidence of an edge in the form of high ncg (CO_2) concentrations, which is understandable if the steam zone is expanding into cooler ground so that the steam condenses leaving the gas. More recently Zarrouk et al. [2007] made use of the variable load operation of the station to examine the type of permeability model which best fits the well measurements. They show that the station was operated at an output of 29 MWe with a very regular step change to 8 MWe for about 5 h every day, with a minor but also very regular 7 MWe pulse superimposed. By modelling the response of the production wells and a monitor well to the pulsed discharge, they concluded that the steam zone is supplied by a fracture network which provides significant vertical flow. The shallow steam zone pressure has not declined to the extent initially predicted, no doubt because of this previously unknown enhanced permeability.

14.4.3 The First Proposed Tauhara Development

The resistivity boundary of the Wairakei resource is not a closed circle around Wairakei, but narrows then opens out again to surround what is known as the Tauhara resource—the two are connected by a relatively narrow neck in a single resistivity boundary. The town of Taupo sits partly over the neck and partly over what was understood at the time to be the Tauhara resource, which was explored by drilling while Wairakei was being drilled, and the wells mainly left unused thus acting as monitor wells. The effects of Wairakei production were seen as a fall in deep reservoir pressure which was interpreted as a fall in water level in the upper levels. The geology of the area can be imagined as a sandwich of permeable and impermeable layers, so that drainage of the lower layer by a flow towards Wairakei led to drainage of the layer in the sandwich immediately above, and then on to the next. Shallow wells had been drilled in Taupo for domestic hot water supply, and separate water levels at various layers of the sandwich were detected; Brockbank et al. [2011] show three layers and the pressure measurements from a large number of wells of varying depth, so that some are

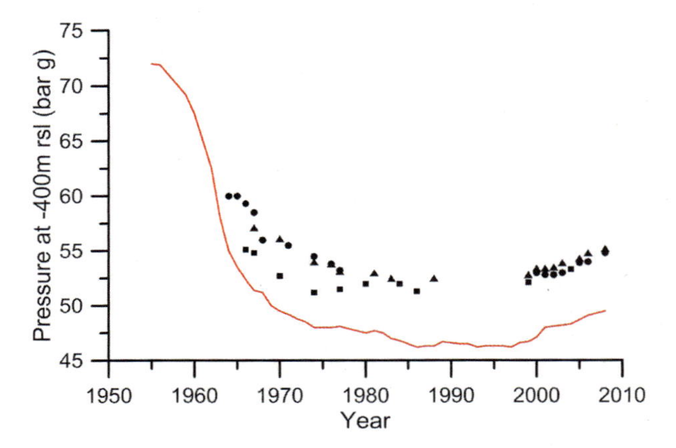

Fig. 14.2 Pressures at production depth in Wairakei (*full line*) and Tauhara (*individual points*) from evidence presented by Bixley to the Tauhara II Board of Inquiry hearing [2010] (*reproduced by permission of Contact Energy*)

in liquid and others in the steam-filled zone above the liquid level. Areas of steaming ground had existed since before the development of Wairakei, but in the 1970s the heat output began to rise; it peaked and later returned to normal and became referred to as a heat pulse. All of this was consistent with the drawdown of the Tauhara part of the resource by Wairakei, and in evidence Bixley produced data from which Fig. 14.2 has been plotted. The continuous line shows the way the mean pressure at production depths fell from the start of production to a minimum in the late 1990s when injection began; it has risen since then. The points are measurements at Tauhara wells—the pressure is higher but follows the same pattern, demonstrating a flow from Tauhara to Wairakei.

The Poihipi development had been prompted by the commercialisation of electricity generation; it began as one man's idea but soon after large organisations became involved and perhaps began to position themselves for the anticipated breakup and sale of ECNZ by the government. The Tauhara area had been regarded as a "poor relation" of Wairakei, but three applications to build power stations using it now appeared. Contact Energy applied for a 50 MWe steam plant combined with a 20 MWe ORC plant, and a Maori group together with a local electricity generator/retailer for a 60 MWe station. A further proposal by the Taupo District Council, which already owned a small hydro-station, combined with Mercury Energy for a 100 MWe was lodged but later withdrawn. After some time all applicants except Contact Energy withdrew, and the outcome was the granting to Contact by the Environment Court [2000] of the rights to take and inject 20,000 tpd for a nominally 15 MWe power station. The court's decision is important here because it addressed the issue of the very slow rate at which some environmental impacts appeared compared to the length of time for which consents were issued. The Resource Management Act by then in force provided a maximum term of 35 years. The debate arose because of subsidence, which was ongoing in the area after almost 50 years of Wairakei production, and was still poorly understood. The Act required the potential effects of a proposed development to be balanced against the potential benefits, community wide. If subsidence from the first consented use of Wairakei was still appearing at the surface, with what should the effects of a new proposal for increased use be compared? Consents are written to require the setting aside of a sum of money from which to compensate the general public from any damage suffered. Were the new holders of the Wairakei consents and owners of the power development to pay for damage only just appearing but resulting from Wairakei operations carried out when it was owned and operated by the NZ Government? Was it possible to analyse the subsidence with sufficient accuracy to associate any future subsidence with particular resource use?

The Tauhara hearing decision (Environment Court [2000]) set out the principles by stating that:

> *We hold that consideration is to be given to the effects on the environment as it actually exists now, including the effects of past abstraction of geothermal fluid from the system, whether by Contact or anyone else. In considering the effects in the future of allowing the proposed abstraction, we hold that we have to consider the environment as it is likely to be from time to time taking into account further effects of past abstraction, and effects of further abstraction authorized by existing consents held by Contact or others, ...*

This part of the decision acknowledged the difficulty in identifying the cause of any particular environmental effect and established the guiding principle.

14.4.4 The Renewed Wairakei Consents and the Te Mihi Power Station

The original Wairakei consents expired in 2001 and Contact sought to renew them. Disposal of separated water to the Waikato River, which was allowed under the original consents, was no longer acceptable—considerable evidence and debate ensued, but Contact submitted proposals to phase the practice out over a 10-year period, injecting the water both inside and outside the resistivity boundary of the resource. "Outfield injection" as the court preferred rather than reinjection raised the issue of contamination of freshwater aquifers, and infield injection became an issue associated with subsidence and the existing environment principle established by the Tauhara decision.

Wairakei subsidence had been quoted for many years as being amongst the greatest in the world—it is, but only in a few areas. Figure 14.3 is a surface plot of subsidence measurements at about 2001 but with the subsidence inverted, so the hollows show up as peaks. A sinuous heavy line leaving the top of the figure represents the Waikato River and the Wairakei power station is on its banks adjacent to the greatest subsidence. The same line, after it bifurcates towards the bottom of the figure, marks the shoreline of Lake Taupo. The dots mark survey points. Generally, the subsidence over the whole area within the resistivity boundary is very small and there is only one place where it is on a world record scale of more than 9 m, not far from the Eastern borefield and near the power station. It is also significant on the outskirts of Taupo (by a loop in the river in the figure). This location is in the area referred to as the neck of the resistivity boundary.

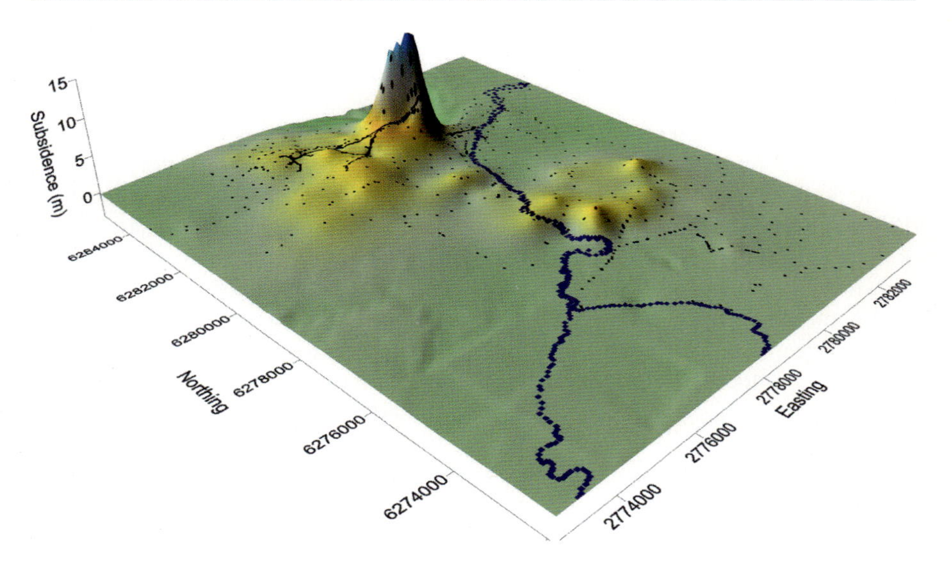

Fig. 14.3 Subsidence distribution over the Wairakei resource, inverted so that maximum subsidence (minimum elevation) appears as a peak (*Data originally provided by T. Hunt, figure created and presented as evidence to the Environment Court by A. Watson*)

At the time of the renewal hearings, attention was drawn to a new subsidence bowl in the built-up part of Taupo known as Crown Road. House damage had been experienced, but it was not clear that it was due to subsidence rather than poor building, and to confuse the matter further, it appeared that some of the houses had been built on gulleys filled with waste from sawmills, which did not offer a firm foundation. The bowl was increasing in depth and extent, however, although it is too small to appear in the figure.

Subsidence had been regularly monitored for many years and a grid of survey points had been established by the original Department of Scientific and Industrial Research (DSIR) and its modern form the Institute of Geological and Nuclear Science (GNS). Allis (see Allis and Zhan [2000]) had developed a method of predicting subsidence using a one-dimensional model. The subsidence was considered to be due to the presence of one or more formations with a very weak matrix which compacted when the fluid pressure decreased as a result of production. The pressure had decreased rather uniformly over the whole Wairakei part of the resource, however, which led to the conclusion that the weak formations were localised and of very small area. It was suggested by the author in evidence that cold downflow from shallow formations could have drained into formations occupied by steam zones, condensing them and reducing the fluid pressure to produce almost a vacuum. The heat capacity of a cold downflow in a well is very large, and if it entered a steam zone, the water flow rate through the permeable medium may not be fast enough to relieve the vacuum (the saturated vapour pressure of CO_2) and the formation would collapse. The Eastern borefield had many wells with broken casings, and the greatest subsidence was in that general locality. Hunt [1970] had demonstrated that changes in saturation of formations in response to production from the Wairakei resource could be detected by gravity measurements; the presence of steam reduced the material density and hence its gravitational attraction and vice versa if a large body of cold water filled the formation. At about the same time as the hearings were in progress, Hunt et al. [2003] independently associated measured gravity changes at Tauhara with a known downflow in a well. In any event, at Wairakei the area of maximum subsidence is very localised, suggesting the presence of either a very localised and particularly weak geological formation or precisely the right conditions for a very particular and as yet unconfirmed physical process to take place (in the same way as hydrothermal eruptions require a rare set of circumstances). The more recent publication by Allis and Bromley [2009] offers no firm conclusions on the mechanism.

The area of maximum subsidence coincided with an unpopulated, localised and very steep sided gulley, where changes in elevation and slope only became evident as a result of secondary effects—a stream backed up and formed a pond, a concrete canal carrying separated water from the field to the river changed slope and the wires between telegraph poles became tight.

For the hearings, and in opposition to the renewal of the consents, Taupo District Council (TDC) commissioned the preparation of expert witness statements which included predictions using a two-dimensional subsidence model based on a soil mechanics programme, described by White et al. [2005]. The choice of a two-dimensional model appears to accept an unnecessary restriction since numerical software was used. Geologists frequently draw two-dimensional cross sections, which convey accurate information if the material is solid. Many fluid mechanics problems were studied in two dimensions, but the solutions were compared with two-dimensional flows. If two-dimensional equations are written to represent a cross section through a three-dimensional flow field, the gradients in the third dimension are effectively being set to zero. Some of the formations are clearly uniform in thickness and could have a two-dimensional flow pattern, but there was interest in areas where the geometry appeared to be more three-dimensional. However, the main difference between these subsidence studies and those of GNS—Allis and Zhan [2000], for example—was that the TDC evidence was based on the idea that the compaction occurred in a deep formation, the upper layer of the main producing formation for Wairakei, and the GNS evidence that it occurred in shallow formations. It was clear to all that the result of the compaction was a changing pressure distribution throughout the resource which produced vertical movement that moved slowly upwards, eventually depressing the ground surface. The court's interest was in proposals as to what to do to reduce the impact of subsidence, in the light of the earlier Tauhara decision which recognised the ongoing problem. TDC's proposal was to inject beneath Taupo township, to bring the pressure distribution back to what it might have been before Wairakei production began in the 1950s; however, the court adopted the approach offered by Environment Waikato, the regional government responsible for resource management. This proposal was to inject into the neck of the resistivity boundary with the aim of maintaining the pressure distribution in that area as it then was. The pressure distribution beneath Taupo could then be regarded as the existing environment—it was a distribution due to the expiring consents. Provision was made in the renewed Wairakei consents for Contact to provide a bond to be drawn on in the event of damage to property by subsidence due to Wairakei operations. Within the hearing there was considerable detailed scientific and engineering discussion between the parties about the combination of deviated injection wells and new pipeline routes required to deliver the fluid to the area and how the resource pressure should be defined and measured. The court's decision was to renew the consents under a set of conditions described by Daysh and Chrisp [2009].

Shortly after the renewal of consents, Contact announced its intention to build a new power station on the Wairakei resource, to be called the Te Mihi power station and to phase out at least part of the original Wairakei plant. The Resource Management Act was constructed on the idea that consents to use natural resources would be granted by regional authorities, but provision was made for an application directly to central government if it was in the national interest, and in this case the application was heard by a Board of Inquiry selected by the central government and chaired by an Environment Court judge. Both the Te Mihi power station and an application for a new development at Tauhara called Tauhara II were considered by such Boards.

14.4.5 The Tauhara II Proposal

The Tauhara II proposal, for which Contact has received consent, is of interest here as an example of planning for a new resource. The proposal is illustrated by Fig. 14.4, based on a map in the evidence of Bixley for Contact Energy (the complete figure is contained in the written Board of Inquiry Decision [2010]). The resistivity boundary of the combined Wairakei and Tauhara resource, with its waist or neck, is shown in the figure.

It had been conjectured for decades that Wairakei and Tauhara had separate deep upflows of hot water, because although the effect of Wairakei production had spread measurably into Tauhara wells, some surface activity had remained unaffected. The decision records that further drilling has revealed possibly two separate upflows. The resistivity boundary comes close to the Rotokawa resource, consistent with the rather densely packed but apparently individual resources in the TVZ as shown in Fig. 2.8. Te Mihi, Wairakei, Poihipi and the two stations at Rotokawa all lie within 10 km of the proposed new station, which is to be a double flash system with a direct-contact condenser and a total generating capacity of 240 MWe.

Production and injection areas are shown on the figure; injection outside the resistivity boundary is permitted. The targetted injection area between the production area and Taupo to maintain the pressure constant beneath the urban area is indicated only approximately. The numerical reservoir model for Wairakei–Tauhara has been extended to cover the new development area and includes the data from

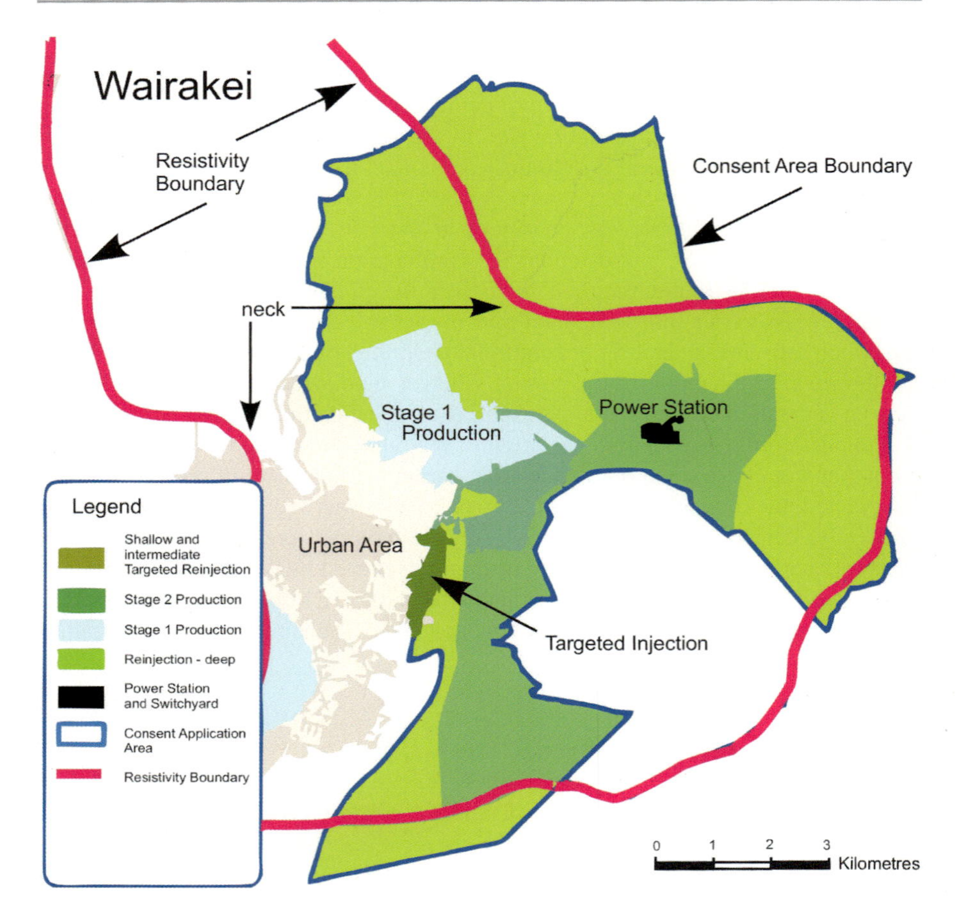

Fig. 14.4 Map of proposed Tauhara II development based on Board of Inquiry report and final decision [2010] Tauhara II geothermal development project, Appendix 8. The neck in the resistivity boundary previously mentioned is shown. (*reproduced by permission of Contact Energy Ltd*)

some new wells drilled. The model assumes that the heat is supplied by thermal conduction through the greywacke basement; from a heat transfer perspective, it is noteworthy that this has been found to be sufficient to provide for the high-temperature resource found, without the addition of a convective supply (upflow).

14.5 Ngawha

Ngawha is an example of a resource developed with a great deal of care for the surface activity and with a successful outcome so far as electricity generation is concerned. If the entire expenditure on its exploration and development was taken into account, its economic success would be questionable, but this is because the wells were drilled by the government and left as an idle investment for 15 years or more. There can be little doubt that the existing operation is economically successful. Ngawha lies several hundred kilometres north of the TVZ and is apparently an isolated, anomalous resource that would have been regarded as a risky exploration project. Nevertheless, it has a maximum measured well temperature of 320 °C, but 220 °C is a more representative average resource temperature.

The tectonic history of northern New Zealand is the subject of active debate (see Schellart [2012] who provides an examination of possible subduction modes viewed in a manner similar to that of Sect. 2.2.2). The geology of the resource is very simple from an engineering perspective; a layer of marine sediments approximately 600 m thick lies over greywacke to below the drilled depth of 2,300 m. The sediments are allochthonous, that is, they were deposited somewhere else and the entire block was moved into its current position by forces resulting from tectonic processes. The greywacke is heavily fractured but appear mainly sealed by mineral deposition. The well measurements show permeability at the contact between the sediments and the greywacke and also at localised zones within the body of the greywacke which suggests that the permeability is provided by faults.

The area has relatively minor surface springs; they are not the vigorous localised discharges of the TVZ. The hot water was known to Maori and used for therapeutic bathing by digging shallow pits into the ground and allowing the hot water to fill them by seepage. The water carries minerals collected from the sediments and the baths are still in use today. The water also has a high mercury content; cinnabar (mercury ore) was mined at Ngawha in the early part of the twentieth century (Mongillo [1985]). To the geothermal engineer, perhaps the most significant features are the amount of CO_2 in solution in the liquid water resource fluid and the geometrical simplicity of the resource—which does not make reservoir simulation simple however. There are high concentrations of boron, and calcite scaling occurs in the wells. Although the resource fluid appears at the surface only at a very few places, CO_2 emerges over wider areas, although not sufficient to prevent vegetation. In particular a concentrated discharge of CO_2 is thought to be the cause of a lake having formed at one place over the resource where the gas flow rate has been enough to create a deep depression into which surface water has drained (Simmons et al. [2005]).

Mongillo [1985] collected all the scientific information on the resource, which was drilled by the New Zealand Ministry of Works and Development in approximately 1980. Although the population north of Auckland was low and industry sparse at that time, any increase in electrical load would have caused problems because generation was several hundred kilometres away and there were few acceptable transmission line routes through the Auckland isthmus. The development was deferred, the structural changes to the electricity industry already mentioned above took place, and a decision to obtain resource consents and build a power station finally resulted in an application in 1992 by Top Energy, the former local electricity distribution company. The wells were the property of the Ministry of Energy and had sat idle since being drilled in 1980–1982. The application was to take and inject up to 40,000 tpd, an amount for which there was no detailed scientific basis as the resource had not been discharged significantly. There was considerable concern by local Maori that the surface springs might disappear as they had at Wairakei, and a claim was made to the Waitangi Tribunal for Maori ownership of the entire resource. This was rejected but it was recommended that the springs be preserved, and this became a consent condition when the Northland Regional Council (NRC), the regional government, eventually gave consents, which it did for 10,000 tpd for a period of 10 years. The first year was to be used for the collection of baseline environmental data, since the area includes wetland with conservation value as well as the springs.

The geothermal water content of the spring water was identifiable by the chloride concentration, and spring protection was provided in the consents by a condition stating that if the mean chloride concentration in the spring water fell outside its normal variation, then the power station must cease to use the resource. The chloride concentration in the springs had a fairly large random variation presumably due to the long passage through the 600 m thick sediments to the surface and then mixing with rainwater near the surface. Nevertheless, this condition posed a risk for the developer, which was added to by the term of consent being only 9 years, a very short time to pay off a loan. The wells were purchased however and an Ormat ORC plant of 10 MWe capacity was commissioned in 1998, with production and injection at opposite sides of the resource. Two production wells and two injection wells were used, and the water was injected at 90 °C.

After 5 years of operation, Top Energy advised NRC that it wished to renew the consents when they expired and to increase production to 25,000 tpd and station capacity to 25 MWe. The porosity of greywacke is only a few % so the volume of water in the resource is small, and downhole pressures had begun to decrease almost from the start of production. One of the original wells, Ng13, was situated close to the springs and was reserved to monitor resource pressure; its wellhead pressure remained high due to the gas content. By 2004 measurement from all wells showed a consistent trend of falling resource pressure, a fall of about 0.3 bars per annum. Chloride concentration in the springs showed the same random variation as before production, with no certain evidence of any decline in their output. Nevertheless, the company took a conservative view, and in the new application, it suggested injecting extra freshwater at a rate sufficient to keep the wellhead pressure of NG13 steady; up to 1,600 tpd was estimated—this was a novel idea for a resource of this type. A 4,000 block TOUGH2 reservoir simulation model was developed by the University of Auckland (Prof O'Sullivan) and improved as time passed. At the time of the application (2004) extra water injection had not been tested, so in case it did not work as planned, reliance had to be placed on the model predictions, which were based on field data collected without the extra injection. The prediction was that the reservoir pressure would continue to fall until 2017 after

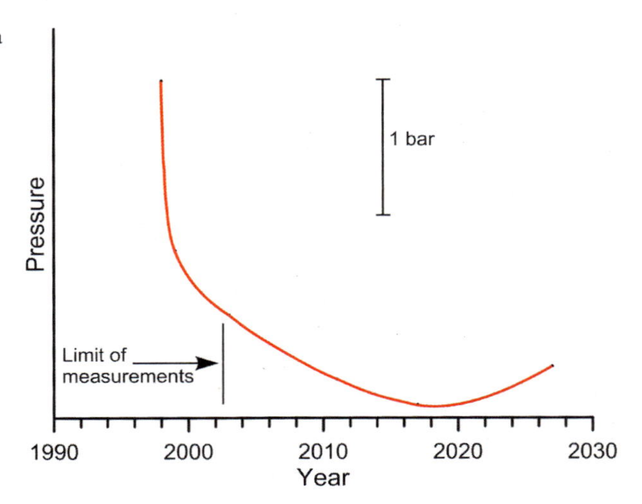

Fig. 14.5 Sketch of Ngawha resource pressure variation predictions made in 2003, showing the extent of the measurements at the time; measurements and predictions were coincident up to the marked "limit of measurements"

which it would rise slowly. The predicted and measured pressure variation is sketched in Fig. 14.5.

The concentration of CO_2 in the reservoir fluid was changing as production fluid was degassed, injected, reheated and recirculated back throughout the resource, and this effect was included in the model. The level of confidence in the prediction was said to be high until 2008, with lower confidence after this—the resource did not have a long history of use with which to refine the model. Figure 14.5 suggested that an important balancing of effects occurred in 2018 to produce a minimum in pressure, after which it would increase slowly. This was encouraging in terms of protection of the springs, but the minimum appeared long after the 2008 high confidence limit. NRC therefore declined the application to increase production but gave consents for the existing 10,000 tpd for a further 15 years, including the injection of up to 10 % more water than discharged. Top Energy appealed but eventually an unusual course of action was taken, with NRC support. Using provisions of the RMA, the Environment Court granted consent for a 6-month delay in the Appeal, with approval to carry out water injection trials, the results of which would be presented to the court at the Appeal hearing. The trial consisted of injecting nominally 1,000 tpd of extra water into the resource via an unused well about 500 m from the monitor well while measuring the pressure increase at the monitor well. The daily rate of injection could not be kept constant but the variations were recorded and used as input to the simulation model; the output was a graph of pressure increase to compare with the measurements. The monitor well pressure responded measurably about 10 h after the start of injection. The model predictions showed a much more rapid rate of pressure rise than was measured, but agreement after the initial rise was good. For the same hearing, a simple numerical model of the springs was produced which suggested that if the reservoir pressure was maintained to within ± 1 bar, the variations in supply of reservoir water to the springs would result in a chloride concentration variation within the range of the natural variations on record. The court granted the renewal and increase in production to 25,000 tpd and approved an increase in installed capacity to 25 MWe. The 25-page decision provides detail (Environment Court [2006]). The amount of extra water to be injected was increased to 3,000 tpd to provide some margin above the 1,600 tpd which was anticipated to be sufficient.

References

Allis R, Bromley C (2009) Unraveling the subsidence at Wairakei, New Zealand. GRC Trans 33:299–302

Allis RJ, Zhan X (2000) Predicting subsidence at Wairakei and Ohaaki fields. Geothermics 29 (4–5):479–497

Board of Inquiry report and final decision (2010) Tauhara II geothermal development project. http://www.epa.govt.nz/applications/tauhara-ii/

Boast RP (1995) Geothermal resources in New Zealand: a legal history. Canterbury Law Review 6:1–24

Bolton RS (2009) The early history of Wairakei (with some brief notes on unforeseen outcomes). Geothermics 38(1):11–29

Brockbank K, Bromley CJ, Glynn-Morris T (2011) Overview of the Wairakei-Tauhara subsidence investigation program. In: Proceedings of the 36th workshop on geothermal reservoir engineering, Stanford

Court of Appeal (1982) Keam v Minister of Works and Development [1982], 1, NZLR 319

Daysh S, Chrisp M (2009) Environmental planning and consenting for Wairakei 1953–2008. Geothermics 38(1):192–199

Environment Court (2000) Contact Energy Ltd v Waikato Regional Council and Taupo District Council, Decision NO A 04/2000

Environment Court (2006) Ngawha Geothermal Resource Company Ltd v Northland Regional Council, Environment Court, Decision A 117/2006

Glover RB (1998) Changes in the chemistry of Wairakei fluids 1929 to 1997. In: Proceedings of the 20th New Zealand geothermal workshop, University of Auckland, Auckland

Healey J (1980) The geothermal story. Chapter in a book published by the Rotorua Historical Society, New Zealand

Houghton BF, Lloyd EF, Keam RF (1980) The preservation of hydrothermal system features of scientific and other interest. Report to the New Zealand geological survey on behalf of the Nature Conservation Council, Wellington

Hunt TM (1970) Net mass loss from the Wairakei geothermal field, New Zealand. Geothermics 2 Pt 1:487–491

Hunt T, Graham D, Kuroda T (2003) Gravity changes at Tauhara geothermal field. In: Proceeding of the New Zealand geothermal workshop

King M (2003) The Penguin history of New Zealand. Penguin, Auckland

Martin, JE (ed) (1998) People politics and power stations: electric power generation in NZ 1880-1998. Electricity Corporation of NZ and the Historical Branch, Department of Internal Affairs

Mongillo MA (ed) (1985) The Ngawha geothermal field: new and updated scientific investigations. DSIR geothermal report no. 8

NZGS (1974) Minerals of New Zealand (Part D Geothermal), New Zealand Geological Survey, 38D

RGUA (1987) Rotorua Geothermal Users Association Inc v Minister of Energy – Attorney General, High Court Wellington Registry, 13 May 1987, CP543/86 (Heron J)

Schellart WP (2012) Comments on "Geochemistry of the Early Miocene volcanic succession of Northland", New Zealand, and implications for the evolution of subduction in

the SW Pacific" by Booden, M.A., Smith, I.E.M., Black, P.M. and Mauk, J.L. J Volcanal Geoth Res 211–212:112–117
Simmons SF, Harris SP, Cassidy J (2005) Lake filled depressions resulting from cold gas discharge in the Ngawha geothermal field. J Volcanal Geoth Res 147(3–4):329–341
Thain, I.A. and Carey, B. (2009) 50 years of power generation at Wairakei. Geothermics 38(1):48–63
White PJ, Lawless JV, Terzaghi S, Okada W (2005) Advances in subsidence modeling of exploited geothermal fields. In: Proceedings of the world geothermal congress
Zarrouk S, O'Sullivan MJ, Croucher A, Mannington W (2007) Numerical modelling of production from the Poihipi dry steam zone: Wairakei geothermal system, New Zealand. Geothermics 36:289–303

Appendix A: Saturation Properties of Water from the Triple Point to the Critical Point

The IAPWS Revised Supplementary Release on Saturation Properties of Ordinary Water Substance, September 1992, sets out the following equations, which provide values for saturation pressure, density, specific enthalpy and specific entropy of saturated liquid and saturated steam. The values are said to be identical to the values tabulated in the IAPS Skeleton Tables, 1985, and this release supercedes a release dated 1986. The equations rely on values of the triple point and the critical point, and these values were slightly modified as a result of a modification to the absolute temperature scale in 1990, referred to as ITS-90.

The IAPWS adopts ‘ to indicate saturated liquid and this has been replaced below by suffix f, which is used throughout this book. It also uses “ to indicate saturated vapour, which has been replaced here by suffix g (use of the term vapour is discussed in Sect. 3.1).

The release has a brief discussion of the importance of retaining the number of digits in the constants given.

Nomenclature (Specific to This Appendix)

h = specific enthalpy kJ/kgK
p = saturation pressure Mpa—note well
s = specific entropy kJ/kgK
T = temperature K
u = specific internal energy kJ/kg
ρ = density kg/m^3
α = auxiliary quantity for specific enthalpy
ϕ = auxiliary quantity for specific entropy
$\Theta = T/T_c$
$\tau = 1 - \Theta$

A. Watson, *Geothermal Engineering: Fundamentals and Applications*,
DOI 10.1007/978-1-4614-8569-8, © Springer Science+Business Media New York 2013

Subscripts:
c—at the critical point
f—saturated liquid water
t—at the triple point
g—saturated steam

Reference Constants

$T_c = 647.096$ K
$p_c = 22.064$ MPa
$\rho_c = 322$ kg/m^3
$\alpha_0 = 1{,}000$ J/kg
$\phi_0 = \alpha_0/T_c$

Saturation Pressure

$$\ln\left(\frac{p}{p_c}\right) = \frac{T_c}{T}\left(a_1\tau + a_2\tau^{1.5} + a_3\tau^3 + a_4\tau^{3.5} + a_5\tau^4 + a_6\tau^{7.5}\right) \quad (A.1)$$

where $a_1 = -7.85951783$; $a_2 = 1.84408259$; $a_3 = -11.7866497$; $a_4 = 22.6807411$; $a_5 = -15.9618719$; $a_6 = 1.80122502$

Density of Saturated Liquid

$$\frac{\rho_f}{\rho_c} = 1 + b_1\tau^{1/3} + b_2\tau^{2/3} + b_3\tau^{5/3} + b_4\tau^{16/3} + b_5\tau^{43/3} + b_6\tau^{110/3} \quad (A.2)$$

where $b_1 = 1.99274064$; $b_2 = 1.09965342$; $b_3 = -0.510839303$; $b_4 = -1.75493479$; $b_5 = -45.5170352$; $b_6 = -6.7469445E5$

Density of Saturated Steam

$$\ln\left(\frac{\rho_g}{\rho_c}\right) = c_1\tau^{2/6} + c_2\tau^{4/6} + c_3\tau^{8/6} + c_4\tau^{18/6} + c_5\tau^{37/6} + c_6\tau^{71/6} \quad (A.3)$$

where $c_1 = -2.03150240$; $c_2 = -2.68302940$; $c_3 = -5.38626492$; $c_4 = -17.2991605$; $c_5 = -44.7586581$; $c_6 = -63.9201063$

Specific Enthalpy and Specific Entropy Auxiliary Equations

Specific enthalpy and specific entropy both depend on the following auxiliary equations:

$$\frac{\alpha}{\alpha_0} = d_\alpha + d_1\Theta^{-19} + d_2\Theta + d_3\Theta^{4.5} + d_4\Theta^5 + d_5\Theta^{54.5} \tag{A.4}$$

$$\frac{\phi}{\phi_0} = d_\phi + \frac{19}{20}d_1\Theta^{-20} + d_2 \ln\Theta + \frac{9}{7}d_3\Theta^{3.5} + \frac{5}{4}d_4\Theta^4 + \frac{109}{107}d_5\Theta^{53.5} \tag{A.5}$$

where $d_1 = -5.65134998\text{E-}8$; $d_2 = 2690.66631$; $d_3 = 127.287297$; $d_4 = -135.003439$; $d_5 = 0.981825814$; $d_\alpha = -1135.905627715$; $d_\phi = 2319.5246$

Specific Enthalpies of Saturated Liquid and Saturated Steam

$$h_f = \alpha + \frac{T}{\rho_f}\left(\frac{dp}{dT}\right) \tag{A.6}$$

$$h_g = \alpha + \frac{T}{\rho_g}\left(\frac{dp}{dT}\right) \tag{A.7}$$

where the differential is found from Eq. (A.1), Eq. (A.2) or (A.3) and α from Eq. (A.4).

Specific Entropies of Saturated Liquid and Saturated Steam

$$s_f = \phi + \frac{1}{\rho_f}\left(\frac{dp}{dT}\right) \tag{A.8}$$

$$s_g = \phi + \frac{1}{\rho_g}\left(\frac{dp}{dT}\right) \tag{A.9}$$

where the differential is found from Eq. (A.1), Eq. (A.2) or (A.3) and ϕ from Eq. (A.5).

Appendix B: Compressibility of Water from 0 to 100 °C and 0–1,000 bar

The following is given by Fine RA, Millero FJ (1973) Compressibility of water as a function of pressure and temperature. J Chem Phys 59(10).

The compressibility is defined as

$$c = \frac{1}{\rho}\left(\frac{\partial \rho}{\partial P}\right)_T \tag{B.1}$$

$$= \frac{V^0\left(B - A_2P^2\right)}{V^P\left(B + A_1P + A_2P^2\right)^2} \tag{B.2}$$

where

$$\mathrm{B = 19654.320 + 147.037T - 2.21554T^2 + 1.0478\ E\text{-}2\ T^3 - 2.2789E\text{-}5T^4} \tag{B.3}$$

$$\mathrm{A_1 = 3.2891 - 2.3910E\text{-}3T + 2.8446E\text{-}4T^2 - 2.8200E\text{-}6T^3 + 8.477E\text{-}9T^4} \tag{B.4}$$

$$\mathrm{A_2 = 6.245E\text{-}5 - 3.913E\text{-}6T - 3.499E\text{-}8T^2 + 7.942E\text{-}10T^3 - 3.299E\text{-}12T^4} \tag{B.5}$$

$$\mathrm{V^0 = (1 + 18.159725E\text{-}3T)/\Big(0.9998396 + 18.224944E\text{-}3T - 7.922210E\text{-}6T^2 - 55.44846E\text{-}9T^3 + 149.7562E\text{-}12T^4 - 393.2952E\text{-}15T^5} \tag{B.6}$$

$$\mathrm{V^P = V^0 - V^0P/\left(B + A_1P + A_2P^2\right)} \tag{B.7}$$

A. Watson, *Geothermal Engineering: Fundamentals and Applications*,
DOI 10.1007/978-1-4614-8569-8,

Nomenclature

T = temperature (°C)
P = pressure (bar abs)
c = compressibility (1/bar)
V^P = specific volume (cm^3/g) at pressure P
V^0 = a specific volume given by Eq. (B.6)
Note: the specific volumes V^P and V^0 are determined from the equations and their units are irrelevant to the calculation of c.

Appendix C: The Boiling-Point-for-Depth Curve

Depth (m)	Tsat (C)	Psat (bars abs)	Depth (m)	Tsat (C)	Psat (bars abs)
0.0	100.00	1.01	150.0	195.99	14.29
5.0	110.99	1.48	155.0	197.37	14.71
10.0	119.36	1.95	160.0	198.72	15.14
15.0	126.18	2.41	165.0	200.03	15.56
20.0	131.98	2.87	170.0	201.32	15.98
25.0	137.04	3.32	175.0	202.58	16.41
30.0	141.56	3.78	180.0	203.01	16.83
35.0	145.63	4.23	185.0	205.02	17.25
40.0	149.36	4.68	190.0	206.20	17.67
45.0	152.80	5.13	195.0	207.36	18.09
50.0	155.99	5.58	200.0	208.50	18.51
55.0	158.98	6.02	205.0	209.01	18.93
60.0	161.78	6.47	210.0	210.70	19.35
65.0	164.43	6.91	215.0	211.78	19.76
70.0	166.94	7.35	220.0	212.83	20.18
75.0	169.33	7.79	225.0	213.87	20.60
80.0	171.60	8.23	230.0	214.89	21.01
85.0	173.78	8.67	235.0	215.89	21.43
90.0	175.87	9.11	240.0	216.87	21.84
95.0	177.87	9.55	245.0	217.84	22.26
100.0	179.80	9.98	250.0	218.80	22.67
105.0	181.66	10.42	255.0	219.74	23.08
110.0	183.45	10.85	260.0	220.66	23.49
115.0	185.19	11.28	265.0	221.57	23.91
120.0	186.87	11.71	270.0	222.47	24.32
125.0	188.50	12.10	275.0	223.36	24.73
130.0	190.08	12.57	280.0	224.23	25.14
135.0	191.62	13.00	285.0	225.09	25.54
140.0	193.11	13.43	290.0	225.94	25.95
145.0	194.57	13.86	295.0	226.78	26.36

(continued)

A. Watson, *Geothermal Engineering: Fundamentals and Applications*,
DOI 10.1007/978-1-4614-8569-8,

(continued)

Depth (m)	Tsat (C)	Psat (bars abs)	Depth (m)	Tsat (C)	Psat (bars abs)
300.0	227.61	26.77	540.0	258.45	45.77
305.0	228.42	27.18	545.0	258.96	46.16
310.0	229.23	27.58	550.0	259.47	46.54
315.0	230.02	27.99	555.0	259.98	46.93
320.0	230.81	28.39	560.0	260.48	47.31
325.0	231.59	28.80	565.0	260.98	47.69
330.0	232.35	29.20	570.0	261.47	48.08
335.0	233.11	29.61	575.0	261.96	48.46
340.0	233.86	30.01	580.0	262.45	48.84
345.0	234.60	30.41	585.0	262.94	49.22
350.0	235.33	30.81	590.0	263.42	49.61
355.0	236.06	31.22	595.0	263.90	49.99
360.0	236.77	31.62	600.0	264.37	50.37
365.0	237.48	32.02	605.0	264.84	50.75
370.0	238.18	32.42	610.0	265.31	51.13
375.0	238.82	32.82	615.0	265.78	51.51
380.0	239.56	33.22	620.0	266.24	51.89
385.0	240.24	33.62	625.0	266.70	52.27
390.0	240.91	34.02	630.0	267.15	52.65
395.0	241.50	34.42	635.0	267.61	53.03
400.0	242.23	34.81	640.0	268.06	53.41
405.0	242.88	35.21	645.0	268.51	53.78
410.0	243.53	35.61	650.0	268.95	54.16
415.0	244.17	36.00	655.0	269.39	54.54
420.0	244.80	36.40	660.0	269.83	54.91
425.0	245.43	36.79	665.0	270.27	55.29
430.0	246.05	37.19	670.0	270.70	55.67
435.0	246.67	37.58	675.0	271.13	56.04
440.0	247.28	37.98	680.0	271.56	56.42
445.0	247.88	38.37	685.0	271.99	56.79
450.0	248.48	38.76	690.0	272.41	57.17
455.0	249.07	39.16	695.0	272.83	57.54
460.0	249.66	39.55	700.0	273.25	57.92
465.0	250.25	39.94	705.0	273.67	53.29
470.0	250.82	40.33	710.0	274.08	58.66
475.0	251.40	40.72	715.0	274.49	59.04
480.0	251.97	41.11	720.0	274.90	59.41
485.0	252.53	41.50	725.0	275.31	59.78
490.0	253.09	41.89	730.0	275.71	60.15
495.0	253.64	42.28	735.0	276.12	60.52
500.0	254.20	42.67	740.0	276.52	60.89
505.0	254.74	43.06	745.0	276.91	61.27
510.0	255.28	43.45	750.0	277.31	61.64
515.0	255.32	43.84	755.0	277.70	62.01
520.0	256.35	44.22	760.0	278.09	62.38
525.0	256.88	44.61	765.0	278.48	62.75
530.0	257.41	45.00	770.0	278.87	63.11
535.0	257.93	45.38	775.0	279.26	63.48

(continued)

(continued)

Depth (m)	Tsat (C)	Psat (bars abs)	Depth (m)	Tsat (C)	Psat (bars abs)
780.0	279.64	63.85	1040	297.2	82.5
785.0	280.02	64.22	1050	297.8	83.3
790.0	280.40	64.59	1060	298.4	84.0
795.0	280.77	64.96	1070	299.0	84.7
800.0	281.15	65.32	1080	299.5	85.4
805.0	281.52	65.69	1090	300.1	86.1
810.0	281.89	66.06	1100	300.7	86.8
815.0	282.26	66.42	1110	301.3	87.5
820.0	282.63	66.79	1120	301.8	88.2
825.0	283.00	67.15	1130	302.4	88.9
830.0	283.36	67.52	1140	303.0	89.6
835.0	283.72	67.80	1150	303.5	90.2
840.0	284.08	68.25	1160	304.1	90.9
845.0	284.44	68.61	1170	304.6	91.6
850.0	284.80	68.98	1180	305.1	92.3
855.0	285.15	69.34	1190	305.7	93.0
860.0	285.50	69.70	1200	306.2	93.7
865.0	285.85	70.07	1210	306.7	94.4
870.0	286.20	70.43	1220	307.2	95.1
875.0	286.55	70.79	1230	307.8	95.7
880.0	286.90	71.15	1240	308.3	96.4
885.0	287.24	71.52	1250	308.8	97.1
890.0	287.59	71.88	1250	309.3	97.8
895.0	287.93	72.24	1270	309.8	98.5
900.0	288.27	72.60	1280	310.3	99.1
905.0	288.60	72.96	1290	310.8	99.8
910.0	288.94	73.32	1300	311.3	100.5
915.0	289.28	73.68	1310	311.8	101.2
920	289.6	74.0	1320	312.3	101.8
925	289.9	74.4	1330	312.8	102.5
930	290.3	74.8	1340	313.3	103.2
935	290.6	75.1	1350	313.7	103.8
940	290.9	75.5	1360	314.2	104.5
945	291.3	75.8	1370	314.7	105.2
950	291.6	76.2	1380	315.2	105.6
955	291.9	76.5	1390	315.6	106.5
960	292.2	76.9	1400	316.1	107.2
965	292.5	77.3	1410	316.6	107.8
970	292.9	77.6	1420	317.0	108.5
975	293.2	78.0	1430	317.5	109.2
980	293.5	78.3	1440	317.9	109.8
985	293.8	78.7	1450	318.4	110.5
990	294.1	79.0	1460	318.8	111.1
995	294.4	79.4	1470	319.3	111.8
1000	294.7	79.7	1480	319.7	112.4
1010	295.4	80.5	1490	320.1	113.1
1020	296.0	81.2	1500	320.6	113.8
1030	295.6	81.9	1510	321.0	114.4

(continued)

(continued)

Depth (m)	Tsat (C)	Psat (bars abs)	Depth (m)	Tsat (C)	Psat (bars abs)
1520	321.4	115.1	1940	337.5	141.4
1530	321.9	115.7	1960	338.1	142.7
1540	322.3	116.4	1980	338.8	143.9
1550	322.7	117.0	2000	339.5	145.1
1560	323.1	117.7	2020	340.1	146.3
1570	323.6	118.3	2040	340.8	147.5
1580	324.0	118.9	2060	341.4	148.7
1590	324.4	119.6	2080	342.0	149.8
1600	324.8	120.2	2100	342.7	151.0
1610	325.2	120.9	2120	343.3	152.2
1620	325.6	121.5	2140	343.9	153.4
1630	326.0	122.2	2160	344.5	154.5
1640	326.4	122.8	2180	345.1	155.7
1650	326.8	123.4	2200	345.7	156.9
1660	327.2	124.1	2220	346.3	158.0
1670	327.6	124.7	2240	346.9	159.2
1680	328.0	125.3	2260	347.5	160.3
1700	328.8	126.6	2280	348.1	161.5
1720	329.5	127.9	2300	348.6	162.6
1740	330.3	129.1	2320	349.2	163.7
1760	331.0	130.4	2340	349.8	164.9
1780	331.8	131.6	2360	350.3	166.0
1800	332.5	132.9	2380	350.9	167.1
1820	333.3	134.1	2400	351.4	168.2
1840	334.0	135.3	2420	351.9	169.3
1860	334.7	136.6	2440	352.5	170.5
1880	335.4	137.8	2460	303.0	171.6
1900	336.1	139.0	2480	353.5	172.7
1920	336.8	140.2	2500	354.1	173.8

Index

A
Accumulator, 123, 137, 247
Acts of Parliament
  Geothermal Energy, 299–301, 303
  Geothermal Steam, 299
  Resource Management, 300, 301, 309, 313
  Thermal Springs District, 299, 302
  Water and Soil Conservation, 300, 301
Adiabatic, 30, 33–34, 129, 135, 163, 184, 261
Advection, 28
Airlifting, 118
Alternator, 5, 6, 214, 217, 218, 227, 228, 232, 233, 243
Ammonia, 245, 246
Asthenosphere, 13
Axial thrust, 228, 233

B
Back-arc rifting, 15
Barometer, 27, 28, 238
Base load, 211–212, 214, 234, 265, 308
Bellows, 256, 257
Bentonite, 81, 85
Bernoulli's equation, 57–58, 130, 152, 239
Bit, 48, 81–87
Bleeding, 105–107, 109, 111, 243
Blowout, 80, 86, 87, 92–94, 304
Boast, R.P., 297, 299
Boiling
  film, 126
  micro-layer, 127
  nucleate, 123, 126–128
Boiling-point-for-depth, 25, 42–45, 327–330
Bourdon tube, 28
Brianca, G., 7, 227
Bubble, 44, 45, 118, 124–128, 130, 132, 138, 165, 267
Buckling, 14, 90
Bursting, 94, 95, 257, 265, 267, 269

C
Calcite, 155, 160, 315
Capillary tubing, 195
Carbon dioxide, 20, 286
Carnot, 25, 34–37, 217, 218, 245, 247
Cash flow
  annual, 202, 205, 208
  cumulative, 202–204, 206
  discounted, 204–205
  net, 204
Casing
  anchor, 78, 80, 84, 87, 94–96
  head flange, 80, 96, 102, 105
  heating of, 117, 121
  production, 31, 78–80, 87, 93–96, 108, 109, 116, 143, 148, 157, 160, 161, 163, 262
  surface, 78
Cavitation, 130, 266
Cellar, 80, 95, 291
Celsius, 26, 29, 36
Centigrade, 26
Cerro Prieto, 215, 255, 260, 293
Choked flow, 163
Clapeyron, 38–39, 137
Compressibility, 42, 73–74, 137, 172, 182, 185, 192, 196, 266, 292, 325–326
Condensation
  dropwise, 130
  film, 130
Condenser
  air-cooled, 239, 243–245
  barometric leg, 239
  jet, 238–240
  tubular, 238, 263, 308
  vacuum, 39, 239
Conservation, 4, 26, 29, 48, 51–55, 132, 165, 282, 297–318
Control volume, 30, 31, 48–50, 52, 53, 55, 72, 172, 282

A. Watson, *Geothermal Engineering: Fundamentals and Applications*,
DOI 10.1007/978-1-4614-8569-8, © Springer Science+Business Media New York 2013

Cooling, 16, 17, 20, 52, 66, 67, 86, 101, 113, 157, 238–240, 243, 263, 291, 292, 294, 307
Cost
capital, 211, 215, 220
drilling, 203, 206, 210, 263
levelised, 208
marginal, 211, 213
operations and maintenance, 202, 206
Cuttings, 14, 81, 83, 85, 86
Cycle
Carnot, 34, 217, 218, 220–227, 237, 240, 242, 243, 245, 246
Kalina, 245–246
organic Rankine, 4, 217, 220, 235, 240–244
Rankine, 4, 217, 218, 220, 222, 224, 226, 227, 235, 240, 247
tri-lateral flash, 246–248

D
Dalton's law of partial pressures, 138
Darcy's law, 47, 61, 71–73, 171, 174, 190, 192, 193, 278, 282
Debt, 201, 203–205, 207
Della-Porta, 7
Depreciation, 200, 201
Dimensionless
pressure, 62, 175, 186
radius, 174–176
variables, 58, 62, 69, 101, 174–176, 185
Discharge characteristic, 141–151, 160–162, 192, 235, 236, 254, 256, 260
Discount
factor, 205, 206
rate, 205, 207, 208
Discounted cash flow, 204–205, 208
Dissipation function, 56, 57, 59
Double porosity, 190, 191
Downflow, in wells, 107–108
Drilling
mud, 83, 85–87, 180, 293
rig, 81–86, 91, 118, 294
Drill string, 81–85, 87
Dryness fraction, 40, 44, 117, 134, 138, 143, 148, 149, 151, 159, 214, 222, 223, 237, 253

E
Earth
core, 11, 12
crust, 3
Energy
conservation, 29, 55
equation, 25, 29–34, 48, 55–58, 62, 67, 69, 135, 165, 166, 223, 232, 282, 283
internal, 26, 29, 31, 33, 38, 40, 56, 58, 283
per capita consumption, 9
Enthalpy
excess, 144, 158, 253
flowing, 144, 193, 194, 279, 286
specific, 25, 29–34, 40, 56, 58, 117, 121, 127, 141, 143–149, 151, 154, 158, 159, 192, 222, 223, 232, 233, 236, 237, 260, 321, 323
Entropy, 25, 34–37, 40, 44, 56, 220, 223, 226, 247, 321, 323
Environment Waikato, 312
Equilibrium, 26, 28, 37, 38, 44, 71, 100, 101, 108, 109, 114, 124, 128, 133, 138, 139, 253
Equivalent diameter, 65, 66, 160
Exergy, 236, 248
Exhaust, 30, 33, 220, 226–228, 232, 236, 237, 244

F
Fahrenheit, 26, 29, 36
Fault
open, 189
sealed, 189
Fauna, 5, 6, 86, 292
Financial
accounting, 200, 201, 203
analysis, 200, 201
appraisal, 201
Finite difference, 70, 270, 278, 281, 284
First law of thermodynamics, 25–29
Flash evaporator, multi-stage, 129, 251
Flashing, 120, 123, 128–129, 158–159, 235–237, 267–270
Flora, 5, 6, 86, 262, 292
Flow
annular, 132
flashing, 128–129
homogenous, 133, 135–137, 161, 165
horizontal, 261, 262
laminar, 58–61, 65, 71, 137, 258
separated, 132, 137, 161
slug, 251, 261, 262
turbulent, 47, 58, 61, 64, 65, 136, 258
two-phase, 4, 58, 118, 123–139, 145, 148, 154, 160, 161, 192–194, 247, 258, 260, 261, 263–265, 279
vertical, 163, 308

Fossil fuel, 1, 34, 37, 209, 213, 217, 240
Fourier number, 69, 70, 281
Fractured media, 190
Fractures, 12, 20, 49, 61, 91–94, 103, 109, 189–191, 308
Fumarole, 20, 289
Function of state, 26, 29, 30, 32, 36, 37, 39

G
Generator, 200, 218, 220, 244, 309
Geotherm Energy Ltd, 307
Geothermometry, 252
Geyser, 17, 21, 141, 164, 166, 292, 303–305, 307
Gibbs function, 224, 225
Grashof number, 68
Groundwater, 72, 74, 158, 169, 171, 173, 183, 278, 291–293
Guericke, von, 7
Guglielmini, 47

H
Heat
- demand for, 1, 8
- engine, 8, 29, 34–36, 218, 248
- flux, 12, 14, 20, 44, 45, 55, 59, 64–66, 70, 109, 124, 126, 127, 131, 163, 259, 260, 293
- internal, 12, 14, 20, 57, 59, 67, 70, 71, 101, 259
- rejection, 226, 238–240, 244, 246
- specific, 40, 66, 70, 110, 246

Heating, 7, 15, 55, 108, 111, 118, 120–121, 127–129, 136, 220, 224, 225, 242, 243, 245–247, 285, 290, 299, 302, 303, 307
Heavy water, 213, 305
Henry's law, 138, 139, 165
Hook, 7, 83, 93
Horner, D.R., 178, 182
Houghton, B.F., 300, 301
Hydro-electric, 265, 266, 298, 308
Hydrothermal eruptions, 123, 128, 131–132, 305, 312

I
Implicit, 146, 281, 283
Inflow, 50, 109–111, 127
Injection, 5, 6, 67, 78, 86, 102, 110, 131, 160, 171, 179, 196, 203, 251, 254, 256, 262–264, 268–270, 278, 279, 282, 284, 286–288, 290, 293, 302, 307, 309, 310, 313, 316, 317
Insulation, thermal, 260
Interference tests, 171, 194, 196
Internal rate of return (IRR), 207–208
Intrusion, 16–18, 78
Isopentane, 241–245
Isopropanol, 159

K
Kakkonda, 17, 18, 77
Kamojang, 114, 257
Kawerau, 108, 109, 111, 291
Keam, R.F., 301
Kelly, 83–85
Kelvin, 36
Kinetic
- energy, 26, 27, 31, 33, 62, 64, 128, 239
- head, 136, 162, 258

KRTA, 304, 305, 307

L
Laminar, 58–61, 64, 65, 71, 137, 258
Laplace equation, 73
Lip pressure pipe, James, 145–147, 150–151
Lithosphere, 13, 15, 20
Load following, 211, 212

M
Magma, 11, 15–20, 67, 131
Manifold, 231, 266
Mantle, 11, 12, 14–16, 20, 67
Maori, 298, 302, 307, 309, 315, 316
Mariotte, 47
Mass
- balance, 110, 149, 150, 159, 183, 270, 275
- quality, 134, 137
- velocity, 134, 135, 145, 161

Master valve, 80, 96, 97, 142
Mercury, 220, 241, 315
Mercury energy, 308, 309
Merz and McLellan, 274
Micro-seismicity, 292
Mises, von, 89, 90
Model
- conceptual, 285, 286
- lumped parameter, 277–279
- mixing, 158
- preliminary, 284
- reservoir, 197, 287, 313

Multi-stage flash evaporator, 251

N
Nappe, 155
Navier-Stokes equations, 51, 52, 55
Network, 132, 199, 210–212, 255, 260, 308
Newcomen, 8, 9
Newton, 27, 42, 47, 51, 55, 66, 232, 284
Ngawha, 4, 188, 297, 298, 314–318
Nozzle, 81, 85, 156, 228, 230–233, 237, 239, 263
Nuclear, 1, 13, 29, 70, 124, 127, 131, 161, 163, 164, 212, 213, 219, 228, 239, 245, 305, 311
Nucleation
  homogenous, 38, 123, 126–128, 132
  site, 126–129
Nusselt, 129
Nusselt number, 64, 66, 69

O
Ohaaki, 196, 240, 291
Oil shock, 219
Orifice plate, 151–154
Ormat, 220, 243, 244, 305

P
Payback period, 204, 207, 208, 213, 307
Pentane, iso-and *n*-, 241
Permeability
  relative, 144, 192
  thickness, 178
Phreato-magmatic, 131
Piezometer, 28
Pipeline
  condensate, 267
  injection, 262, 293
  steam, 213, 254–260, 262, 263
  two-phase, 5, 254, 260–262
Pivot point, 111–114, 188
Plastic deformation, 14
Plate boundaries
  collision, 13, 14
  subduction, 14–17
Pluton, 16, 78, 289
Pohutu, 303–305
Poihipi, 306–309, 313
Poiseuille, 61, 71, 72
Polynesians, 297, 298
Porosity, 72, 73, 176, 190–192, 196, 276, 277, 284, 316
Prandtl number, 63
Present value, 205, 208
Pressure
  absolute, 28, 128, 151, 196, 268
  buildup test, 177–179, 185
  dimensionless, 62, 175, 186
  fall-off test, 179
  gauge, 27, 28, 104, 148, 195
  hydrostatic, 25, 42–45, 53, 67, 86, 87, 91–94, 104, 108, 113, 115, 129, 292
  tapping, 145, 152, 154
Prime mover, 29, 217, 218, 244, 246–248, 273, 287, 288
Project life, 201, 204, 208, 209
Pulse, pressure, 170, 171, 196, 308, 309

R
Radioactive isotopes, 12, 20, 55, 67, 69
Ramey, H.J. Jr., 173, 176, 183, 185, 186, 275
Raoult's law, 138
Recovery factor, 275, 277
Reheat, 226
Reservoir simulation, numerical, 4, 188, 273, 274, 278–287
Resistivity, 17, 293
Resistivity boundary, 17, 19, 275, 284, 306–308, 310, 313, 314
Revenue, 200–208, 303
Reversibility, 36
Reynolds number, 61–63, 65, 136, 137, 160, 258, 259, 270
Risk, 78, 87, 94, 207–210, 273, 274, 278, 287, 307, 316
Rotorua, 4, 161–164, 291, 297, 298, 300–305

S
Sandface, 74, 101, 110, 116, 117, 119, 143, 144, 160, 171, 172, 174–183, 185, 187, 188, 193, 195
Saturation
  conditions, 32, 38, 39, 43, 44, 117, 123, 128, 137, 221, 262
  definition, 38, 44
  envelope, 38, 39, 226, 241, 242
  line, 25, 38–39, 41, 138, 191
Savery, 6–8
Scale, 1, 3, 4, 6, 26, 28, 29, 36, 100, 157, 189, 199, 200, 218, 242, 251, 265, 310, 321
Scaling, 61, 251, 253, 257, 262, 315
Seismicity, 292, 293
Separated water, 5, 6, 78, 143, 148, 157, 215, 235–237, 241, 244, 245, 251–256, 262, 264, 267–270, 290, 293, 306, 307, 310, 312

Separator, 5, 6, 47, 78, 85, 147–150, 155–156, 237, 252, 254, 256, 257, 262–265, 268–270
Silencer, 86, 121, 147–151, 156, 157, 263–265
Silica
amorphous, 252, 253, 257
quartz, 252, 253
Skin effect, 180–182, 195
Slab, 12, 16, 69, 71, 276
Slotted liner, 65, 79, 100, 101, 105, 107, 109, 160, 161, 163, 171, 188, 265
Smith, I.K., 9, 247, 248
Species concentration, 158, 159
Springs, 14, 28, 106, 109, 128, 158, 289, 291, 293, 298, 299, 302–306, 315–318
Static formation temperature test (SFTT), 93
Steady flow energy equation, 25, 29–34, 58, 135
Steam
engine, 8, 26, 219, 227, 247
explosions, 131
formations, 114, 144
interface, 126
jet ejectors, 215, 239
sales contract, 199, 213–215
saturated, 32, 33, 40, 93, 106, 137, 143, 148, 221, 222, 226, 258, 267, 321–323
superheated, 38, 40
system, 215
wet, 237, 239, 257
zone, 100, 144, 164, 192, 306–308, 311
Stored heat, 4, 5, 12, 247, 274–278
Strain, 13, 54, 74, 88
Stream tube, 57, 58
Stress
axial, 88, 89, 256
direct, 53, 56
failure, 89, 90
hoop, 89, 90
proof, 88
shear, 48, 49, 51–60, 62, 64–66, 89, 135, 136, 258
yield, 88–90, 94
Subsidence, 14, 15, 291–292, 309–313
Sulphur hexafluoride (SF_6), 159
Sumikawa, 196
Superficial velocity, 135
Superposition, 169, 177–179, 190, 304
Supersonic, 145, 146
Surface
discharge, 11, 20–21, 115, 146, 274, 285, 289–291, 301
features, 21, 289–290, 300, 301, 303
springs, 305, 315, 316
subsidence, 14
tension, 20, 38, 123–125

T
Tauhara, 308–310, 312, 313
Taupo District Council, 309, 312
Te Aroha, 164
Tectonic plates, 11, 12, 14, 64
Te Mihi, 310–311, 313
Theis solution, 173, 177, 181, 182, 185, 187, 188, 196, 262, 304
Thermal
conductivity, 20, 44, 48, 49, 55, 58, 61, 67, 69, 70, 184, 258
diffusivity, 16, 63, 70
efficiency, 36, 226, 244
insulation, 36, 251, 258, 260
Throttling, 30, 33–34, 233
Tongonan, 17, 158, 164, 180
Top energy, 316, 317
TOUGH2, 279, 284, 285, 316
Tracers, 158–160, 265
Transient pressure testing, 50, 93, 137, 172, 191–192, 197
Turbine
axial flow, 228, 235
back pressure, 228, 244
condensing, 227, 237
impulse, 230, 231
mixed pressure, 228, 229, 235
multi-stage, 227
pass-in, 235
radial flow, 235
reaction, 227, 230, 231
seals, 231
stage, 237, 243
steam, 1, 4–6, 33, 40, 211, 217, 218, 227–235, 237, 241, 243, 244, 247, 256, 269, 307
Willans line, 234, 270
Turbo-generator, 218, 305
Type curves, 175, 182–188, 196

V
Vacuum pump, 240
Velocity diagram, 232, 233
Viscosity
dynamic, 49, 73
kinematic, 49, 63, 193
Void fraction, 134, 164

Volcanism, 13–15, 20, 128
Volumetric flux, 135

W
Wairakei, 4, 105, 107, 209, 235, 236, 240, 254, 257, 261, 263, 264, 274, 279, 284–286, 290, 293, 297, 299, 300, 303, 305–314, 316
Waitangi Tribunal, 298, 302, 316
Water
  International Association for the properties of, 25, 40
  saturated, 16, 32, 104, 108, 111, 115, 118, 123, 131, 137, 148, 149, 221, 224, 225, 237, 246, 251, 253, 267, 268, 275
Water level, 99, 104–107, 113, 114, 116–120, 155, 238, 262, 264, 289, 303, 304, 308
Weir, 147, 149, 150, 154–157
Well
  injection, 5, 6, 78, 196, 251, 254, 262–263, 268, 288, 313, 316
  production, 4–6, 78, 116, 121, 157, 197, 251, 254–256, 262, 288, 293, 307, 308, 316
Wellbore storage, 169, 180, 182–188, 195
Wellhead
  pressure, 95, 104–108, 111, 113, 121, 142–144, 148, 157, 161, 164, 236, 255, 256, 260, 316
  valve, 78, 95, 96, 171, 178, 182
WELSIM, 160
Willans line, 233, 234, 270
Work, thermodynamic, 248

Y
Young's modulus, 88

Printed by Printforce, the Netherlands